The Information Retrieval Series

Series Editor

W. Bruce Croft

Sándor Dominich

The Modern Algebra of Information Retrieval

Sándor Dominich
Computer Science Department
University of Pannonia
Egyetem u. 10.
8200 Veszprém, Hungary
dominich@dcs.vein.hu

ISBN: 978-3-642-09643-3 e-ISBN: 978-3-540-77659-8

ACM Computing Classification (1998): H.3, G.1, G.3

Cover Design: KünkelLopka, Heidelberg

Printed on acid-free paper

9 8 7 6 5 4 3 2 1

springer.com

To my parents Jolán and Sándor

To my wife Emőke and our daughter Emőke

Acknowledgments

Special thanks, first, to my family, the two Emőke, who were very understanding during the entire time I was involved in writing this book.

Next, I would like to thank Ferenc Friedler, the head of the Department of Computer Science, University of Pannonia (Veszprém, Hungary), for providing a highly supportive environment.

At some stages I derived much benefit from discussions with the following colleagues: Rozália Piglerné Lakner, Tamás Kiezer, Júlia Góth (all from the University of Pannonia, Veszprém, Hungary), Iadh Ounis (Glasgow University, Glasgow, Scotland, U.K.) and I enjoyed their active support.

Parts of this book were included in a B.Sc. curriculum on Information Retrieval (Faculty of Information Technology, University of Pannonia, Hungary). The experience thus gained as well as students' feedback proved very helpful, especially for elaborating examples and proposed problems, and in presenting methods.

May all find here my expression of gratitude.

Last but not least, I am indebted to Springer-Verlag Gmbh for making the publication of this book possible.

Sándor Dominich

Contents

1 Introduction

Problems worthy of attack prove
their worth by fighting back.
(Pál Erdős)

Information retrieval (IR) is concerned with finding and returning information stored in computers that is relevant to a user's needs (materialized in a request or query). With the advent of the Internet and World Wide Web (Web for short), IR has acquired tremendous practical impact as well as theoretical importance.

The number of IR books that have appeared in the last ten years is 70% of the total number (approximately 30) published on the subject in all thus far. This is a clear sign that interest in learning, teaching, researching, and applying IR methods and theory has grown and is increasing rapidly (most probably owing to the Internet and the World Wide Web, which are "invading" practically every aspect of human activity and life).

Most of the books published thus far are concerned with describing IR methods and theories, and they range from classical texts (Hays 1966, Salton 1971, van Rijsbergen 1979, Salton and McGill 1983, Korfhage 1997, Kowalski 1997, Baeza-Yates and Ribeiro-Neto 1999) to ones that are based on linear algebra (Berry and Browne 1999, Langville and Meyer 2006), concept lattice (Koester 2006b), geometry (e.g., van Rijsbergen 2004, Widdows 2004), user modeling and context (Belew 2000, Spink and Cole 2005, Ingwersen and Järvelin 2005), natural language processing (Tait 2005), algorithms (Grossman and Frieder 2004), logic (Crestani et al. 1998), language modeling (Croft and Lafferty 2003), and the mathematical axiomatic method (Dominich 2001).

The present volume differs from all of the books that have appeared thus far in both approach and style. Retrieval methods (major proven models and ranking techniques) and information retrieval in general are treated in a unified manner within the one formal framework of modern algebra, namely abstract algebraic structures (primarily lattices, but also linear space, clans, and algebras), while keeping traditional algebraic tools (equation solving, matrix). This approach has some clear advantages:

- It sheds new light on the very mechanism of retrieval methods (at a conceptual-mathematical level).
- New properties are revealed.
- New and efficient retrieval methods can be developed.
- It allows for a very elegant treatment of IR.
- Connections with modern algebra are created for further research.

The book should be helpful for a wide range of readers from newcomers through students and educators to researchers and system developers coming from a variety of fields such as computer science, mathematics, information science, engineering, logics, linguistics, and physics. A precise description of every method is given in detail. For every method, a complete example is provided to enhance understanding. Every chapter, apart from Chapters 1 and 7, ends with exercises and problems designed to convey a deeper understanding of concepts, methods, and applications. Hints for solving the exercises are given at the end of the book.

1.1 Information Retrieval

1.1.1 Brief History of Information Retrieval

In this section, an attempt is made to give a concise summary of the history (Wellish 1991, Lesk 2007) of IR, which may help one to understand exactly what is meant by IR (and perhaps also to clear up some misunderstandings and confusion according to which IR is a synonym for data mining, or data retrieval, or library science, albeit that it stems from them and shares some aspects with them even today).

Very briefly, and yet very broadly, but very exactly, *information retrieval* means *finding relevant information in a store of information.* In other words, in principle, IR does not mean finding any information that we happen to come across or information we are fortunate enough to discover by chance without having anything particular in mind. IR means that we already have a need for information that we are able to formulate, and then find relevant items in a store (collection) of items.

1.1.1.1 Table of Contents

Try to imagine that our scientific books, journals, or teletexts lack tables of contents. Then imagine how you find out when a particular TV program begins or on which page a specific article in, say, *Scientific American*, begins. It would be possible, but is it not easier to use a table of contents? And yet, there were times when there was no such thing as a table of contents: e.g., there were none for the clay boards produced in ancient Babylon that contained arithmetical calculations.

In ancient Greece and the Roman Empire papyrus scrolls were used to record data in a written format, and scholars found it useful to devise a means of organizing the material to make locating certain sections of text easier. For example, Pliny the Elder (around 70 CE) produced a table of subjects (similar to what we call today a table of contents) to his work *The Natural History in 37 Books*. The method of tables of contents was first used by Valerius Soranus in the second century BCE. Thus, we may assume it had been employed earlier by the Greeks.

The method of tables of contents to retrieve information is an ancestor of what we call today inverted file structure, a most important data structure that is used extensively today in computer science as well as IR.

1.1.1.2 Alphabet

We all take it for granted that, e.g., people's names can be put in alphabetical order. But try to imagine that you are looking for someone's telephone number in a London telephone directory in which the names are not ordered. How would you go about searching? How long would it take you to find the phone number?

However, alphabetization did not always exist. This special type of ordering strings of characters was probably invented, or rather devised, by Greek scholars in the third century BCE at the Library of Alexandria (Egypt) in order to better organize the large body of Greek literary work. Today alphabetization serves as the basis for, and is an ancestor of, one of the most important and widely used algorithms in computer science (as well as in IR), namely sorting.

1.1.1.3 Hierarchy

Try to imagine that you are looking for soup recipes in a cookbook that contains all kinds of recipes for French and Hungarian food in general, but is not divided into sections nor organized into headings or groups of foods. How would you find the recipes you are looking for? How long would it take?

Organizing written material into a hierarchy of groups (chapters, sections, headings, paragraphs) is a useful means by which it is possible to consult the material more easily. In the first and second century CE, Roman scholars (e.g., Valerius Maximus, Marcus Julius Frontinus, and Aulus Gellius) used to group and organize their written work into books and divide the books into chapters with headings.

The hierarchical organization of written material to ease the retrieval of information is an ancestor of one of the most important data structures in computer science in general and in information retrieval in particular, and one of the most important concepts in graph theory, namely tree.

1.1.1.4 Index

Nowadays, if someone is looking for a particular author (and his works) in many libraries around the world, then he/she naturally uses a computer network. However, there may be readers who remember the days when this was not the case—a time when there were no computers. In the precomputer era (and still in many of today's libraries), one way was to simply ask the librarian. But there was another way as well: the use of indexes. This meant going to the big index board, pulling out index shelves, and browsing through paper cards containing authors' names in alphabetical order.

The use of indexes goes back to ancient Rome, when indexes meant slips attached to papyrus scrolls that contained the title of the work. Thus, each scroll on the shelf could be identified without having to pull it out. With time, indexes could also include an abstract of the work.

Papyrus scrolls did not have page numbers, leaf numbers, or line numbers. (*Note*: The modern successor of papyrus scrolls is/was the microfilm.) On the other hand, even if some works were produced in several copies (sometimes on the order of hundreds), no two copies were exactly the same.

The invention of printing (in the fifteenth century) made it possible to have page numbers and any number of exact copies. This, in turn, enabled the compilation of reliable indexes (e.g., in books on herbals). The first indexes were not fully alphabetized (the words were ordered only on their first letter). Full alphabetization only became the rule in the eighteenth century.

Indexing is a method by which the exact location (chapter, page, line, record, sector, etc.) of an identifier (word, term, name, subject, code, etc.) in a unit (book, file, database, etc.) can be given, and it served as a basis of and is an ancestor of one of the most important data structures, used, e.g., in computer science, database systems, Web search engines, and IR, namely inverted file structure.

1.1.1.5 Inception of Information Retrieval

In the late 1940s, the United States military carried out an indexing of wartime scientific research documents captured from Germany. In the 1950s, the Soviet Union sent up the first artificial Earth satellite. This exploit was interpreted a sign of the "science gap" between the United States and the Soviet Union and led to the realization that little was known in the United States about Russian science and technology, and it served to motivate American funding of research in mechanized literature search.

During this period, among other results:

- The citation indexing method was invented (E. Garfield), which is at the basis of link analysis methods used in today's Web search engines.
- The terms "information retrieval" and "descriptor" were coined (C. Mooers).
- Term occurrences were introduced to represent a piece of text as a sequence of pairs (t_i, f_i), $i = 1,...,n$, where t_i denotes a term occurring in text, and f_i is the number of times t_i occurs in the text (H. P. Luhn, KWIC index).

1.1.1.6 Models

The 1960s and 1970s witnessed a boom in IR experimentation, which yielded the most important measures as well as measurement principles (known as Cranfield paradigm) used to evaluate the relevance effectiveness of a retrieval method (i.e., how well a retrieval method or system performs) under laboratory conditions: precision and recall (C. Cleverdon). They have been in use ever since.

The recognition that computing methodology (methods, algorithms, software, programming) and technology (hardware) made it possible to identify terms and to perform Boolean operations on sets of data automatically yielded systems that allowed full text searches. As a result, computerized commercial retrieval systems (e.g., Dialog and BRS) were developed, which implemented what we call today the Boolean model of IR (based on mathematical logic and set theory).

Later, more sophisticated theoretical frameworks, or models, were elaborated. They served and serve as bases for the development of retrieval methods and have motivated an enormous body of research:

- Probabilistic model (based on probability theory: M. E. Maron, J. L. Kuhns, S. Robertson, C. J. van Rijsberegen, C. Spärck-Jones).
- Vector space model (based on linear algebra: G. Salton, M. Lesk, J. Sammon),

Later, in the 1980s and 1990s, further models were proposed:

- Information logic (based on logical imaging: C. J. van Rijsbergen).
- Fuzzy model (based on fuzzy set theory: D. H. Kraft, B. E. Boyce, D. A. Buell).
- Language model (based on linguistics and probability theory: W. B. Croft, J. Ponte).
- Inference network model (H. Turtle, W. B. Croft).
- Associative interaction model (based on the Copenhagen interpretation in quantum mechanics: S. Dominich).

1.1.1.7 World Wide Web

At the end of 1980s, the World Wide Web (Web or WWW for short) was proposed (T. Berners-Lee). The Web is a worldwide network of electronic documents stored in computers belonging to the Internet (which is a worldwide computer network).

By the 1990s, many IR models, methods, and algorithms were already known, a huge amount of research had gone into IR, many experimental retrieval systems had been used and tested in IR laboratories, and a large body of experimental results had accumulated. However, their application in practice, in large and real retrieval systems meant to be used by groups of real people at large was yet to come.

And the Web offered just that opportunity.

In the late 1990s, retrieval systems, called Web search engines, appeared. They implement many features and results obtained in IR and enable many people around the world to search for information in an ever-growing collection of electronic documents.

The Internet and the Web have made it possible to more easily design and use intranet systems, i.e., dedicated retrieval systems for a specific company, university, or organization.

At the same time, the Internet, the Web, and search engines have definitely changed the way we (should) think about IR. IR is, like most nascent fields, interdisciplinary. The Web has taught us that if we want to search for information in data stored in computer memories successfully, then we should considerably enlarge our understanding of IR so as to encompass:

- Fields other than those that traditionally belong to IR (information science, library science, linguistics, etc.).
- Fields that would not have been previously considered to be in the mainstream of IR, such as mathematics, algorithms, computational complexity, physics, neural networks, etc.
- Completely new technologies developed, run, and maintained by important information companies (search engine companies).

1.1.2 "Definition" of Information Retrieval

Let us start this section by giving a widely accepted formulation for the meaning of the term information retrieval. This may seem superfluous, or unnecessarily meticulous (first of all for specialists). However, it will prove useful to review different definitions given over a time span of more than 40 years. Thus, the meaning of the term IR can be made as precise as possible inline with our present understanding of the field.

- Salton (1966) defines IR as follows: "*The SMART retrieval system takes both documents and search requests in unrestricted English, performs a complete content analysis automatically, and retrieves those documents which most nearly match the given request.*"
- Van Rijsbergen (1979) gives the following definition: "*In principle, information storage and retrieval is simple. Suppose there is a store of documents and a person (user of the store) formulates a question (request or query) to which the answer is a set of documents satisfying the information need expressed by this question.*"
- Some years later, Salton (1986) phrased it as follows: "*An automatic text retrieval system is designed to search a file of natural language documents and retrieve certain stored items in response to queries submitted by the user.*"
- Meadow et al. (1999) defined IR as follows: "*IR involves finding some desired information in a store of information or database. Implicit in this view is the concept of selectivity; to exercise selectivity usually requires that a price be paid in effort, time, money, or all three. Information recovery is not the same as IR...Copying a complete disk file is not retrieval in our sense. Watching news on CNN...is not retrieval either...Is information retrieval a computer activity? It is not necessary that it be, but as a practical matter that is what we usually imply by the term.*"
- Berry and Browne (1999) formulated it as follows: "*We expect a lot from our search engines. We ask them vague questions ... and in turn*

anticipate a concise, organised response. ... Basically we are asking the computer to supply the information we want, instead of the information we asked for. ... In the computerised world of searchable databases this same strategy (i.e., that of an experienced reference librarian) is being developed, but it has a long way to go before being perfected."

- Baeza-Yates and Ribeiro-Neto (1999) wrote: "*In fact, the primary goal of an IR system is to retrieve all the documents which are relevant to a user query while retrieving as few non-relevant documents as possible.*"
- Belew (2000), within his cognitive FOA (finding out about) framework, formulated retrieval in a pragmatic way: "*We will assume that the search engine has available to it a set of preexisting, 'canned' passages of text and that its response is limited to identifying one or more of these passages and presenting them to the users.*
- A few years later, Baeza-Yates (2003) formulated a definition similar to his earlier view: "*IR aims at modelling, designing, and implementing systems able to provide fast and effective content-based access to large amounts of information. The aim of an IR system is to estimate the relevance of information items to a user's information expressed in a query.*"

What can be seen form the above definitions? The answer is that, in essence, the meaning of the term IR has remained the same over the last 40 years (practically from its inception). Thus, we may say that IR is concerned with (typically using computer systems) the organization, storage, retrieval, and evaluation of information relevant to a user's information need.

The user (researcher, tourist, etc.) has an information need (e.g., articles published on a certain subject, travel agencies with last-minute offers, etc.). The information need is expressed in the form of a query, i.e., in a form that is required by a computer program (e.g., according to the syntax of some query language). The program then retrieves information (journal articles, Web pages, etc.) in response to the query.

Thus, the meaning of the term IR may be formulated formally as the following mapping:

$$IR : (U, IN, Q, O) \rightarrow R,$$

where

- U = user
- IN = information need
- Q = query
- O = collection of objects to be searched
- R = collection of retrieved objects in response to Q

The information need IN is, obviously, more than its expression in a query Q. IN comprises the query Q plus additional information about the user U. The additional information is specific to the user (e.g., spoken languages, fields of interest, preferred journals, etc.). The importance of additional information is found in that it is one factor in relevance judgment, i.e., when the user is judging whether a retrieved object is relevant or not to his/her specific IN. The additional information is obvious for the user (he/she implicitly assumes it) but not for the computerized retrieval system. Thus, we may say that the additional information is an implicit (i.e., not expressed in Q) information I specific to the user U, and we may write $IN = (Q, I)$.

With this, a stricter reformulation of the meaning of IR is the following: IR is concerned with finding a relevance relationship $\Re$ between object O and information need IN; formally:

$$IR = \Re(O, IN) = \Re(O, (Q, I)).$$

In order for an IR system to find such a relationship $\Re$ it should be possible to also take into account the implicit information I, and ideally the information that can be inferred from I to obtain as complete a picture of user U as possible. Finding an appropriate relationship $\Re$ would mean obtaining (deriving, inferring) those objects O that match the query Q and satisfy the implicit information I. With these, the notion of IR is formally rewritten as follows:

$$IR = \Re(O, (Q, \langle I, \vdash \rangle)),$$

where $\langle I, \vdash \rangle$ means I together with information inferred (e.g., in some formal language or logic) from I. Relationship $\Re$ is established with some (un)certainty m, and thus we may write that (Dominich 2001):

$$IR = m[\Re(O, (Q, \langle I, \vdash \rangle))].$$

Thus, in essence, *information retrieval is a kind of measurement* in that it is concerned with measuring the relevance of an item stored in computer memory to a user's information request (and then returning the items sorted in descending order based on their measure of relevance). All IR frameworks, methods, and algorithms aim at as good a measurement as possible.

Note: A few words on the meaning of the term "information" are in order.[1] The notion of energy was separated from the notion of matter when man was able to construct and use equipment that converted one type of energy into another type (e.g., the steam engine). Likewise, with the advent of computer systems, the notion of information is being used, more and more, as a distinct idea, separated from the notions of energy and matter. One may distinguish different forms of information: genetic information (carried by DNA), human information (generated in our brains), computer information, information in crystals, etc. There is no absolute consensus on the meaning of the term "information." In everyday life, information is usually used as a synonym for data, knowledge, news, experience, or facts. Indeed, human information may be conceived as that which humans perceive, create, or convey, without checking its validity or reliability. In one word, information is that which is susceptible to be known. In contrast, human knowledge is information that possesses order. Meaning should be distinguished from information. It may be viewed as being information that is interpreted in a given context. For example, a book contains information, regardless of whether someone is reading it or not, or whether we understand its language or not. But it gains meaning only when we are able to read it and place it into some context. In IR, whether it is data, meaning, knowledge, or information that is being retrieved constitutes a subject for debate. Generally and pragmatically speaking, however, IR deals with human information, and is concerned with measuring the degree of similarity between information need and items in a collection of information (practically regardless of whether the information need and the items are, at a philosophical level, more information, or rather meaning, or perhaps knowledge.).

[1] Stonier, T. (1990). *Information and the Internal Structure of the Universe.* Springer Verlag, London.

1.2 Retrieval Methods

Many retrieval methods have been elaborated since the inception, about half a century ago, of the field of IR. These can be categorized in several ways. For example:

- Classical methods (Boolean, vector space, probabilistic).
- Nonclassical methods (information logic, situation theory, associative).
- Alternative (or hybrid) methods (e.g., cluster, fuzzy, latent semantic, artificial neural network, genetic algorithm, natural language processing, knowledge base).
- Web methods (link analysis, browsing, visualization, etc.).

Albeit that these methods differ from one another, they nonetheless have properties in common. Thus, in general:

- They are typically based on text in the sense that terms are identified (associated with) in the documents (objects) to be searched, and the number of occurrences of terms is used to represent numerically (using so-called weights) the content of documents (objects).
- The query is conceived as being a piece of text. (Hence, it too is represented by weights of terms.)
- Apart from their content, the importance of Web pages stems from another source as well: from the fact that they can be linked to each other (and thus form a network).
- The relevance degree (or similarity) between a query and a document (or a Web page) is established by a numeric computation based on weights and/or link importance.

The ways in which the document weights, the importance of Web pages, and the similarity values are computed depend on the mathematical framework in which the retrieval method used is being based. Typically, different mathematical frameworks are used in different methods.

For an example: The Boolean retrieval method is based on set theory and mathematical logic in that:

- Documents are represented as sets of terms.
- The query is conceived as a Boolean expression of terms.
- The retrieved document set is obtained by performing the set operations corresponding to the logical operators occurring in the query.

As another example, in the vector space retrieval method,

- Both the documents and the query are represented as vectors in the linear space of terms.
- The similarity values are based on the scalar product between the query vector and document vectors.

In this book, after reviewing (in Chapter 2) the fundamental mathematical notions (mathematical logic, set theory, relations theory) used in IR, the well-known major retrieval methods are described, namely:

- Boolean method
- Vector space method
- Generalized vector space method
- Probabilistic method
- Language model method
- Inference model method
- Impact factor method
- Connectivity method
- Mutual citation method
- PageRank method
- HITS method
- SALSA method
- Associative interaction method
- Bayesian methods

The major retrieval methods are particularly important in that many other retrieval methods are based on them. The methods are described in such a way that they can be understood and applicable themselves as well. Thus, they are helpful for a wide range of readers from newcomers through students and educators to researchers and system developers coming from a variety of fields such as information science, computer science, mathematics, engineering, linguistics, etc. The exact description of every method is given in detail. For every method, a complete example is also given to help enhance understanding.

However important the knowledge of the major retrieval methods, a clear understanding of their usefulness and application possibilities is equally important. For this reason, in Chapter 4, the basics of IR technology are presented:

- Identification of terms.
- Power law.
- Stoplisting.

- Stemming.
- Weighting schemes.
- Term-document matrix.
- Inverted file structure.
- Typical architecture of a retrieval system.
- Web characteristics.
- General architecture of a Web search engine.
- Architecture of a Web metasearch engine.
- Measures of retrieval effectiveness.
- Laboratory measurement of retrieval effectiveness (precision-recall graph).
- Measurement of relevance effectiveness of Web search engines.

The basics of IR technology are useful for practitioners as well for more theoretically minded readers, and they also enable a better understanding of the experimental results reported throughout the book.

1.3 Modern Algebra

In order to acquire a better and a more complete understanding of the approach as well as of the results described in this book, it is useful to know exactly what is meant by the term "modern algebra."

First, in just a few words, it all started many thousands of years ago with solving equations and has ended up nowadays with structures and mapping between them. But this may be just a new beginning. In order to delineate the meaning of the term modern algebra, we note very succinctly a few of the milestones in its long history.

1.3.1 Equations

Everyday life (agriculture, land surveying, financial operations, commerce, etc.) has always generated equations—even in ancient times in Babylon, Egypt, and Greece. For instance, in ancient Babylon, about 4000 years ago, agriculture and land surveying gave rise to problems such as: *The length and width of a piece of land is equal to 30, while its area is equal to 221. What are the length and the width of the piece of land?* In today's notation, this problem is rewritten as the following (system of) equations:

$$xy = 221, \ x + y = 30,$$

where x denotes length and y denotes width. Finding solutions to equations was, in general, a difficult task. Today, we would solve this problem as follows:

- Obtain x from the second equation: $x = 30 - y$.
- Substitute this into the first equation: $(30 - y)y = 30y - y^2 = 221$.
- Solve the equation $30y - y^2 - 221 = 0$ to obtain the roots y_1 and y_2.
- Use the relationship $x = 30 - y$ to obtain the value x_1 corresponding to y_1 (and x_2 corresponding to y_2).

However, it is very instructive to see how this problem was solved in ancient Babylon. In today's notation, the solution was obtained as follows (symbols to denote unknowns and operations were not used; computation procedures were expressed as a sequence of sentences):

1. Let $x = 30/2 + u$, $y = 30/2 - u$. (Note that, with this notation, $x + y = 30$).
2. Then, $xy = 30^2/4 - u^2 = 900/4 - u^2 = 225 - u^2 = 221$.
3. From this, $u = \sqrt{4} = 2$. [The computation of the square root of a number was known in ancient Babylon and has been used ever since. In today's notation, in order to compute $\sqrt{a}$, guess a value x_0 such that $1 < x_0 < \sqrt{a}$. Then, calculate $x_1 = 0.5(x_0 + a/x_0)$ to obtain a better approximation of $\sqrt{a}$. An even better approximation is $x_2 = 0.5(x_1 + a/x_1)$, and so on.]
4. Hence, $x = 30/2 + u = 15 + 2 = 17$, and $y = 30/2 - u = 15 - 2 = 13$.

We can see that this is a very ingenious technique.

However, there were equations for which a formula was known and was used to compute the roots. For example (again, in today's notation), the second-degree equation $13x^2 + 60x = 13{,}500$ was solved using the formula that we know and use today:

$$\frac{-b \pm \sqrt{b^2 - 4ac}}{2a} = \frac{-60 \pm \sqrt{60^2 - 4 \times 13 \times (-13500)}}{2 \times 13}.$$

Unfortunately, we do not know how this formula was arrived at.

1.3.2 Solving by Radicals

Having a formula to calculate the roots of an equation has some clear theoretical and practical advantages:

- The formula is an expression containing the coefficients of the equation, and no other variable.
- The formula uses some or all of the well-known operations on numbers: addition, subtraction, multiplication, division, and radicals. No other operation occurs. This is referred to as *solving by radicals*.
- There is no need to look for tricks to solve the equation.
- Thus, root computation becomes a mechanical process.

Legend tells us that in ancient Greece, during a plague epidemic, the gods asked that the size of the altar stone to be doubled in order to stop the epidemic. The architects were faced with the problem of carving a new stone altar having its height (and length and width) equal to the root of the equation $x^3 = 2$. They did not succeed. This problem, known today as "doubling the cube," has since become very famous. (More probably, however, the problem originates in the geometrical interpretation of the root of the equation $x^3 = 2$ coming from ancient Babylon. It has since been shown that this equation does not have rational roots; hence, doubling the cube cannot be solved.) Even in ancient Greece, many tried to find a solution. One solution involved solving the third-degree equation $2 = (d - 2)^3$.

The computation of the roots of an equation of third degree was done in different ways. For example, in the Middle Ages:

- In China, a method based on polynomial division was used.
- In Arabian mathematics, it meant finding the intersection points of a parabola with a hyperbola.

The formula that allowed one to solve a third-degree equation by radicals is the result of the efforts of Italian mathematicians (Scipione del Ferro, Ludovico Tartaglia, Girolamo Cardano) in the sixteenth century. Cardano showed that (in today's notation) any third-degree equation $ax^3 + bx^2 + cx + d = 0$ can be transformed into a simpler form $y^3 + py + q = 0$ with the substitution $x = y - b/(3a)$, where $p = c/a - b^2/(3a^2)$, $q = 2b^3/(27a^3) - (bc)/(3a^2) + d/a$. Tartaglia gave the following formula, using radicals for the equation $y^3 + py + q = 0$:

$$y_1 = u + v,\ y_{2,3} = -\frac{u+v}{2} \pm i\frac{u-v}{2}\sqrt{3},$$

where

$$u, v = \sqrt[3]{-\frac{q}{2} \pm \sqrt{\left(\frac{q}{2}\right)^2 + \left(\frac{p}{3}\right)^3}}\ .$$

Around 1540, the Italian mathematician Ludovico Ferrari offered a solution by radicals (which looked even more complicated than the one above) to the fourth-degree equation $x^4 + ax^3 + bx^2 + cx + d = 0$.

Also in the sixteenth century, owing to the work by the English mathematician Thomas Harriot, the Dutch mathematician Albert Girard, and the French mathematician François Viète, the relationships (known today as Viète's relationships) between the roots and the coefficients of an n-degree equation $a_1x^n + a_2x^{n-1} + \ldots + a_nx + a_{n+1} = 0$ were established. For example, when $n = 2$, $x_1 + x_2 = -a_2/a_1$, $x_1x_2 = a_3/a_1$.

At the end of the eighteenth century, the German mathematician Carl F. Gauss proved that the n-degree equation $a_1x^n + a_2x^{n-1} + \ldots + a_nx + a_{n+1} = 0$ has exactly n roots in the set $\mathbb{C}$ of complex numbers (a result known as the fundamental theorem of algebra today).

By the eighteenth century, as a result of the work by Viète, the French mathematician René Descartes, the English mathematician Isaac Newton, the Swiss mathematician Leonhard Euler, and Gauss, algebra became an established branch of mathematics concerned with solving equations.

Many methods for computing roots were known, and many properties were discovered. And yet, the big question remained unanswered:

Can a fifth-degree or higher-degree equation be solved by radicals?

1.3.3 Birth of Modern Algebra

Euler studied the rational functions (i.e., radical expressions that contain only addition, subtraction, multiplication, division, radicals) of the roots of the n-degree equation. He noted that if a rational function of the roots is invariant with respect to all the permutations of the roots, then this function is a rational function of the coefficients of the equation. Joseph L. Lagrange, a French mathematician, believed that the answer to this question would be found by studying the group of the permutations of roots. Around 1771 he conjectured that the answer to the "big" question was negative. Following in the footsteps of Euler and Lagrange, the Italian mathematician Paolo Ruffini gave a proof of this conjecture at the end of the eighteenth century. Unfortunately, the proof was incomplete. The Norwegian mathematician Niels H. Abel gave another (alas, also incomplete) proof in 1826.

It was and is clear that many n-degree equations can be solved by radicals. The problem of solvability by radicals is a general one; more specifically: When can an n-degree equation be solved by radicals? Is there a formula that allows one to compute the roots of any n-degree equation?

Complete and correct answers to these questions were given by the French mathematician Évariste Galois around 1830. It is worth sketching the line he followed, which helps us to understand how modern algebra started and the essence of the idea behind it.

For simplicity, let us consider a third-degree equation. By the fundamental theorem of algebra, it has exactly three roots: x_1, x_2, and x_3. A permutation of roots is represented as a table with two rows and as many columns as the number of roots. For example,

$$\begin{pmatrix} 1 & 2 & 3 \\ 2 & 3 & 1 \end{pmatrix}$$

means that 1 stands for x_1, 2 stands for x_2, 3 stands for x_3, and any element in the first row is replaced by the element below it in the second row (e.g., instead of 3 we write 1). The number of all permutations of n elements is equal to $n! = 1 \times 2 \times 3 \times \ldots \times n$. Thus, the number of permutations of all the roots of a third-degree equation is $3! = 6$. The set P of all permutations can be arranged as a matrix:

$$p_{11} = \begin{pmatrix} 1 & 2 & 3 \\ 1 & 2 & 3 \end{pmatrix} \quad p_{12} = \begin{pmatrix} 1 & 2 & 3 \\ 2 & 1 & 3 \end{pmatrix} \quad p_{31} = \begin{pmatrix} 1 & 2 & 3 \\ 3 & 1 & 2 \end{pmatrix}$$

$$p_{21} = \begin{pmatrix} 1 & 2 & 3 \\ 1 & 3 & 2 \end{pmatrix} \quad p_{22} = \begin{pmatrix} 1 & 2 & 3 \\ 3 & 2 & 1 \end{pmatrix} \quad p_{23} = \begin{pmatrix} 1 & 2 & 3 \\ 2 & 3 & 1 \end{pmatrix}$$

Let the consecutive application of two permutations be denoted by $\otimes$, e.g., $p_{12} \otimes p_{22} = p_{23}$ (i.e., 1 transforms into 2 in p_{12}, 2 transforms into 2 in p_{22}, and so 1 is transformed into 2 as in p_{23}, and so on). It can be shown that the operation $\otimes$ has the following properties:

- It is an *internal operation*, i.e., for any two permutations, it yields another existing permutation.
- The permutation p_{11} leaves every permutation in place, i.e., $p_{11} \otimes p_{ij} = p_{ij}$, for every permutation. (It behaves like the number 1 with respect to the multiplication of numbers.) p_{11} is the *neutral element* of $\otimes$.
- For every permutation p_{ij}, there is a permutation p_{uv} such that $p_{ij} \otimes p_{uv} = p_{11}$. In other words, for every permutation there is an *inverse permutation* (just like, e.g., $1/a$ for the number a with respect to multiplication).
- It can be shown that $\otimes$ is *associative*.

Galois called the set P of permutations together with the operation $\otimes$, i.e., the structure $(P, \otimes)$, a *group*, and this concept has been in use ever since.

The coefficients $a_1, a_2, \ldots, a_{n+1}$ of the n-degree equation belong to a set C of rational, real, or complex numbers. In this set, addition (+) and multiplication ($\times$) are defined and satisfy well-known properties. In other words, the triple $(C, +, \times)$ forms a well-defined structure (namely, field) given by the properties of + and $\times$.

Galois showed that there is a connection between the following two structures:

- The group $(P, \otimes)$ of permutations of the roots of the n-degree equation.
- The field $(C, +, \times)$ to which the coefficients of the equation belong.

He also delineated the conditions under which the n-degree equation can be solved by radicals.

Galois's work is very important from two points of view:

- Practical point of view: He solved the problem of solvability of equations by radicals.
- Theoretical point of view: He laid down the foundations of a new type of reasoning and approach in science, namely by considering structures (i.e., sets endowed with operations satisfying certain properties) and mappings between them.

1.3.4 Modern Algebra

In the nineteenth century, approaches based on structures multiplied. In 1854, the English mathematician George Boole showed that Aristotle's laws of human thought can be expressed using mathematical symbols and operations that form a structure. This is known today as Boolean algebra, and is usually represented as the sets of all subsets of a set together with the set union, intersection, and complementation.

In 1888, the Italian mathematician Giuseppe Peano defined an axiomatic foundation of natural numbers as a structure of elements generated by certain operations. The German mathematician David Hilbert did the same for geometry in 1899. (Of course, we should not forget Euclid, who, as far as we know, was the first to provide an axiomatic foundation for geometry more than 2000 years ago in ancient Greece.)

Around 1890, the German mathematician Richard J. W. Dedekind observed that the greatest common divisor and the least common multiple are in number theory what logical conjunction and logical disjunction are in logic (i.e., they form Boolean algebras). This led him to create what is called today "lattice theory": the theory of a very general and abstract structure (see the next section). He showed that a Boolean algebra is a special type of lattice.

We will not say more regarding the development of structure-based approaches, as what we have noted is, we think, enough for the reader to see that since the nineteenth century, when algebra meant the study of solving equations, the meaning of the term "algebra" has been changed (more exactly, has been expanded). Today it means *the study of abstract structures and the mappings between them*, i.e., the study of collections of objects in general, endowed with operations or relations satisfying certain properties. As the properties can be studied without regard to the particular elements of the sets involved, the terms "abstract algebra" or "modern algebra" are now current. In order to emphasize the importance of modern algebra, it is sufficient to note that entire fields of science, e.g., the various geometries, different areas in physics, functional analysis, and tensor calculus all deal with algebraic structures (at an abstract level).

We end this section by noting that one of the goals of this book is to show that there is a strong connection between IR and modern algebra, provided primarily through one of the basic notions in the latter—the lattice—in the sense that it will be shown that the major retrieval methods can all be based upon this concept.

1.4 Lattice

"Never in the history of mathematics has a mathematical theory been the object of such vociferous vituperation as lattice theory." So begins one of the papers[2] Gian-Carlo Rota, one of the prominent figures in lattice theory.

The hostility toward lattices began when Dedekind published the papers that first gave birth to the theory. Kronecker, one of his contemporaries, wrote in a letter: Dedekind lost "his mind in abstraction."

So, what is a lattice? A lattice is a "well-behaved" set (i.e., a certain collection of objects). It is well behaved for three reasons:

- There is an order among its objects (to be exact, the order is partial, but this is irrelevant at this point).
- For any two objects, there is one that is 'greater' than (or 'equal' to) both of them.
- For any two objects, there is one that is 'smaller' than (or 'equal' to) both of them.

[2] The Many Lives of Lattice Theory. *Notices of the AMS*, vol. 44, no. 11, pp: 1440–1445.

The drawing below is a (typical) visual representation of a network of points that form a lattice. In the middle row of three points, any two points 'converge' (follow the links) into one point (one upward and one downward):

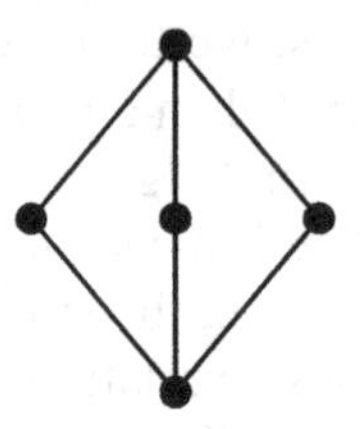

As opposed to the drawing above, the one below shows a network of points that do not form a lattice. In the two middle rows of the two pairs of points, any two points 'converge' to both points in the other row, which is not permissible in a lattice:

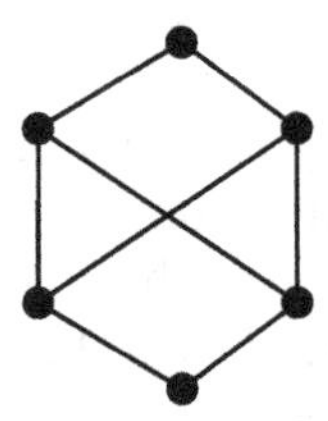

In order to make the concept of lattice clearer, let us consider, as an example, possible prices for bread (prices are given in Hungarian forints, though this is now actually irrelevant):

$$17, 32, 110, 164, 210, 255, 280, 320.$$

The prices are in an ascending order (from left to right) to better visualize the order that characterizes the collection of prices. It can be seen that whichever two prices we take, there is a price that is higher and one that is lower. Mathematically, bread prices, when gathered into one ordered collection (i.e., in a set), may be viewed as forming a lattice.

This may sound trivial. Yet it is very important from a practical point of view, because it makes it possible:

- To answer questions like "Which type of bread is cheaper?," or "Which is the most expensive bread?"
- To introduce prices into the database of an accounting system and to answer a question like "Give me the names of bread types whose prices are between 100 and 400."

1.5 Importance of Lattices

What we said about lattices in the preceding section is common to every lattice.

In theoretical as well as practical applications, lattices having certain additional properties are of real interest and use. Let us consider an example. It is known that the position and velocity of a car (or ship, etc.) can be measured simultaneously and with any desired precision. However, in quantum mechanics, e.g., the position and velocity of an electron cannot be measured simultaneously with any degree of accuracy. Thus, it turns out that there are quantities in nature that cannot always be measured simultaneously with any precision.

What is to be done?

First, both situations should be accepted as being aspects of reality. Second, if that is the case, then we may not think about (or describe) all the parts of reality using the same "scheme" (reasoning structure). It seems that different structures (ways of reasoning) are needed for different parts of reality (situations).

It was shown that what is common in the two situations above (car and electron) is that in both cases the structure of our thinking (the way in which we may logically combine propositions expressing measurements) can be formally expressed as a lattice.

What is different in the two situations, however, is that the lattices used are not exactly the same: they differ in certain properties (to be somewhat more exact, in the distributive law). The two drawings below are visual representations of the two kinds of lattices that show the structure of the thinking (reasoning) that we have to follow in these two situations. The lattice on the left represents the structure of our thinking (i.e., the structure of the corresponding logic) in situations of the first type (e.g., in everyday life), while that lattice on the right corresponds to the structure of our thinking (i.e., the structure of the corresponding logic) in situations of the second type (e.g., the subatomic world).

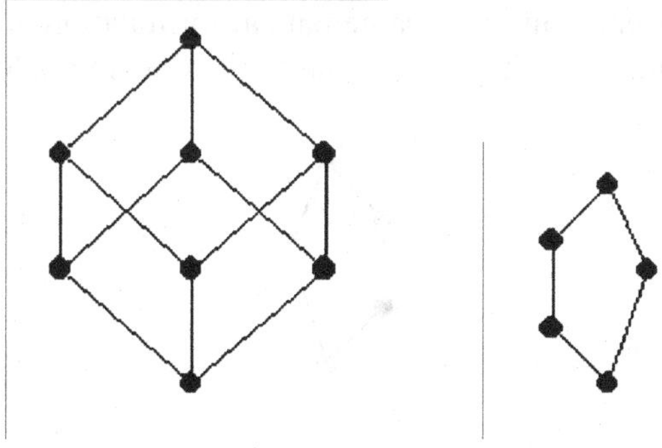

Once we have the formal models (lattice), we can enjoy some advantages, e.g.:

- A better understanding of phenomena.
- The ability to make predictions (e.g., for future measurements, we can be sure that the position and velocity of a car (or ship, etc.) can always be measured, whereas we cannot even hope—at our current level of understanding of nature—that we will ever be able to measure the position and velocity of subatomic particles simultaneously with any precision).

In Chapter 3, the concept of lattice as well as certain properties (that prove useful in IR) are presented in detail (and in such a way that they can be understood even by beginners). Every notion and property is illustrated by clarifying examples.

1.6 Lattices in Information Retrieval

1.6.1 Retrieval Systems

Retrieval systems that apply lattices are described in Chapter 5, namely:

- Moors
- FaIR
- BR-Explorer
- FooCA
- Rajapakse-Denham

A method to transform a term-document matrix into a concept lattice is also described.

Further, a detailed mathematical treatment of the properties of the lattices applied in these retrieval systems is presented. Perhaps the primary advantage of this treatment, in a mathematical formulation is that these lattices are not modular, i.e., they are similar to the drawing below.

This result sheds light upon the structure of the logic that underlies these systems. Moreover, it is far from intuitive. Reasoning with documents and queries (i.e., the structure of the logic applied) may be different than the structure of reasoning with propositions in mathematical logic. In the latter, whatever operation we perform on propositions always and necessarily leads to another proposition. Furthermore, any two propositions are compatible with one another. The study of the lattices applied in retrieval systems reveals that documents, queries, or terms are not always and necessarily compatible with one another: subjecting them to logical operations may yield an entity having a different quality or nature.

1.6.2 Boolean Retrieval

The Boolean retrieval method is a very important one in that it is widely used in database systems (e.g., Oracle, SQL) and World Wide Web search engines. In principle, it is a simple method, but all the more important for that.

The Boolean retrieval method (both formally and using an example) and the application of lattices in Boolean retrieval are described in Chapter 6. An efficient way to answer Boolean queries is presented as well.

1.6.3 Vector Space Retrieval

After the description of the required mathematical concepts and results (Chapter 7), the application of lattices in vector space retrieval is presented and discussed (Chapter 8):

- Calculation of meaning.
- Queries with disjunction.
- Compatibility of relevance assessments.
- Vector space method as lattice-lattice mapping.

It is shown that, unlike the nonmodular character of the lattices applied in retrieval systems (Chapter 5), the lattices used in vector space retrieval are modular but are not generally distributive; i.e., they look like (or are equivalent to) the following drawing:

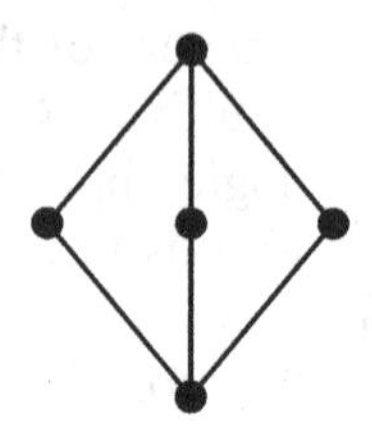

Further, we show that the very character of the underlying mechanism of vector space retrieval is nonsubmodularity. As is shown in a parallel to the lattices applied in quantum mechanics, whose underlying mechanism (i.e., way of reasoning) is characterized by orthomodularity, the application of lattices in vector space retrieval is characterized by nondistributivity and nonsubmodularity.

Quantum mechanics and vector space retrieval have a common "weakest link": distributivity does not always hold (i.e., their entities are not always compatible with one another). In other words, their assertions do not always commute. However, as compared to quantum mechanics, the underlying mechanism of vector space retrieval has an additional ingredient: a nonsubmodular lattice-lattice mapping, which means that the logic of retrieval has a more sophisticated structure than the logic of quantum mechanics. Retrieval is a very special correspondence between two different types of lattices (i.e., between two different types of logic), one for the objects to be searched (which is well-behaved and nicely organized) and another for queries (which is not so well-behaved, this being perhaps an expression of free will).

1.6.4 Fuzzy-Algebra-Based Retrieval Methods

In Chapter 9, after introducing the necessary mathematical concepts and properties (from tensor algebra, fuzzy set theory, fuzzy algebra), three retrieval methods are described: fuzzy cardinality, fuzzy entropy, and fuzzy probability. Experimental results are reported to show that they yield increased retrieval effectiveness [when compared to traditional vector space and latent semantic indexing (LSI) methods].

It has long been known that the linear space is not generally an adequate mathematical framework for IR. In Chapter 9, the principle of invariance (PI) is described. According to PI, documents may or may not preserve their identities when looked at from different points of view. It is shown that PI together with the notion of fuzzy cardinality form a correct mathematical framework for the traditional vector space retrieval method, from which the latter can be formally (and hence correctly) obtained.

1.6.5 Probabilistic Retrieval

Probabilistic retrieval methods are based on the concept of conditional probability. The nonbinary model, language model, and inference model retrieval methods are described in detail, together with examples for each.

It has long been known that whether the documents, queries, relevance, and irrelevance form a σ-algebra or not (for the conditional probabilities to have sense from a mathematical point of view) is questionable. Thus, probabilistic retrieval methods need another mathematical framework to support them. In Chapter 10, we show that the notion of lattice offers such a framework and how the lattice of logical implications from mathematical logic offers a correct mathematical background for probabilistic retrieval.

1.6.6 Web Retrieval and Ranking

After introducing the notion of a Web graph and discussing degree distribution, the basic methods using link structure analysis are presented:

- Impact factor
- Connectivity
- Mutual citation
- PageRank
- HITS
- SALSA
- Associative
- Belief network
- Inference network

Clarifying examples are provided for each one and a connection between HITS and LSI is described.

Then, using the results obtained for lattices in Chapter 9, we present an aggregated method for Web retrieval based on lattices. This method allows one to determine the importance of pages taking into account both their link importance (using link analysis) and their intrinsic importance (stemming from page content). Experimental evidence for the relevance effectiveness of this method is also given in terms of comparisons with commercial search engines (Google, Altavista, Yahoo!).

After introducing the notion of Web lattice and chain, we define Web ranking as a lattice-lattice function between a Web lattice and a chain. We show that ranking is not submodular. Then, global ranking is defined as a lattice-lattice function (i.e., a mapping from the direct product of Web lattices to the chain [0; 1]). It is shown that global ranking is not submodular.

Based on the notion of global ranking, a global ranking method is given that enables one to compute the global importance of a Web page at Web level taking into account the importance of the site that the page belongs to, but without the need to consider the entire Web graph of all pages.

After proving that any tree as well as any document can be transformed into a lattice, we show that the DocBall model and Galois (concept) lattice representations of a document are equivalent to one another.

Based on these results as well as on the fact that the structure of any site is a lattice, we present a method to compute site importance.

1.7 Exercises and Problems

Exercises and problems are found at the end of every chapter (apart from Chapters 1 and 7). These are, of course, IR-oriented, and they are designed to help the reader better understand and deepen his/her knowledge of the concepts and methods discussed as well as their applications in practice. There are hints for solving them at the end of the book.

2 Mathematics Basics

You cannot conceive the many without the one.
(Confucius)

This chapter presents the concepts, operations, and properties of mathematical logic (proposition, negation, conjunction, disjunction, implication, equivalence), set theory (sets, set operations), and relations theory (binary relations, functions, equivalence relations, posets) that are being applied in modern computerized (IR) and which are used in the modern algebra of IR.

Every notion and property is illustrated by several examples, which are designed to enhance understanding. Some are purely mathematical, but the rest are taken from everyday life or have an IR flavor.

Apart from examples, a number of exercises and problems are also proposed at the end of the chapter. They are IR-oriented, and are included to improve understanding and show how logic, sets, and relations are/can be applied in IR. Solutions are given at the end of the book (in Chapter 12).

At the end of the chapter, the literature referred to, as well as recommended, is listed.

2.1 Elements of Mathematical Logic

The main goal of logic has always been the study of reasoning, proof, and truth.

In the eighteenth century, the German mathematician Gottfried W. Leibniz utilized an algebraic language to express logical ideas. At the end of the nineteenth century, thanks to logicians such as, e.g., Gottlob Frege and George Boole, a formal language, consisting of just a few symbols, was developed that allowed for the writing of mathematical assertions.

The ultimate principles of human reasoning (syllogism, excluded middle, etc.) as well as its basic concepts (the notion of number, the axioms of geometry, etc.) have remained—and probably will remain—the same over time. Hardly anyone will doubt them.

Mathematical logic is concerned with the study (using formal means) of the structural properties of the correct deduction of conclusions. Here we present the fundamentals of mathematical logic. They are important in that they constitute the formal mathematical basis of, and are used intensively in, modern computerized IR.

To a reader acquainted with the cold facts of reality, or exclusively technically minded, this part of the book may seem, in places, less friendly, or too abstract, or perhaps bizarre. However, those readers should understand that the symbols and operations of mathematical logic have well-defined meanings and that their power is in their ability to capture and express many different aspects of reasoning in both science and everyday life in a unified formal way.

The reader already familiar with mathematical logic can skip this chapter.

2.1.1 Proposition

A *proposition* is a statement (formulation, assertion) that can be assigned either a value T or a value F (there is no third alternative), where T and F are two different values, i.e., $T \neq F$. For example, T = true, F = false (these values are used throughout this book), or T = yes, F = no, or T = white, F = black, or $T = 1$, $F = 0$. The values T and F are referred to as *truth values*. A proposition cannot be true and false at the same time (*principle of noncontradiction*).

Example 2.1

- "*I am reading this text.*" is a — true — proposition.
- The sentence "*The sun is shining.*" is also a proposition because either the value *T* or *F* can be assigned to it.
- The sentence "*The cooks wearing red hats are playing football at the North Pole*" becomes a proposition if a truth value can be assigned to it. □

In general, it is not a necessary quality of an assertion or proposition that it be true. For example, the proposition "*It is raining*" may be true or false. However, there are propositions that are "absolutely" true (e.g., "*The year 2001 is the first year of the twenty-first century.*")

2.1.2 Negation

The *negation* of a proposition P is a proposition denoted by $\neg P$ and pronounced "not P." If P is true, then $\neg P$ is false, and if P is false, then $\neg P$ is true (**Table 2.1**). Hence, $\neg(\neg P)$ is always P (*law of double negation*).

Table 2.1. Truth Table of Logical Negation

P	$\neg P$
T	F
F	T

Example 2.2

"*I am not reading this text*" is a—false—proposition, and is the negation of the proposition "*I am reading this text.*" □

2.1.3 Conjunction

Given two propositions: P, Q, the proposition denoted by $P \land Q$ (expressed as "P and Q") is called a *conjunction*. The conjunction is true if and only if both P and Q are true, and false otherwise (**Table 2.2**). Thus, $P \land (\neg P)$ is always false (*law of contradiction*).

Table 2.2. Truth Table of Logical Conjunction

P	Q	$P \land Q$
T	T	T
T	F	F
F	T	F
F	F	F

Example 2.3

- "*I am reading this text* ∧ *It is raining*" is a proposition, and its truth value can be assigned by the reader.
- "*I am thinking to myself* ∧ *A bicycle has two wheels*" is a proposition (the reader can assign a truth value to it), albeit that one would rarely link its two constituent propositions into one sentence in everyday speech. □

2.1.4 Disjunction

Given two propositions P, Q, the proposition denoted by $P \vee Q$ (expressed as "P or Q") is called a *disjunction.* The disjunction is false if and only if both P and Q are false, and true otherwise (**Table 2.3**). Thus, $P \vee (\neg P)$ is always true (*law of the excluded third*).

Table 2.3. Truth Table of Logical Disjunction

P	Q	$P \vee Q$
T	T	T
T	F	T
F	T	T
F	F	F

Example 2.4

"*I am reading this text* ∨ *It is raining*" is a true proposition (regardless of whether it is actually raining or not). □

2.1.5 Implication

Given two propositions P, Q, the proposition denoted by $P \Rightarrow Q$ (expressed as "P implies Q") is called an *implication* (alternate notation: $P \rightarrow Q$). The implication is false if and only if P is true and Q is false, and true otherwise (**Table 2.4**). The values of the implication $P \Rightarrow Q$ coincide with the values of the disjunction $\neg P \vee Q$, which can be easily checked using **Tables 2.3** and **2.4**.

Table 2.4. Truth Table of Logical Implication

P	Q	$P \Rightarrow Q$
T	T	T
T	F	F
F	T	T
F	F	T

Example 2.5

- "*I am reading this text* $\Rightarrow$ *It is raining*" is a proposition (its truth value depends on whether it is actually raining or not).
- "*I am reading this text* $\Rightarrow$ *I am not here*" is a false proposition.
- "*I am not reading this text* $\Rightarrow$ *The circle is a square*" is a true proposition (albeit that one would rarely formulate such a sentence in everyday speech).
- "*I am not reading this text* $\Rightarrow$ *The Sun does not exist in 2007*" is a true proposition. □

In Section 10.6, we have more to say about implication (also called a *material conditional*).

2.1.6 Equivalence

Given two propositions P, Q, the proposition denoted by $P \Leftrightarrow Q$ (expressed as "P is equivalent to Q") is called *equivalence* (alternate notation: $P \leftrightarrow Q$). The equivalence is true if and only if both P and Q have the same truth values, and false otherwise (**Table 2.5**).

The truth values of the equivalence $P \Leftrightarrow Q$ coincide with the truth values of the conjunction $(P \Rightarrow Q) \wedge (Q \Rightarrow P)$, which can be easily checked using the respective truth tables.

Table 2.5. Truth Table of Logical Equivalence

P	Q	$P \Leftrightarrow Q$
T	T	T
T	F	F
F	T	F
F	F	T

The following equivalences are very important and useful. (Their proofs can be easily given using the corresponding truth tables.):

- Law of contraposition: $(P \Rightarrow Q) \Leftrightarrow (\neg Q \Rightarrow \neg P)$
- $(P \Rightarrow Q) \Leftrightarrow (\neg P \vee Q)$
- De Morgan's laws: $\neg(P \wedge Q) \Leftrightarrow (\neg P) \vee (\neg Q)$
  $\neg(P \vee Q) \Leftrightarrow (\neg P) \wedge (\neg Q)$

Example 2.6

"*I am not here now* $\Leftrightarrow$ *The circle is a square*" is a true proposition. □

Note: As can be seen, negation, disjunction, and conjunction are enough to express any logical expression (because both implication and equivalence can be written using negation and disjunction).

2.2 Elements of Set Theory

It is widely believed that every mathematician should learn set theory. This is true, but it is also widely believed that only they should learn set theory, but this latter notion is a delusion (see below).

Set theory not only allows one to group and thus talk about entities under consideration and express operations among them in a compact form, but it also represents a formal way of dealing with logical aspects of structures (structures of objects in general). In this respect, it is enough to say that, e.g., the subsets of a set have the same formal structure as the logical propositions. In a word—they are equivalent. Thus, set theory is a very useful formal tool for engineers and information scientists alike. This part of the chapter presents the concepts and operations in set theory that are the most useful in IR.

2.2.1 Set

The notion of set is a fundamental one, but it does not have a mathematical definition. A *set* is a collection of distinct objects. The objects in a set are called *elements*. If an object x is an *element of* a set S (equivalent formulation: x *belongs to* S), this is denoted as $x \in S$. The term $x \notin S$ means that x *does not belong* to S.

It is very important to note that:

- An element can occur at most once in a set.
- The order of the elements in a set is unimportant.

A set can be given by enumerating its elements between brackets, e.g., $A = \{a_1, a_2,...,a_n\}$, or by giving a property $P(x)$ [e.g., using a predicate $P(x)$; see Section 2.3.3 for the notion of predicate] that all elements must share: $A = \{x \mid P(x)\}$. A set having a fixed number of elements is *finite*, and *infinite* otherwise. An *empty set* contains no elements and is denoted by $\varnothing$.

Example 2.7

- $\mathbb{N} = \{1, 2,\ldots,n,\ldots\}$ denotes the set of natural numbers.
- $\mathbb{Z} = \{..., -2, -1, 0, 1, 2, ...\}$ denotes the set of integer numbers.

- $\mathbb{Q}$ denotes the set of rational numbers.
- $\mathbb{R}$ denotes the set of real numbers.
- $\mathbb{C}$ denotes the set of complex numbers.
- {thought, ape, quantum, Rembrandt} is a set.
- {mammal | water content of mammal's milk is less than 20%} is a set. □

There are two important *quantifiers* that are used extensively in set theory (as well as in mathematics, logic, and formal disciplines in general):

1. *Universal quantifier,* which is denoted by $\forall$ and means *for every*, *for any*.
2. *Existential quantifier,* which is denoted by $\exists$ and means *there exists* (at least one), *there is* (at least one).

Example 2.8

- $\exists\, x \in \{1, 2, 3\}$ such that x is an even number.
- $\forall$ set A $\exists$ set B such that $A = B$. □

2.2.2 Subset

If all the elements of a set B belong to a set A, then B is called a *subset* of A (**Fig. 2.1**); this is denoted by $B \subseteq A$, i.e.,

$$B \subseteq A \Leftrightarrow (\forall x \in B \Rightarrow x \in A). \tag{2.1}$$

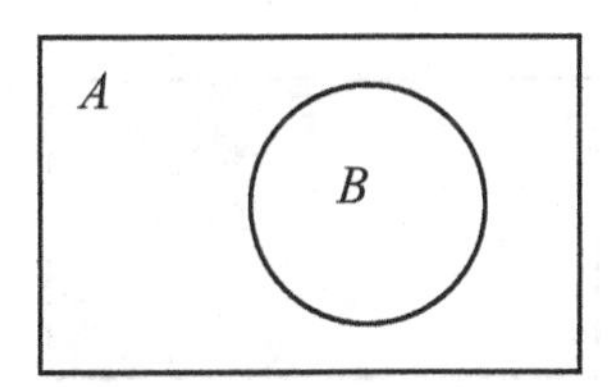

Fig. 2.1. Visualization of the notion of subset, $B \subset A$.

$B \subset A$ denotes the fact that B is a *proper subset* of A, i.e., all the elements of B belong to A, but A also has other elements:

$$B \subset A \Leftrightarrow ((\forall x \in B \Rightarrow x \in A) \wedge (\exists y \in A \Rightarrow y \notin B)). \tag{2.2}$$

Note that the empty set $\varnothing$ is a subset of any set A, i.e., $\varnothing \subseteq A$.

2.2.3 Equality of Sets

The *equality* of sets A and B is denoted by the symbol = and defined as

$$A = B \Leftrightarrow ((A \subset B) \wedge (B \subset A)), \tag{2.3}$$

i.e., A and B have exactly the same elements.

Example 2.9

{thought, ape, quantum, Rembrandt} = {thought, Rembrandt, quantum, ape}. Note that the order of elements in a set does not matter. □

2.2.4 Set Union

The *union* of sets A and B is denoted by the symbol $\cup$ and defined as (**Fig. 2.2**)

$$A \cup B = \{x \mid (x \in A) \vee (x \in B)\}. \tag{2.4}$$

Example 2.10

{thought, ape, quantum, Rembrandt} $\cup$ {1, 2} = {thought, ape, quantum, Rembrandt, 1, 2}. Note that the operation of union is a purely formal one (just like the other set operations); it does not require that the elements of the sets be compatible with each other or have the same nature in any way. □

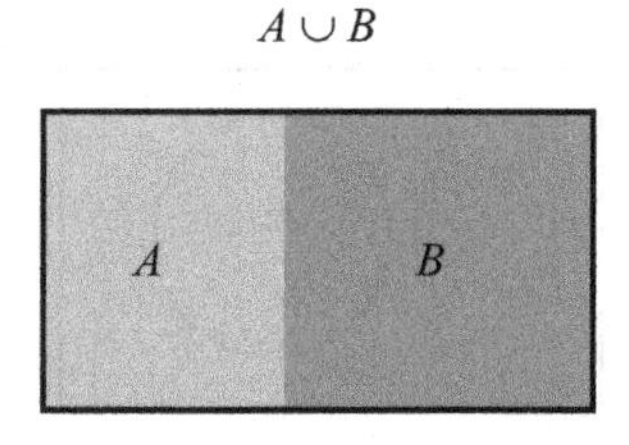

Fig. 2.2. Visualization of set union $A \cup B$.

Set union satisfies the following properties (as can be easily checked using the definitions of sets equality and union):

- Commutativity: $A \cup B = B \cup A$, for any two sets A, B.
- Associativity: $A \cup (B \cup C) = (A \cup B) \cup C$, for any three sets A, B, C.
- Idempotency: $A \cup A = A$, for any set A.

2.2.5 Set Intersection

The *intersection* of sets A and B is denoted by the symbol $\cap$ and defined as **(Fig. 2.3)**

$$A \cap B = \{x \mid (x \in A) \wedge (x \in B)\}. \tag{2.5}$$

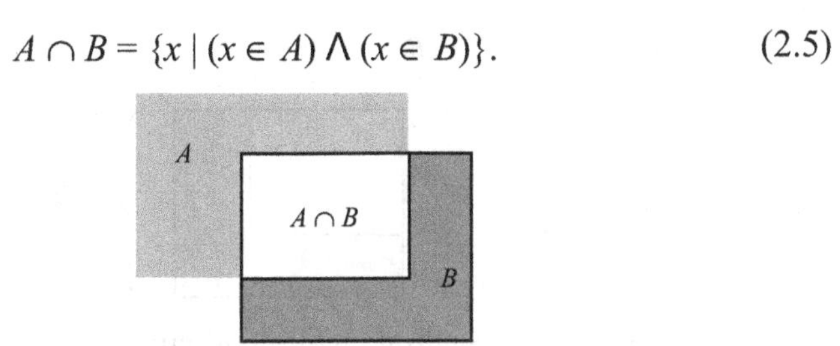

Fig. 2.3. Visualization of set intersection $A \cap B$.

If $A \cap B = \varnothing$, A and B are said to be *disjoint* sets **(Fig. 2.4)**.

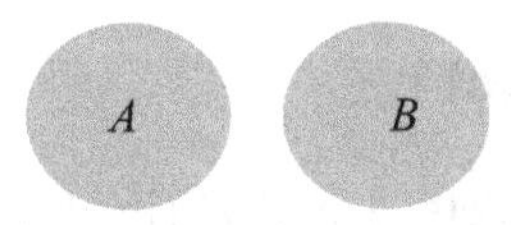

Fig. 2.4. Visualization of the disjoint sets A and B.

Example 2.11

{thought, ape, quantum, Rembrandt} $\cap$ {thought, Rembrandt, 1, 2} = {thought, Rembrandt}. Note that the result of the intersection consists of the elements that are exactly the same. □

Set intersection satisfies the following properties (as can be easily checked using the definitions of set equality and intersection):

- Commutativity: $A \cap B = B \cap A$, for any two sets A, B.
- Associativity: $A \cap (B \cap C) = (A \cap B) \cap C$, for any three sets A, B, C.
- Idempotency: $A \cap A = A$, for any set A.

2.2.6 Set Difference

The *difference* of sets A and B (in this order) is denoted by the symbol \, and is defined as **(Fig. 2.5)**

$$A \setminus B = \{x \mid (x \in A) \wedge (x \notin B)\}. \tag{2.6}$$

Example 2.12

{thought, ape, quantum, Rembrandt} \ {thought, Rembrandt, 1, 2} = {ape, quantum}. □

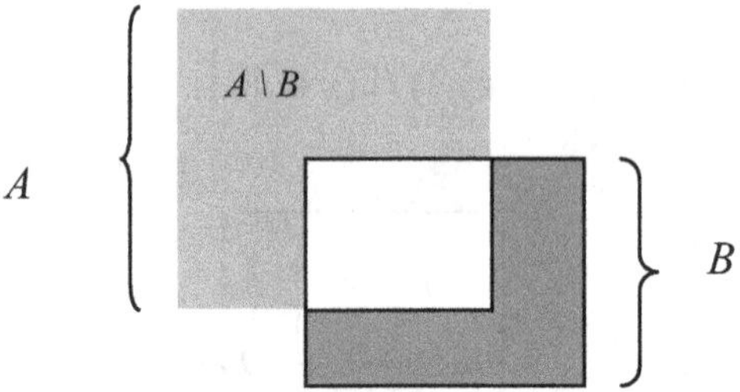

Fig. 2.5. Visualization of set difference $A \setminus B$.

We note that, in general, $A \setminus B \neq B \setminus A$ (i.e., set difference does not commute, just like, e.g., the subtraction of numbers).

2.2.7 Cartesian Product

The *Cartesian product* of sets A and B (in this order) is denoted by the symbol $\times$ and defined as (**Fig. 2.6**)

$$A \times B = \{(a, b) \mid (a \in A) \wedge (b \in B)\}. \tag{2.7}$$

We note that $A \times B \neq B \times A$ if $A \neq B$.

Example 2.13

{thought} × {1, 2} = {(thought, 1), ((thought, 2)}. Note that the pairs of a Cartesian product are "ordered" pairs, i.e., the pair (thought, 1) is not the same as the pair (1, thought), and thus the latter pair is not an element of this Cartesian product. □

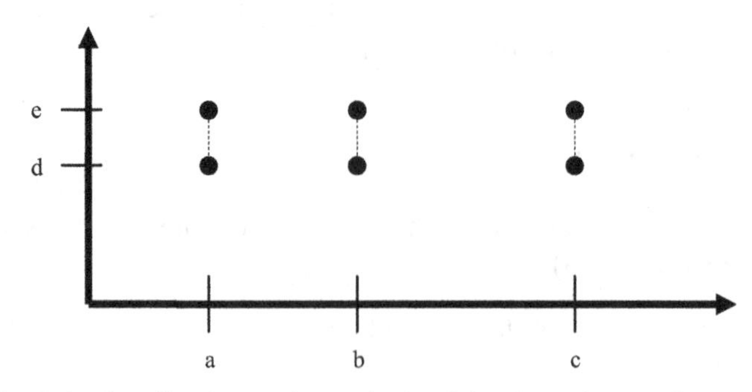

Fig. 2.6. Visualization (using points) of the Cartesian product $A \times B = \{a, b, c\} \times \{d, e\} = \{(a, d), (a, e), (b, d), (b, e), (c, d), (c, e)\}$.

2.2.8 Set Complement

Let $A \subseteq B$. The *complement* $\complement_B A$ of set A relative to set B is defined as (**Fig. 2.7**)

$$\complement_B A = \{x \mid (x \in B) \wedge (x \notin A)\} = B \setminus A. \tag{2.8}$$

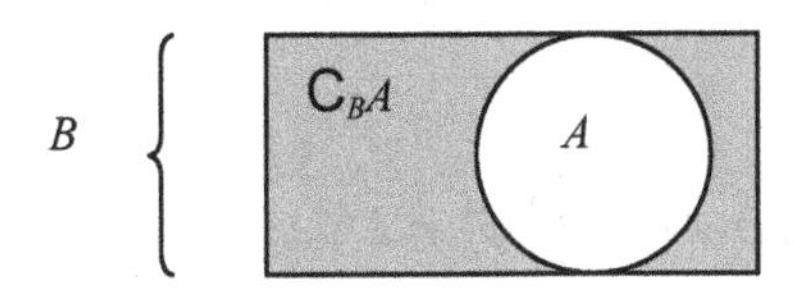

Fig. 2.7. Visualization of set complement $\complement_B A$.

Example 2.14

$\complement_{\{\text{thought, ape, quantum}\}}\{\text{thought}\} = \{\text{ape, quantum}\}$. □

2.2.9 Powerset

The *powerset* $\wp(A)$ of a set A is defined as $\wp(A) = \{X \mid X \subseteq A\}$, i.e., the set of all subsets of A. The empty set $\varnothing$ is a member of the powerset of any set A, i.e., $\varnothing \in \wp(A)$.

Example 2.15

$\wp(\{\text{thought, ape, quantum}\}) = \{\varnothing, \{\text{thought}\}, \{\text{ape}\}, \{\text{quantum}\}, \{\text{thought, ape}\}, \{\text{thought, quantum}\}, \{\text{ape, quantum}\}, \{\text{thought, ape, quantum}\}\}$. □

2.2.10 Cardinality of Set

The *cardinality* of a set A is denoted by $|A|$ and defined (from a practical point of view) as the number of elements it contains. The cardinality of a finite set A having n elements is denoted as $|A| = n$, whereas that of an infinite set B is equal to infinity, i.e., $|B| = +\infty$.

The cardinality of powerset $\wp(A)$ is equal to $|\wp(A)| = 2^n$, where $|A| = n$. Indeed, the elements of $\wp(A)$ are:

- The empty set $\varnothing$.
- The subsets of A containing the elements of A one by one, two by two, and so on.
- The set A itself.

That is, if $A = \{a_1, a_2,\ldots,a_n\}$, then

$\wp(A) = \{\varnothing, \{a_1\}, \{a_2\},\ldots,\{a_n\}, \{a_1, a_2\},\ldots,\{a_1, a_n\},\ldots,\{a_1, a_2, a_3\},\ldots, A\}$.

Thus,

$$|\wp(A)| = \sum_{k=0}^{n} C_n^k = 2^n, \tag{2.9}$$

where C_n^k denotes the combinations of n taken by k.

Example 2.16

From Example 2.15: $|\{\varnothing$, {thought}, {ape}, {quantum}, {thought, ape}, {thought, quantum}, {ape, quantum}, {thought, ape, quantum}$\}| = 2^3 = 8$. □

2.2.11 Properties of Set Operations

Apart from the properties given thus far, the following properties (holding for any set A, B, C) are also important and applied in IR:

- Distributivity: $A \cap (B \cup C) = (A \cap B) \cup (A \cap C)$,
  $A \cup (B \cap C) = (A \cup B) \cap (A \cup C)$,
- Absorption: $A \cap (A \cup B) = A$, $A \cup (A \cap B) = A$,
- De Morgan's laws (see Section 2.1.6 for an analogy in mathematical logic):

$$\complement_A(B \cap C) = \complement_A(B) \cup \complement_A(C).$$
$$\complement_A(B \cup C) = \complement_A(B) \cap \complement_A(C).$$

2.3 Elements of Relations Theory

The aim of this section is to present the major concepts of relations theory, with an emphasis on ordering and equivalence relations (as applied in IR).

However trivial the word "order" might sound, it is all important in theoretical sciences as well in practical applications. In order to illustrate this, let us consider the following situation. We assume that we are given the following objects:

$$O_1 = \text{ball}, O_2 = \text{photograph}, O_3 = \text{building},$$

$$O_4 = \text{spaceship}, O_5 = \text{blouse}.$$

At first glance, we may have the impression that considering (or even mentioning) these objects together (i.e., in one set) is strange and that it hardly makes any sense. However, order can be introduced among them. For example, it is possible to calculate a price P_i for every object O_i:

$$O_1 \text{ has price } P_1, O_2 \text{ has price } P_2, O_3 \text{ has price } P_3,$$

$$O_4 \text{ has price } P_4, O_5 \text{ has price } P_5.$$

Prices are real numbers, so they can be ordered, e.g., as follows:

$$P_2 \leq P_5 \leq P_1 \leq P_3 \leq P_4.$$

Using the ordering of prices, we can define an order among the objects themselves :

$$\text{photograph} \leq \text{blouse} \leq \text{ball} \leq \text{building} \leq \text{spaceship}.$$

In this way, a structure (namely, an order) has been introduced among our objects. This structure makes it possible to deal with and talk about them from a different perspective. For example:

- It has now become possible to answer the question "Which is the cheapest object?"
- It has also become possible to incorporate these objects into an accounting system.

2.3.1 Binary Relations

Given two sets A, B, a *binary relation* R is a subset of the Cartesian product $A \times B$, i.e., $R \subseteq A \times B$. A is called the *domain* and B is called the *codomain* of R. The fact that $(x, y) \in R$ can also be denoted by xRy (which should be read as "x is in relation R with y").

Example 2.17

- {(thought, 1)} is a relation of the Cartesian product {thought} × {1, 2} = {(thought, 1), ((thought, 2)}.
- Let A denote the set of words of language L_1 (e.g., Hungarian) and B the set of words of language L_2 (e.g., English). Then, the structure of a

bilingual dictionary D (Hungarian-English) can be modeled as a binary relation $D \subseteq A \times B$, i.e., as pairs of corresponding words. □

2.3.2 Function

Let A and B denote two sets. A *function* f defined over set A with values in set B is a binary relation $f \subseteq A \times B$ for which $\forall a \in A\ \exists b \in B$ such that afb. The function is usually denoted as f: $A \to B$, $f(x) = y$, where $x \in A$, $y \in B$. The way in which the relation f (i.e., the mapping of x onto y) is performed, or constructed, falls outside our scope (this generally depends on the application or problem being considered).

Example 2.18

- f: $\mathbb{R} \to \mathbb{R}_+$, $f(x) = x^2$ is a function (its graphic representation is a parabola).
- g: $\{1, 2, 3\} \to \{4, 9, 11\}$, $g(2) = 4$, $g(3) = 9$ is not a function because it does not assign any value to 1.
- $f = \{(1, 4), (2, 4), (3, 9)\} \subseteq A \times B = \{1, 2, 3\} \times \{4, 9, 11\}$ is a function. In the usual notation, one writes the following:

$$f\text{: } \{1, 2, 3\} \to \{4, 9, 11\}, f(1) = 4, f(2) = 4, f(3) = 9.$$

$$f: A \longrightarrow B$$

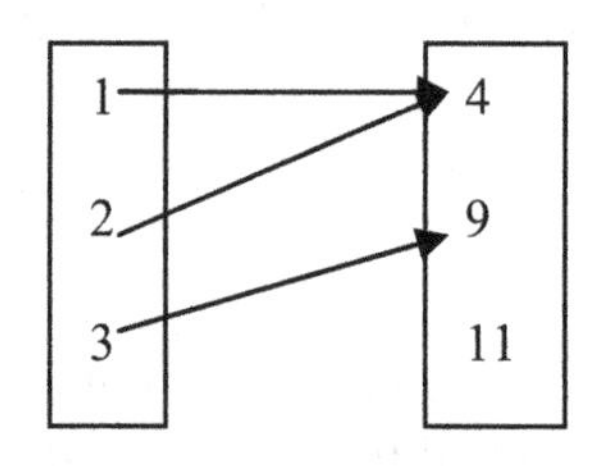

□

A function f: $A \to B$ is

- Surjective if $\forall b \in B\ \exists a \in A$ such that $f(a) = b$.
- Injective if $\forall a_1, a_2 \in A$, $a_1 \neq a_2$, it follows that $f(a_1) \neq f(a_2)$.
- Bijective if it is surjective and injective.

2.3.3 Predicate

Let A denote an arbitrary set. A *predicate* is a function $Pred: A \rightarrow \{T, F\}$ such that $Pred(a)$ is a proposition for every a from A, i.e., $Pred(a)$ is assigned exactly one of the values T or F.

Example 2.19

Let A be the set of the names of all members of a family. For example, A = {William, Anne, Edward, John, Eve, Deirdre}. Then, the function assigning a marital status to everybody in the family is a predicate $Pred: A \rightarrow \{married, not\ married\}$. For example:

- *Pred*(William) = *married* (i.e., in words: William is married.),
- *Pred*(Deirdre) = *not married* (i.e., in words: Deirdre is not married.). □

2.3.4 Equivalence Relation

A binary relation $R \subseteq A \times A$ is an *equivalence relation* if it satisfies the following conditions:

- Reflexivity: $xRx, \forall x \in A$.
- Symmetry: $xRy \Rightarrow yRx, \forall x, y \in A$.
- Transitivity: $((xRy) \land (yRz)) \Rightarrow xRz, \forall x, y, z \in A$.

Example 2.20

- Let x denote a positive rational number: $x \in \mathbb{Q}_+$. Let us define the following rounding rule: if number x has a nonzero fractional part, then the number is rounded to the integer number immediately greater than x. For example, 6.17 = 7. The rounding rule is an equivalence relation in the set $\mathbb{Q}_+$. The elements of the set $\{x \mid n < x \leq n + 1, n \in \mathbb{N}\}$ are equivalent to each other.

- The relation "brother/sister of" may be viewed as an equivalence relation between people (if we are allowed to say, for the sake of reflexivity, that anyone is a brother/sister of him/herself). □

2.3.5 Ordering Relation

A binary relation $R \subseteq A \times A$ is an *ordering relation* if it satisfies the following conditions:

- Reflexivity: $xRx, \forall x \in A$.
- Transitivity: $((xRy) \wedge (yRz)) \Rightarrow xRz, \forall x, y, z \in A$.
- Antisymmetry: $((xRy) \wedge (yRx)) \Rightarrow x = y, \forall x, y \in A$.

Example 2.21

- The relation $\leq$ (meaning: "less than or equal to") is an ordering relation in the set $\mathbb{R}$ of real numbers.

- In the set $\mathbb{N}$ of natural numbers, the relation "divides" (e.g., 3 divides 12), denoted by $|$, is an ordering relation. □

2.3.6 Partially Ordered Set

A set A with an ordering relation R defined on it is called a *partially ordered set* (or *poset*, for short). Notation: (A, R).

Example 2.22

From Example 2.21, the following structures are posets:

- $(\mathbb{R}, \leq)$

- $(\mathbb{N}, |)$

Let A denote a set. Then, the structure $(\wp(A), \subseteq)$ is a poset. Indeed, one can easily check, using the corresponding definitions, that $\subseteq$ is reflexive (i.e., $A \subseteq A$), transitive (i.e., if $A \subseteq B$ and $B \subseteq C$, then $A \subseteq C$), and antisymmetric (i.e., if $A \subseteq B$ and $B \subseteq A$, then $A = B$). □

2.3.7 Partition

The *partition* of a set A is given by mutually disjoint subsets $X_1, X_2,...,X_n$ (of A):

$$A = X_1 \cup X_2 \cup ... \cup X_n \qquad (2.10)$$
$$X_i \cap X_j = \varnothing,\ i = 1, 2,...,n, j = 1, 2,...,n,\ i \neq j.$$

Any equivalence relation R on a set A provides a partitioning of A into mutually disjoint equivalence classes (all the elements belonging to the same class are equivalent to each other).

Example 2.23

- The sets $\{x \mid n < x \leq n + 1, n \in \mathbb{N}\}$ in Example 2.20 are equivalence classes, and they provide a partition of the set $\mathbb{Q}_+$ of positive rational numbers.
- A relation R is referred to as a *preordering* relation if it is reflexive and transitive. For example, transportation priorities in a logistics system form a preordering relation. Such a system can become very complicated in practice, but its complexity can be reduced by decomposing it into equivalence classes. □

2.4 Exercises and Problems

1. Is the sentence "*The sun will shine over the Niagara waterfall on the April 10, 6045*" a proposition?
2. What is the negation of the proposition "*The Sun is shining*"? Is it "*The Sun is not shining*? Is it "*It is not the Sun that is shining*"?
3. When is the sentence "*If John is a liar, then Peter is a truth teller*" true?
4. Let P denote the following proposition P = "*I wear a necktie,*" and Q denote the following proposition Q = "*I am elegant.*" Write the sentence "*I am elegant because I wear a necktie*" using the formalism of mathematical logic.
5. Let us assume that you go to library (or you search the World Wide Web) in order to read newspaper articles on the art of dancing. For example, waltz, rock and roll, tango, salsa, but you are not interested in salsa as a Web ranking method. Express the information you using the formalism of mathematical logic.

6. Let us assume that $J_1,...,J_n$ denote Web pages (or journal papers). Let the URL (title) of the page (paper) J_i be T_i, and its posting date (publication date) be D_i ($i = 1,...,n$). Turn the set of pages (papers) into a poset on their posting date, and then on the length (length is equal to number of words) of their URLs. What do you observe?
7. Let us assume that you are given the task to design an ordered structure for toys in a kindergarten. How would you order the toys? Is there just one way to order them? Can you find more than one way to create order among the toys, i.e., to define posets of toys?
8. Let us assume that $B = \{B_1,...,B_n\}$ denotes a set of Web pages (or books). Let $T_1,...,T_m$ denote the terms appearing in them. Create a partition P of set B if all the pages (books) in which the term T_i has the same number of occurrences are equivalent to each other. Do this for every $i = 1, 2,...,m$. What do you observe?
9. Let $B_1,...,B_n$ denote all the Web pages of the World Wide Web, or all the books held in a library. Do they form a set?

2.5 Bibliography

Dumitriu, A.: *History of Logic* (Abacus Press, Kent, 1977)

Enderton, H. B.: *A Mathematical Introduction to Logic* (Academic Press, New York, 1972)

Halmos, P. R.: *Naive Mengenlehre (Naïve Sets Theory)* (Vandenhoek and Ruprecht, Goettingen, 1968)

Kneale, W., and Kneale, M.: *The Development of Logic* (Oxford University Press, 1962)

Krivine, J. L.: *Introduction to Axiomatic Set Theory* (D. Reidel, 1971)

Kurtz, M.: *Handbook of Applied Mathematics for Engineers and Scientists* (McGraw-Hill, New York, 1991)

Makowsky, J. A.: Model theory and computer science: an appetizer. In: *Handbook of Logic in Computer Science vol 1 Background and Mathematical Structures*, ed by Abramsky, S., Gabbay, D. M., and Maibaum, T. S. E. (Clarenden Press, Oxford1992)

Mendelson, E.: *An Introduction to Mathematical Logic* (Van Nostrand Reinhold, Princeton, NJ, 1964)

3 Elements of Lattice Theory

Lattice theory will play a leading role in the mathematics of the twenty-first century.
(Gian-Carlo Rota)

In this chapter, we discuss concepts and properties that pertain to lattice theory (lattice, poset, duality, Hasse diagrams, atomicity, modularity, distributivity, complementation, orthocomplementation, orthomodularity, Boolean algebra, important lattices) that are applied in the theory of information retrieval (IR), in the development of IR systems, and in the major retrieval methods. Further concepts and properties of lattices are introduced in subsequent chapters, as they become relevant.

By important lattices we mean those lattices that are widely applied in IR, namely: the Boolean algebra of the powerset, of the logical propositions, and of the logical predicates, and further the distributive lattice of logical implications.

Every property is discussed and proved in detail. Many examples and figures are given in order to help the reader grasp the concepts and properties presented.

Apart from examples, there are exercises and problems that are IR-oriented and are designed to improve understanding and show how basic properties of lattices are/can be applied in IR. Solutions are given at the end of the book.

A bibliography on the theory of lattices is included at the end of the chapter.

3.1 Lattice

The notion of lattice represents one of the basic structures in the modern theory of algebraic structures (next to, e.g., groupoid, group, factor, ring, field, linear space, poset, sequence, clan, incline, etc.). The concept of lattice has important applications in several mathematical disciplines (e.g., topology, functional analysis). It is, at the same time, an interesting notion in that it is the only concept that denotes both a relational and an algebraic (i.e., operations-based) structure.

The German mathematician Richard Dedekind wrote a book based on the notes he took of Dirichlet's lectures, which was first published in 1863. In the later 1893 edition, there is a section about lattices (axioms, modular law, duality, distributive law, free lattices). Apparently, Dedekind used lattice terminology in an 1877 paper well before he published the second edition of his book. Dedekind noted that lattices were discussed by Ernst Schröder in an 1880 volume, and that that led him to consider nonmodular lattices. However, the founder of modern lattice theory was Garrett Birkhoff, who proposed it in a book first published 1940, which went through several editions.

Given a set L, two operations (i.e., functions), denoted by $\wedge$, called *meet*, and $\vee$, called *join*, are expressed as

$$\wedge: L \times L \to L, \tag{3.1}$$

$$\vee: L \times L \to L.$$

The structure $(L, \wedge, \vee)$ is called a *lattice* if the following properties hold:

- Commutativity: $A \wedge B = B \wedge A$, $A \vee B = B \vee A$, $\forall A, B \in L$.
- Associativity: $A \wedge (B \wedge C) = (A \wedge B) \wedge C$, $\forall A, B, C \in L$.
  $A \vee (B \vee C) = (A \vee B) \vee C$, $\forall A, B, C \in L$.
- Absorption: $A \wedge (A \vee B) = A$, $A \vee (A \wedge B) = A$, $\forall A, B \in L$.

It is worth noting that:

1. First, absorption is the only property that connects the meet and the join.
2. Second, any lattice L is at the same time a commutative semigroup with respect to the meet and the join (a semigroup is a structure $(G, *)$ in which the operation $*$ is associative).
3. Third, a property called *idempotency* holds in any lattice: $A \wedge A = A$, $\forall A \in L$. Indeed, taking $B = A \wedge X$, the absorption property $A \wedge (A \vee B) = A$ becomes $A \wedge (A \vee (A \wedge X)) = A \wedge A = A$. Idempotency

also holds for the join: $A \vee A = A, \forall A \in L$ (which can be shown in a similar manner).

3.2 Lattice and Poset

In any lattice $(L, \wedge, \vee)$, an ordering relation $\leq$ can be defined as

$$(A \leq B) \Leftrightarrow (A \wedge B = A). \tag{3.2}$$

The structure $(L, \leq)$ is a poset. The relation $\leq$ is

- Reflexive because the meet $\wedge$ is idempotent.
- Transitive because from $A \wedge B = A$ and $B \wedge C = B$ it follows that $A \wedge B = A \wedge (B \wedge C) = (A \wedge B) \wedge C = A \wedge C = A$.
- Antisymmetric because $A \wedge B = A = B \wedge A = B$.

In a poset P, an element $A \in P$ is an *upper bound* of subset $H \subset P$ if and only if $X \leq A, \forall X \in H$. An upper bound A of H is the *least upper bound* (also called the *supremum*) of H if and only if for any upper bound U we have $A \leq U$. The notions of *lower bound* and *greatest lower bound* (also called the *infimum*) are similarly defined.

In a lattice L, any two elements A and B have a supremum $\sup\{A, B\}$ and an infimum $\inf\{A, B\}$:

- A *supremum*: $\sup\{A, B\} = A \vee B$.
- An *infimum:* $\inf\{A, B\} = A \wedge B$.

Example 3.1

- The powerset $\wp(A)$ of a set A ordered by set inclusion is a lattice. For every X and Y element of $\wp(A)$ we have (see also Sections 2.2.4, 2.2.5, 2.2.11)

$$\sup\{X, Y\} = X \vee Y = X \cup Y,$$

$$\inf\{X, Y\} = X \wedge Y = X \cap Y.$$

- The set $\mathbb{N}$ of natural numbers ordered by the relation "divides" is a lattice, and $\sup\{a, b\}$ = l.c.m.(a, b), $\inf\{a, b\}$= g.c.d.(a, b), where l.c.m. = least common multiple and g.c.d. = greatest common divisor.

- Let $\mathcal{E}$ denote all the equivalence relations R on a set X. The structure $(\mathcal{E}, \leq)$ is a lattice, where $R_1 \leq R_2 \Leftrightarrow (x\, R_1\, y \Rightarrow x\, R_2\, y)$. □

3.3 Duality

If we examine the properties of the meet and join operations defining a lattice, we can observe that they are in pairs in the sense that the meet and join are swapped. It follows that any other property valid in a lattice (and being a consequence of the axioms) will also be valid if the meet and the join are interchanged. This phenomenon is referred to as the *principle of duality*. Duality is a useful tool for obtaining new results from given ones simply. The dual of a given property is obtained as follows:

- The relation $\leq$ is replaced by the relation $\geq$.
- The meet is replaced by the join, and vice-versa.

The dual of a valid property is also valid. For example, $A \wedge B$ is a lower bound of A and B, i.e., $A \wedge B \leq A$ and $A \wedge B \leq B$. Taking the dual, we obtain the following valid properties: $A \vee B \geq A$ and $A \vee B \geq B$.

3.4 Hasse Diagram

The fact that any lattice can be turned into a poset allows us to visualize (i.e., create a drawing of) the lattice. Lattices are represented graphically using a *Hasse diagram*:

- Any element of the lattice is usually represented by a point or circle (depending on which yields a better visualization of the problem or application being considered), or by a square (or rectangle) in which data can be written (if so required by the application).
- Two elements A and B are connected by a nonhorizontal (usually straight) line if $A \leq B$ and there does not exist any element C ($\neq A, B$) such that $A \leq C \leq B$ (in the diagram, A is situated below B).

For example, the Hasse diagram of the lattice $A \leq B \leq C$ is shown in **Fig. 3.1**.

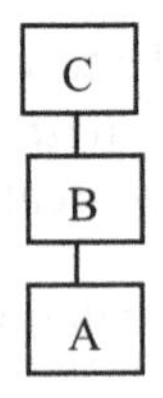

Fig. 3.1. Hasse diagram representation of the lattice $A \leq B \leq C$.

The Hasse diagrams of the lattices that can be formed using $n = 1, 2, 3, 4, 5$ elements are shown below.

Any 1-element set is a lattice; its Hasse diagram is a single point: •. Using $n = 2$ elements, one can form just one lattice. The Hasse diagram of the one 2-element lattice is seen in **Fig. 3.2**.

Fig. 3.2. The one 2-element lattice.

With $n = 3$ elements, one can also form just one lattice. The Hasse diagram of the one 3-element lattice is shown above in **Fig. 3.1**.

Using $n = 4$ elements, one can form two lattices. The Hasse diagrams of the two 4-element lattices are pictured in **Fig. 3.3**.

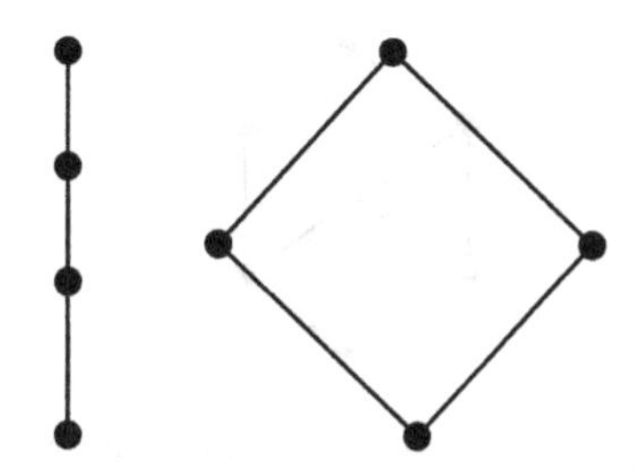

Fig. 3.3. The two 4-element lattices.

The Hasse diagrams of the five 5-element lattices are illustrated in **Fig. 3.4**.

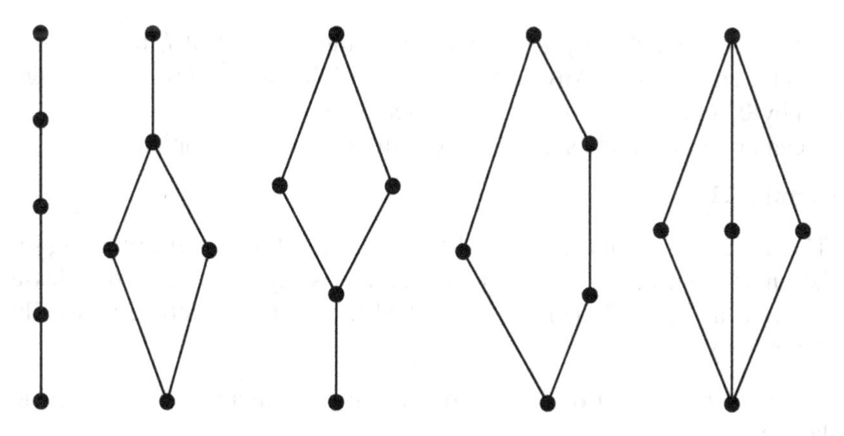

Fig. 3.4. The five 5-element lattices.

We note that while every lattice is a poset, not every poset is a lattice. A poset is a lattice if and only if any two of its elements have a supremum and an infimum. For example, the poset $\{A, B, C\}$ with $A \leq B$ and $C \leq B$ is not a lattice because A and C do not have an infimum (**Fig. 3.5**).

Fig. 3.5. A 3-element poset that is not a lattice (the bottom elements do not have an infimum).

Figure 3.6 shows a 6-element poset that is not a lattice because there are multiple infima (maxima).

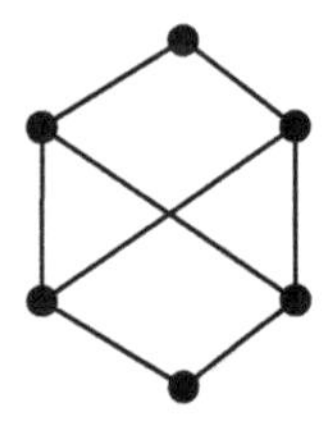

Fig. 3.6. A poset that is not a lattice.

3.5 Complete, Atomic Lattice

A lattice L is called *complete* if every nonempty subset of L has a supremum and an infimum. Any complete lattice L has a *smallest element*, denoted by $\mathbf{0}$, and a *largest element*, denoted by $\mathbf{1}$.

It can be immediately seen that every finite lattice is complete.

Example 3.2

- The lattice $(L, \leq)$ on the set $L = \{\mathbf{0}, x, y, z, \mathbf{1}\}$ defined by the ordering relation $\{(\mathbf{0}, x), (\mathbf{0}, z), (x, y), (y, \mathbf{1}), (z, \mathbf{1})\}$ is represented by the Hasse diagram in **Fig. 3.7**. This lattice, called the *pentagon* lattice, is usually denoted by N_5.

- The powerset $\wp(A)$ of a set A ordered by set inclusion is a complete lattice.

- The set ℕ of natural numbers ordered by the relation $\leq$ is not complete because there is no greatest natural number.
- The set ℤ of integer numbers ordered by the relation $\leq$ is not complete because there is no greatest or least integer number. □

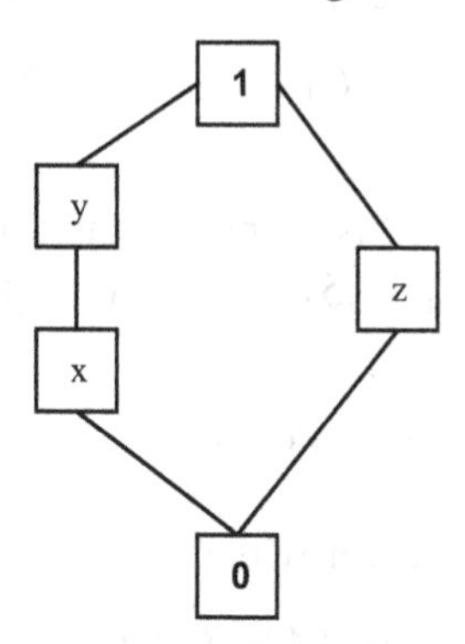

Fig. 3.7. The pentagon lattice (N_5).

An *atom* of a lattice L is an element A of L such that

$$(\mathbf{0} \leq B \leq A) \Rightarrow (B = \mathbf{0} \text{ or } B = A). \tag{3.3}$$

A lattice with $\mathbf{0}$ is *atomistic* if every one of its elements is a join of atoms, and it is referred to as *atomic* if

$$\forall\, x \in L, x \neq \mathbf{0} \Rightarrow \exists \text{ atom } a \neq \mathbf{0} \text{ such that } a \leq x. \tag{3.4}$$

3.6 Modular Lattice

In any lattice L, the following property, referred to as *weak distributivity*, holds:

$$A \vee (B \wedge C) \leq (A \vee B) \wedge (A \vee C), \forall\, A, B, C \in L. \tag{3.5}$$

Indeed, in any lattice, we have $A \wedge X \leq X$ and $A \wedge X \leq A$. If $A \leq B$, then $A \wedge X \leq B \wedge X$. Also, we have $B \leq B \vee C$. Then, taking $X = A$, we obtain

$$B \wedge A \leq (B \vee C) \wedge A, \tag{3.6}$$

and so

$$A \wedge B \leq A \wedge (B \vee C). \tag{3.7}$$

Similarly,

$$A \wedge C \leq A \wedge (B \vee C). \tag{3.8}$$

In other words, $A \wedge (B \vee C)$ is an upper bound for both $A \wedge B$ and $A \wedge C$, and thus $A \wedge (B \vee C)$ is an upper bound for their supremum, i.e.,

$$(A \wedge B) \vee (A \wedge C) \leq A \wedge (B \vee C), \tag{3.9}$$

whose dual is Eq. (3.5).

If $A \leq C$, then $A \vee C = C$ [this being the dual of Eq. (3.2)], so $A \vee (B \wedge C) \leq (A \vee B) \wedge C$ [using Eq. (3.5)]. Thus, the following definition can be introduced:

Definition 3.1. A lattice L is *modular* if

$$(\forall A, B, C \in L \text{ for which } A \leq C) \Rightarrow \tag{3.10}$$
$$A \vee (B \wedge C) = (A \vee B) \wedge C. \; \square$$

There are lattices that are not modular, e.g., the pentagon lattice N_5 (**Fig. 3.7**). It is clear that $x \leq y$. However,

$$x \vee (z \wedge y) = x \vee \mathbf{0} = x, \tag{3.11}$$

which is different from

$$(x \vee z) \wedge y = \mathbf{1} \wedge y = y. \tag{3.12}$$

Modularity is thus not a consequence of the axioms that define the notion of lattice.

Example 3.3

The set $\mathbb{N}$ of natural numbers ordered by the relation "divides" is a modular lattice. It is known from arithmetic that l.c.m. [a, g.c.d.(b, c)] = g.c.d. [l.c.m.(a, b), c]. □

In a modular lattice, the following property holds:

Theorem 3.1. *Let L denote a modular lattice, and let $A \leq B$, A, $B \in L$. Then,*

$$A \wedge C = B \wedge C \text{ and } A \vee C = B \vee C$$

imply $A = B$.

Proof. Indeed, we have:

$A = A \vee (C \wedge A) =$	(absorption)
$A \vee (A \wedge C) =$	(commutativity)
$A \vee (B \wedge C) =$	(by assumption)
$A \vee (C \wedge B) =$	(commutativity)
$(A \vee C) \wedge B =$	(modularity)
$(B \vee C) \wedge B =$	(by assumption)
$= B$	(absorption). □

3.7 Sublattice

Let L denote a lattice. A subset S of L is a *sublattice* of L if for any two elements A and B of S we have $\sup\{A, B\} \in S$ and $\inf\{A, B\} \in S$.

It can be shown that any sublattice S of a modular lattice L is modular. Let X and Y be two arbitrary elements of S. Then, we have

$$\begin{aligned} \inf_S\{X, Y\} &= \inf_L\{X, Y\} = X \wedge Y, \\ \sup_S\{X, Y\} &= \sup_L\{X, Y\} = X \vee Y. \end{aligned} \quad (3.13)$$

Let X, Y, and Z be three arbitrary elements of S and let $X \leq Z$. Since L is modular, using Eq. (3.13), we find that it follows that S is modular as well.

3.8 Distributive Lattice

We have seen that the weak distributivity property [Eq. (3.5)] as well as its dual hold in any lattice. We may have a lattice with the relation $<$ (instead of $\leq$). In such a lattice, if the property (3.5) holds for $=$ (i.e., for equality), then nothing can be said about its dual (because modularity is not a consequence of the axioms defining a lattice). However, it can be shown that, in any lattice, the following equivalence holds:

$$\begin{aligned} &A \vee (B \wedge C) = (A \vee B) \wedge (A \vee C) \Leftrightarrow \\ &A \wedge (B \vee C) = (A \wedge B) \vee (A \wedge C). \end{aligned} \quad (3.14)$$

Let us show the $\Rightarrow$ part of this equivalence. We have

$(A \wedge B) \vee (A \wedge C) =$	
$((A \wedge B) \vee A) \wedge ((A \wedge B) \vee C) =$	[left part of Eq. (3.14)]
$A \wedge ((A \wedge B) \vee C) =$	(absorption)

$A \wedge (C \vee (A \wedge B)) =$	(commutativity)
$A \wedge ((C \vee A) \wedge (C \vee B)) =$	[left part of Eq. (3.14)]
$(A \wedge (C \vee A)) \wedge (C \vee B) =$	(associativity)
$A \wedge (C \vee B) =$	(absorption))
$A \wedge (B \vee C).$	(commutativity)

The $\Leftarrow$ can be shown in a similar fashion. Thus, we introduce the following notion of distributivity:

Definition 3.2. A lattice L is *distributive* if

$$A \vee (B \wedge C) = (A \vee B) \wedge (A \vee C), \qquad \forall A, B, C \in L. \square \tag{3.15}$$

The dual of Eq. (3.15) also holds. It can be immediately seen that every distributive lattice is modular. If $A \leq C$, i.e., $A \vee C = C$, then we have $A \vee (B \wedge C) = (A \vee B) \wedge C$.

However, not every modular lattice is distributive. Hence, any nonmodular lattice is nondistributive (according to the law of contraposition), and thus the pentagon lattice (**Fig. 3.7**) is not distributive.

The lattice shown in **Fig. 3.8** is modular but not distributive:

$$x \vee (y \wedge z) = x \vee \mathbf{0} = x, \tag{3.16}$$

which is different from

$$(x \vee y) \wedge (x \vee z) = \mathbf{1} \wedge \mathbf{1} = \mathbf{1}. \tag{3.17}$$

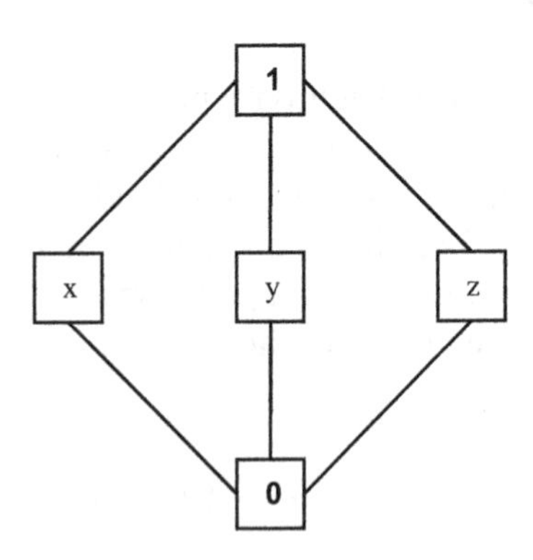

Fig. 3.8. Modular lattice (it is not distributive).

Example 3.4

- The set $\mathbb{Z}$ of integer numbers ordered by the relation $\leq$ is a distributive lattice.
- The powerset $\wp(A)$ of any set A is a distributive lattice (ordered by $\subseteq$).
- The set $\mathbb{N}$ of natural numbers ordered by the relation "divides" is a distributive lattice. □

Figure 3.9 shows the Hasse diagram of a distributive lattice (which can be easily checked).

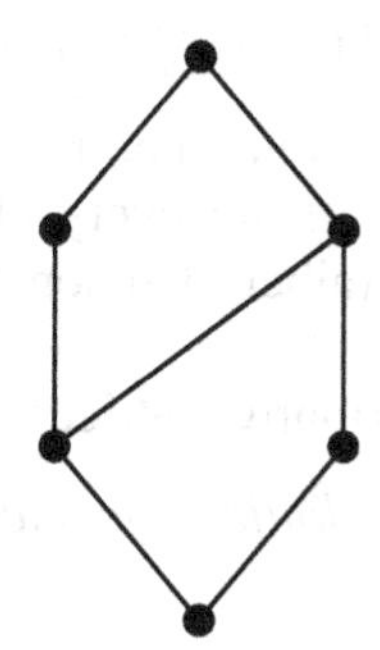

Fig. 3.9. Distributive lattice.

Any sublattice S of a distributive lattice L is distributive. Let A, B, and C be three arbitrary elements of S. Then, because they are also elements of L, we have $A \vee (B \wedge C) = (A \vee B) \wedge (A \vee C)$. However, because S is a sublattice, the distributivity condition is also valid in S.

It can be shown that the following property holds:

Theorem 3.2. *In any distributive lattice, we have*

$$(A \wedge C = B \wedge C \text{ and } A \vee C = B \vee C) \Rightarrow A = B. \qquad (3.18)$$

Proof. Indeed, we can write that

$$\begin{aligned} A = A \vee (C \wedge A) = & \quad \text{(absorption)} \\ A \vee (A \wedge C) = & \quad \text{(commutativity)} \\ A \vee (B \wedge C) = & \quad \text{(by assumption)} \\ (A \vee B) \wedge (A \vee C). & \quad \text{(distributivity)} \end{aligned}$$

In a similar way, it can be shown that

$$B = B \vee (C \wedge B) = (A \vee B) \wedge (A \vee C).$$

Hence, $A = B$. □

3.9 Complemented, Orthomodular Lattice

We first introduce the notion of complementation:

Definition 3.3. A lattice L with $\mathbf{0}$ and $\mathbf{1}$ is *complemented* if there is a mapping from L to L such that

$$A \mapsto A^{\mathrm{C}}, \forall A \in L, \tag{3.19}$$

so that

$$A \wedge A^{\mathrm{C}} = \mathbf{0}, A \vee A^{\mathrm{C}} = \mathbf{1}, \tag{3.20}$$

for every A in L. $A^{\mathrm{C}} \in L$ is called the *complement* of A. □

The complement may not be unique; i.e., it may happen that an element has more than one complement. If not every element has a complement, the lattice is not complemented (albeit that some elements may have complements).

An important property of distributive lattices is the following:

Theorem 3.3. *In any distributive lattice, any element can have at most one complement.*

Proof. If

$$A \wedge C = B \wedge C = \mathbf{0} \text{ and } A \vee C = B \vee C = \mathbf{1},$$

then A and B are complements of C. Because $A = B$ (Theorem 3.2), it follows that the complement is unique (if it exists). □

Example 3.5

The powerset $\wp(A)$ of any set A ordered by set inclusion is a complemented lattice with $\mathbf{0} = \varnothing$ and $\mathbf{1} = A$. □

Figure 3.10 shows a (not uniquely) complemented lattice. The pentagon lattice is also a complemented lattice. In any complemented lattice, we have $A \wedge A^{\mathrm{C}} = \mathbf{0}$, $\forall A \in L$. If we take $A = B^{\mathrm{C}}$, then $B^{\mathrm{C}} \wedge B^{\mathrm{CC}}$ is the same as $B \wedge B^{\mathrm{C}} = B^{\mathrm{C}} \wedge B = \mathbf{0}$. A special case would be to require that $B = B^{\mathrm{CC}}$, $\forall B \in L$. Further, we have

$$(A \leq B^{\mathrm{C}}) \Leftrightarrow (A \wedge B^{\mathrm{C}} = A), \tag{3.21}$$

and

$$(B \leq A^{\mathrm{C}}) \Leftrightarrow (B \wedge A^{\mathrm{C}} = B), \tag{3.22}$$

$\forall A, B \in L$. Then we have

$$(A \leq B) \Leftrightarrow (A \wedge B^{C} \leq B \wedge A^{C}). \tag{3.23}$$

A special case of equivalence [Eq. (3.23)] would be to require that

$$(A \leq B) \Leftrightarrow (B^{C} \leq A^{C}). \tag{3.24}$$

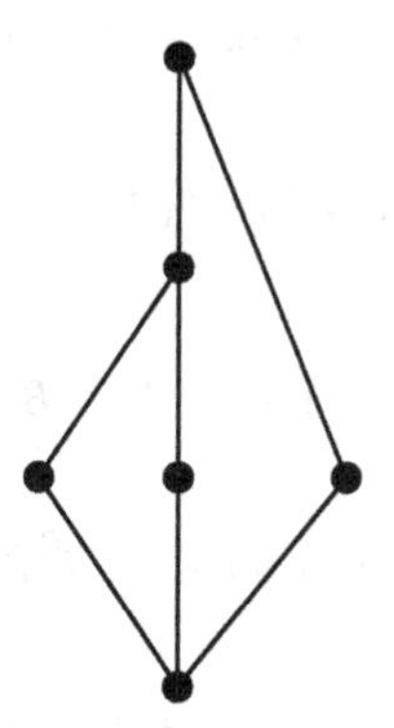

Fig. 3.10. Complemented lattice.

Thus, the following definition may be introduced:

Definition 3.4. A complemented lattice L is *orthocomplemented* if

$$A^{CC} = A, \tag{3.25}$$

$$A \leq B \Leftrightarrow B^{C} \leq A^{C}. \;\square$$

Figure 3.11 shows the Hasse diagram of an orthocomplemented lattice.

Fig. 3.11. Orthocomplemented lattice.

The analogues of de Morgan's laws (known from set theory, Section 2.2.11) hold in lattices as follows:

Theorem 3.4. *In any orthocomplemented lattice the following relations (known as De Morgan's laws) hold:*

$$\begin{aligned}(A \vee B)^{\mathrm{C}} &= A^{\mathrm{C}} \wedge B^{\mathrm{C}},\\ (A \wedge B)^{\mathrm{C}} &= A^{\mathrm{C}} \vee B^{\mathrm{C}}.\end{aligned} \tag{3.26}$$

Proof. Let $X = A \wedge B$. We have that $A \wedge B \le A$. Thus by Definition 3.4, $A^{\mathrm{C}} \le (A \wedge B)^{\mathrm{C}}$. As $A \wedge B \le B$, we have $B^{\mathrm{C}} \le (A \wedge B)^{\mathrm{C}}$. Using the idempotency property, we obtain

$$(A \wedge B)^{\mathrm{C}} = (A \wedge B)^{\mathrm{C}} \vee (A \wedge B)^{\mathrm{C}} \ge A^{\mathrm{C}} \vee B^{\mathrm{C}}.$$

From this, we get that

$$(A \wedge B)^{\mathrm{CC}} \le (A^{\mathrm{C}} \vee B^{\mathrm{C}})^{\mathrm{C}},$$

i.e.,

$$A \wedge B \le (A^{\mathrm{C}} \vee B^{\mathrm{C}})^{\mathrm{C}}.$$

The dual of this also holds:

$$A \vee B \ge (A^{\mathrm{C}} \wedge B^{\mathrm{C}})^{\mathrm{C}}.$$

Taking $A \to A^{\mathrm{C}}$ and $B \to B^{\mathrm{C}}$, we obtain

$$A^{\mathrm{C}} \wedge B^{\mathrm{C}} \le (A \vee B)^{\mathrm{C}},$$

from which we get

$$A \vee B \le (A^{\mathrm{C}} \wedge B^{\mathrm{C}})^{\mathrm{C}}.$$

Therefore,

$$A \vee B = (A^{\mathrm{C}} \wedge B^{\mathrm{C}})^{\mathrm{C}},$$

and thus

$$(A \vee B)^{\mathrm{C}} = (A^{\mathrm{C}} \wedge B^{\mathrm{C}})^{\mathrm{CC}} = (A^{\mathrm{C}} \wedge B^{\mathrm{C}}).$$

The other De Morgan law is the dual of this. □

A special case of the modular law (3.10) is the following concept:

Definition 3.5. An orthocomplemented lattice L is called *orthomodular* if the modularity condition (3.10) holds for $B = A^{\mathrm{C}}$:

$$A \le C \Rightarrow A \vee (A^{\mathrm{C}} \wedge C) = C. \; □ \tag{3.27}$$

It can be shown that the following property holds:

Theorem 3.5. *In any orthomodular lattice, we have (absorption of the complement)*

$$(A^C \wedge B^C) \vee ((A \vee B) \wedge A^C) = A^C, \forall A, B \in L.$$

Proof. As $A \leq A \vee B$, we have from Definition 3.4 that $(A \vee B)^C \leq A^C$. By Theorem 3.4, $A^C \wedge B^C \leq A^C$. Using Definition 3.5, we find that

$$(A^C \wedge B^C) \vee ((A^C \wedge B^C)^C \wedge A^C) = A^C.$$

By Theorem 3.4, this can be rewritten as

$$(A^C \wedge B^C) \vee (A \vee B) \wedge A^C) = A^C. \square$$

3.10 Boolean Algebra

A complemented and distributive lattice $(L, \vee, \wedge)$ is called a *Boolean algebra.* Any element of a Boolean algebra has a unique complement. **Figure 3.12** shows the Hasse diagram (in two different versions) of one and the same Boolean algebra.

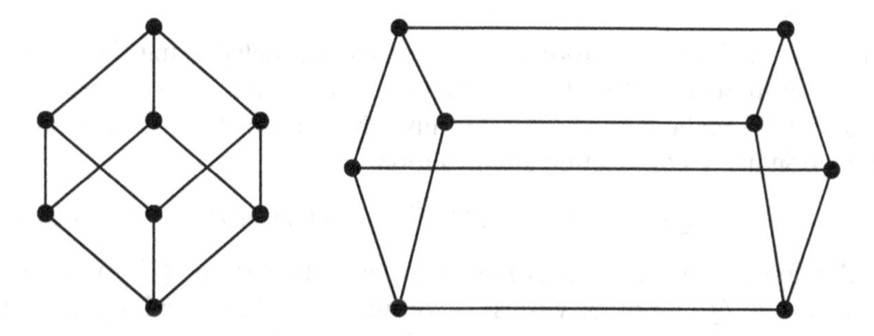

Fig. 3.12. Boolean algebra (two different drawings of the same Boolean algebra).

3.11 Important Lattices

The following lattices are of basic importance in IR theory as well as practical applications and retrieval systems.

3.11.1 Powerset Lattice

The structure ($\wp(X), \cap, \cup, \backslash$), i.e., the set of all subsets of set X, where

- $\mathbf{0} = \varnothing$
- $\mathbf{1} = X$
- $\cap$ = set intersection
- $\cup$ = set union
- $\backslash$ = set complement

is a Boolean algebra (see Example 3.1 for a proof).

3.11.2 Lattice of Logical Propositions

The structure ($\{T, F\}, \wedge, \vee, \neg$) of propositions in mathematical logic, where

- $\mathbf{0} = \mathit{False}$
- $\mathbf{1} = \mathit{True}$
- $\wedge$ = conjunction
- $\vee$ = disjunction
- $\neg$ = negation

is a Boolean algebra (a proof can be easily constructed using the truth tables given in Section 2.1). De Morgan's laws (Theorem 3.4) are especially important in mathematical logic because they make it possible to define disjunction using conjunction and negation:

$$\neg(P \vee Q) = \neg P \wedge \neg Q, \; P \vee Q = \neg(\neg P \wedge \neg Q), \tag{3.28}$$

which should be read as "it is not true that neither P nor Q," or "at least one of P and Q." In other words, according to mathematical logic, if one asserts anything about the world around us, then the assertion or its negation should/must be true. It cannot be neither true nor false.

3.11.3 Lattice of Logical Predicates

The structure ($Pred(X), \wedge, \vee, \neg$) of predicates in mathematical logic is a Boolean algebra. $Pred(X)$ denotes the set of all predicates over X; furthermore

$$Pred_1(x) \wedge Pred_2(x) = (Pred_1 \wedge Pred_2)(x),$$

$$Pred_1(x) \vee Pred_2(x) = (Pred_1 \vee Pred_2)(x).$$

The proof is similar to that in Section 3.11.2.

3.11.4 Lattice of Logical Implications

Let P = $\{P_1, P_2, \ldots, P_j, \ldots, P_m\}$ denote a set of propositions. The logical equivalence $\Leftrightarrow$ partitions set P into equivalence classes:

$$\mathsf{P} = \{C_1, C_2, \ldots, C_i, \ldots, C_n\}, \tag{3.29}$$

$$C_i = \{P_{i1}, \ldots, P_{ik}\},\ P_{i1} \Leftrightarrow \ldots \Leftrightarrow P_{ik}.$$

The structure (P, $\Rightarrow$) is a distributive lattice. The logical implication $\Rightarrow$ is to be understood between class representatives (say $c_i \in C_i$ representing class C_i, i.e., at the class level). One can define the following ordering relation $\leq$ in the lattice P:

$$(c_i \leq c_j) \Leftrightarrow (c_i \Rightarrow c_j). \tag{3.30}$$

P becomes a complemented lattice, and thus a Boolean algebra, if we introduce

- The null element **0** = the equivalence class of propositions that are always false (e.g., "*This object is a table and is not a table.*"),
- The unity element **1** = the equivalence class of tautologies (e.g., "*This object is a table or is not a table*").

3.11.5 Lattice Types

A diagram of lattice types is shown in **Fig. 3.13**.

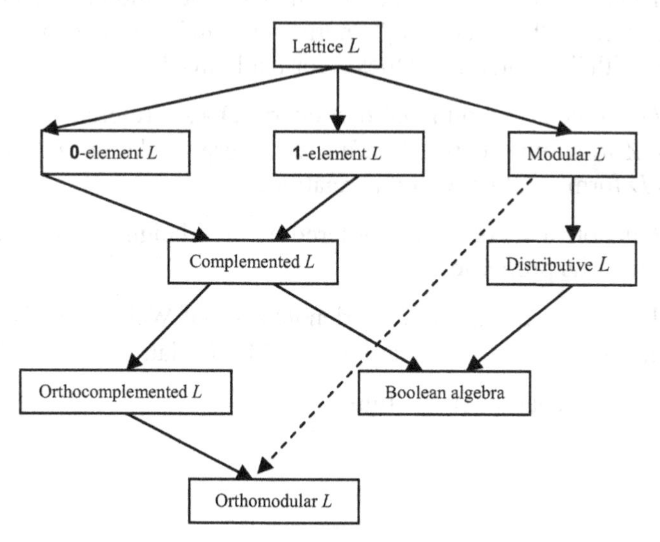

Fig. 3.13. The diagram of the basic types of lattices used in IR.

3.12 Exercises and Problems

1. Prove that the set ℕ of natural numbers ordered by the relation of divisibility is a distributive lattice.

2. Let $T = \{t_1, t_2, t_3\}$ denote a set of properties. Give the lattices corresponding to the following cases: (i) no property is comparable to any other property, (ii) two properties are comparable, (iii) three properties are comparable but two are not, and (iv) all three properties are comparable with each other two by two.

3. Prove that a sufficient and necessary condition for a lattice to be distributive is that $(Z \wedge X = Z \wedge Y, Z \vee X = Z \vee Y) \Rightarrow X = Y$.

4. Let A and B denote two convex figures in a plane (e.g., circle, rectangle). Let $A \cdot B$ denote the largest convex figure that is contained in both A and B, and let $A + B$ denote the smallest convex figure that contains both A and B. Prove that the set of such figures is a nondistributive lattice.

5. Prove that the collection of all distinct publications in a library can be viewed as a Boolean algebra.

6. Let T denote the terms of a thesaurus ordered by the relation "broader than." In general, does T form a lattice? Can you specify cases when T is a lattice and when T is not a lattice?

7. Let T denote the terms of a thesaurus with the following relations between terms: "broader than," "narrower than," "similar to," "synonymous with," "related to." Does T form a lattice?

8. Let D denote a monolingual dictionary. Does a reference given in an entry A to another entry B define an ordering relation $A \leq B$? If yes, does D form a lattice with this relation?

9. Let P denote a set of people ordered by their heights. Is P a lattice? If yes, study its properties.

10. Let $W = \{w_1,...,w_i,...,w_j,...,w_N\}$ denote a set of Web pages. Do the hyperlinks $w_i \rightarrow w_j$ form a relation on W? Is W a lattice?

11. Prove that the lattice in the figure is:

(a) Nonmodular:

(b) Distributive:

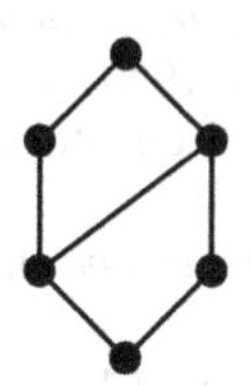

(c) Modular and nondistributive:

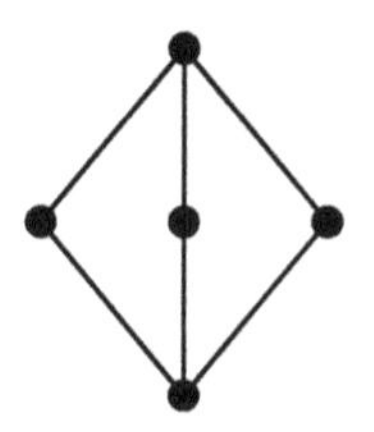

(d) Noncomplemented:

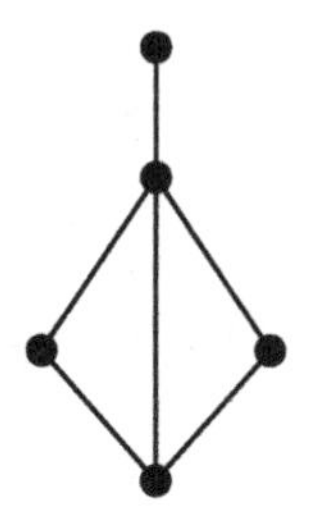

3.13 Bibliography

Birkhoff, G. Birkhoff, G.: *Lattice Theory* (American Mathematical Society Colloquium. Publication, 1948)

Birkhoff, G.: Lattice-ordered demigroups. *Séminaire Dubreil. Algebre et théorie de nombres.* **14**(2), 1–19 (1960)

Birkhoff, G., and von Neumann, J.: The logic of quantum mechanics. *Annals of Mathematics.* **37**(4), 823–843 (1936)

Gratzer, G.: *General Lattice Theory* (Birkhauser Verlag, Basel/Boston/Berlin, 2003)

Kaufmann, A., and Précigout, M. : *Cours de mathématiques nouvelle (A Course in New Mathematics)* (Dunod, Paris, 1966)

Piziak, R.: Orthomodular lattices and quantum physics. *Mathematics Magazine.* **51**(5), 299–303 (1978)

Stern, M.: *Semimodular Lattices: Theory and Applications* (Cambridge University Press, 1999) pp. 1–18

Szász, G.: *Introduction to Lattice Theory* (Academic, New York, 1963)

4 Basics of Information Retrieval Technology

I hear and I forget. I see and I remember. I do and I understand.
(Confucius)

This chapter introduces the basics of information retrieval technology (document, stoplist, term, power law, stemming, inverted file structure, weighting schemes, term-document matrix, architecture of retrieval system, architecture of a search engine, relevance effectiveness, measures and measurement, precision-recall graph method, search engine effectiveness measurement). These are presented, first, with an eye toward practitioners, and the material will be useful for those interested in developing practical retrieval systems. However, the material herein may also be helpful for theoretically minded readers as well as it will enable a better understanding of the chapters that follow.

The ways in which a query (expressing a user's information need) can be matched against entities (documents) stored in computers is not dealt with at this point. Matching and ranking constitute the topics that will be discussed further on.

The chapter ends with exercises and problems designed to promote a deeper understanding of the basics of information retrieval.

4.1 Documents

Let $E_1,\ldots,E_j,\ldots,E_m$ denote entities in general. They can be:

- Texts (books, journal articles, newspaper articles, papers, lecture notes, abstracts, titles, etc.),
- Images (photographs, pictures, drawings, etc.),
- Sounds (pieces of music, songs, speeches, etc.),
- Multimedia (a collection of texts, images, and sounds),
- A collection of Web pages,
- And so on.

For retrieval purposes, it is assumed that each entity E_j is described by (is assigned, is characterized, is identified by) a piece of text D_j. Obviously, D_j may coincide with E_j itself (e.g., when E_j is itself a piece of text). D_j is traditionally called a *document*.

This assumption is not as restrictive as it may seem at a first look. It is based on a quite natural hypothesis, according to which we are/should be able to describe in words (of some language) any entity that we want to store in a computer for retrieval purposes. If we accept that describing entities using words is an act of thought, then the hypothesis is all the more acceptable, in that, according to Wittgenstein, "language is a vehicle of thought" (Blair 2006).

This hypothesis seems, indeed, to be very helpful at the present stage of computing and retrieval technology. However, new technologies may eventually offer other possibilities that will grow out of result of research being carried out at present (e.g., retrieval of drawings by comparing them with a sample drawing, without using words).

4.2 Power Law

From a computational point of view (and from the viewpoint of a computer programmer), and thus formally, documents consist of words as automatically identifiable lexical units. Thus,

lexical unit = word =

string of characters preceded and followed by "space" (or some

special character, e.g., ! , . ?).

Thus, words can be recognized automatically (using a computer program). Moreover, word occurrence has a remarkable statistical property that is

- Not at all intuitive.
- Has practical impact.

It has been shown that the number f of occurrences of words in an English text (corpus) obeys a power law (Yule 1924, Dewey 1929, Thorndike 1937, Zipf 1949, Smith and Devine 1985), i.e.,

$$f(r) = Cr^{-\alpha}, \tag{4.1}$$

where C is a corpus-dependent constant, and r is the rank of words; α is referred to as the *exponent* of the power law. The power law $f(r) = Cr^{-1}$ is known as Zipf's law ($\alpha = 1$).

For visualization purposes, the power law is represented in a log-log plot, i.e., as a straight line obtained by taking the logarithm of Eq. (4.1):

$$\log f(r) = \log C - \alpha \times \log r, \tag{4.2}$$

where

- $\log r$ is represented on the horizontal axis.
- $\log f(r)$ is represented on the vertical axis.
- $-\alpha$ is the slope of the line.
- $\log C$ is the intercept of the line.

In practice, the following *regression line method* can be applied to fit a power law to data:

Power Law Fitting Using the Regression Line Method

1. We have a sequence of values $X = (x_1,\ldots,x_i,\ldots,x_n)$ on the horizontal axis and another sequence of corresponding values $Y = (y_1,\ldots,y_i,\ldots,y_n)$ on the vertical axis (y_i corresponds to x_i, $i = 1,\ldots,n$).
2. If the correlation coefficient

$$r(X, Y) = \frac{n\sum_{i=1}^{n} x_i y_i - \sum_{i=1}^{n} x_i \sum_{i=1}^{n} y_i}{\sqrt{\left(n\sum_{i=1}^{n} x_i^2 - \left(\sum_{i=1}^{n} x_i\right)^2\right)\left(\left(n\sum_{i=1}^{n} y_i^2 - \left(\sum_{i=1}^{n} y_i\right)^2\right)\right)}}$$

suggests a fairly strong correlation—i.e., it is close to +1 or −1—between X and Y on a log scale, then a regression line can be drawn to exhibit a relationship between the data X and Y.

3. Using the

$$\text{slope} = \frac{n\sum_{i=1}^{n} x_i y_i - \sum_{i=1}^{n} x_i \sum_{i=1}^{n} y_i}{n\sum_{i=1}^{n} x_i^2 - (\sum_{i=1}^{n} x_i)^2}$$

and the

$$\text{intercept} = \frac{\sum_{i=1}^{n} y_i \sum_{i=1}^{n} x_i^2 - \sum_{i=1}^{n} x_i \sum_{i=1}^{n} x_i y_i}{n\sum_{i=1}^{n} x_i^2 - (\sum_{i=1}^{n} x_i)^2}$$

of the regression line, we can write the corresponding power law .

It should be noted, however, that even a strong correlation of the two quantities X and Y does not mean a necessary cause-effect relationship between them. The power law can be used as an approximation of some behavior (possible connection) between X and Y, especially when no other relationship is known.

The parameters α and C of the power law can be computed (approximated) using the method of least squares, as follows:

Power Law Fitting by Least Squares

1. We have a sequence of values $X = (x_1,\ldots,x_i,\ldots,x_n)$ on the horizontal axis and another sequence of corresponding values $Y = (y_1,\ldots,y_i,\ldots,y_n)$ on the vertical axis (y_i corresponds to x_i, $i = 1,\ldots,n$).
2. The parameters α and C should be so computed as to minimize the squared error

$$\sum_{i=1}^{n} (f(x_i) - y_i)^2 = \sum_{i=1}^{n} (Cx_i^{-\alpha} - y_i)^2 ,$$

i.e., the partial derivatives with respect to C and α should vanish.

The least squares method is, in general, a nonlinear optimization problem. As such, no generally valid method is known that solves it exactly. However, different approximation methods (e.g., Newton's method, gradient descent method, Levenberg-Marquardt method) can be used to find an approximate solution.

In practical applications, the number of data (i.e., n) is very large, so the use of an appropriate mathematical software or other computer program is highly recommended in order to apply the regression line or the least squares method (e.g., MathCAD, Matlab, etc.). In general, we recommend using both methods. The values for the parameters that should be accepted are those for which the approximation error is smaller or which best fit the problem being considered.

Example 4.1

Let us assume that the data we want to approximate by a power law is X and Y, $n = 150$. Fragments of X and Y are shown below. The correlation coefficient is equal to $r(X, Y) = -0.95$, which suggests a fairly strong correlation between X and Y. Using the regression line method, we obtain the following power law : $f(x) = 10^{8.38}x^{-3}$, whereas using the least squares method, we get: $f(x) = 5677733x^{-2.32}$. The approximation error is 2.8×10^8 in the regression line method, and 3.6×10^6 in the least squares method. Thus, we should accept the power law obtained by least squares.

X =

	1
1	1
2	2
3	3
4	4
5	5
6	6
7	7
8	8
9	9
10	10

Y =

	1
1	$5.722975 \cdot 10^6$
2	744343.1
3	510729.7
4	449741.1
5	441213
6	313464.3
7	300948.4
8	235022.1
9	182827.1
10	167201.1

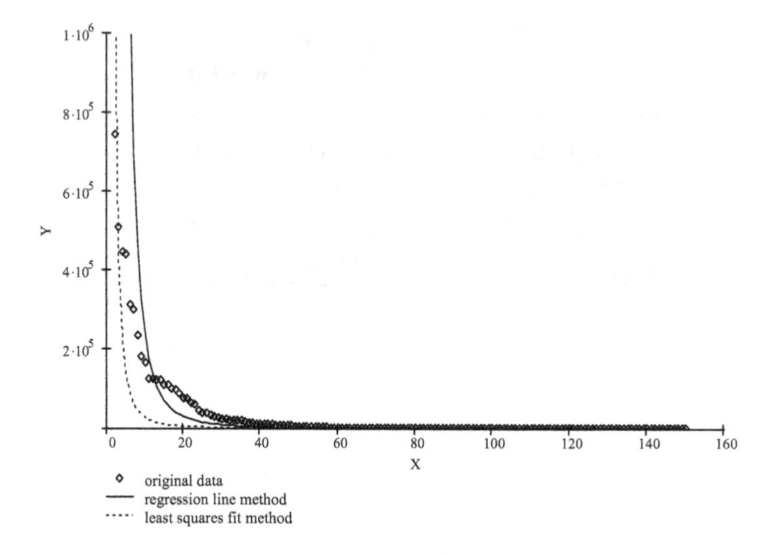

□

Recent experiments have shown that the distribution of *n*-grams in English, Chinese, and Greek texts obey a power law with $\alpha \neq 1$, i.e., different from Zipf's law (Egghe 2000). (*Note:* An *n*-gram is a subsequence $x_{i_1},\ldots,x_{i_n}$ of *n* items from a given sequence $x_1,\ldots,x_i,\ldots,x_m$ of items, $m \geq n$. If the sequence consists of words, then an *n*-gram is a subsequence of *n* consecutive words. For a sequence of characters, an *n*-gram is a subsequence of *n* consecutive characters.) Similarly, Le Quan Ha et al. (2003) showed that using very large English corpora (1987–1989 *Wall Street Journal* articles, 41 million words) as well as very large Chinese text corpora (20 million words TREC database and 250 million syllables in the *Mandarin Daily News* article database), the power law holds with $\alpha = 0.6$.

Dominich and Kiezer (2005) showed that the Hungarian language also obeys a power law, also different from Zipf's law. The following five Hungarian text corpora—having very different style and content and spanning a period of nearly five centuries—were used (Table 4.1):

- ARANY: all the writings by János Arany (Hungarian author).
- JÓKAI: all the writings by Mór Jókai (Hungarian author).
- BIBLE (Hungarian translation of the Holy Bible, the "Károly-féle Biblia").
- PALLAS: Great Lexikon Pallas (all 16 volumes).
- WEB: Hungarian Web corpus.

Table 4.1. Statistics of the Hungarian Corpora Used in Experiments

Corpus	Number of word forms	Number of word stems
ARANY (19th c.)	57,376	31,909
JÓKAI (19th c.)	443,367	200,022
BIBLE (1590)	62,474	29,360
PALLAS (1897)	871,635	605,358
WEB (2003)	11,547,753	7,516,221

Table 4.2 shows fragments of the lists of terms and their frequency.

Table 4.2. The First Ten Most Frequent Words in the Hungarian Corpora Used in Experiments (r = rank of word, f = frequency of word)

	BIBLE		ARANY		JÓKAI		PALLAS		WEB	
r	**word**	***f***	**word**	***f***	**word**	***f***	**word**	***f***	**word**	***f***
1	a	48796	a	17475	a	539612	a	900498	a	113416230
2	és	40658	az	7146	az	290432	és	313237	az	47124033
3	az	35248	van	4974	van	162547	az	311770	és	27129451
4	ő	12191	nem	4772	hogy	110183	van	147165	van	26089040
5	van	9396	s	3814	s	99039	is	90004	hogy	16594469
6	úr	8171	is	3200	nem	95309	mely	83363	nem	16022747
7	hogy	7791	hogy	3102	egy	75891	ez	61737	is	15872013
8	ki	7650	egy	2865	ez	62873	hogy	55998	egy	12018565
9	én	7074	és	2530	is	58486	nagy	49286	ez	12007607
10	te	6470	de	2276	és	56907	nem	47076	the	7534824

Table 4.3 shows the values of the power law exponent obtained in experiments.

Table 4.3. The Values of the Power Law Exponent α for the Corpora Used in Experiments

Corpus	**The value of the exponent α for**	
	word forms	**word stems**
ARANY	0.88	1.1
JÓKAI	1.11	1.36
BIBLE	1.03	1.29
PALLAS	1.09	1.15
WEB	1.59	0.99
Average deviation from $\alpha = 1$	+42.5%	+47.3%

In order to explain the empirical observation of power law in language, Zipf (1949) used the “principle of least effort.” According to this principle, the writer uses as few words as possible to accomplish a job of communication, whereas the reader prefers unique words (and hence different words for different situations) to minimize ambiguity (the effort of interpretation). The power law is the result of a balance between these two opposing and competing tendencies. There are also other explanations; see, e.g., Belew (2000).

4.3 Stoplist

The experimental result according to which word occurrences in a text obey a power law can be exploited in IR.

Typically, there are words in a document that occur many times, and there are words that occur once or just a few times. One may disregard frequently occurring words (i.e., frequency f exceeds some threshold value) on the grounds that they are almost always insignificant, as well as infrequent words (i.e., frequency f is below some threshold value) on the grounds that they are not much on the writer's mind (or else they would occur more often). The list of frequent and infrequent words thus obtained in an entire corpus is called a *stoplist*. For the English language, a widely accepted and used stoplist is the so-called TIME stoplist[1] (a fragment is shown below):

A
ABOUT
ABOVE
ACROSS
...
BACK
BAD
BE

...

Note: Of course, a stoplist is, in practice, dependent of the application context. For example, the word "a" may be in a stoplist in, say, a mechanical engineering context, but it is the name of an important vitamin in medicine.

When we take Table 4.2 into account, a fragment for a Hungarian stoplist is as follows:

a
és
az
van
is
mely
ez
hogy
...

The construction of a stoplist can be automated (using computer programs). Other stoplists can also be used depending on, e.g., the topic of the documents being considered. One usually starts with a general stoplist, and enlarges/modifies it depending on the topic or on experimental results.

---

[1] http://www.dcs.gla.ac.uk/idom/ir_resources/linguistic_utils/stop_words

4.4 Stemming

After excluding stopwords, it is advisable that the remaining words be transformed to their lexical roots. This operation is referred to as *stemming*. The justification for stemming can be shown through an example. Let us assume that the document D reads as follows:

> From an organizational point of view, the structure of the institution is consistent with the principle of hierarchical organization. Albeit that hierarchically structured organizations can be very effective in many cases, it is advisable to consider moving toward a network type of organizational model, at the same time maintaining consistency.

After removing stopwords, among the remaining words there will be, e.g., the words "consistent," "consistency." When the above piece of document D is to be retrieved, some users may use the query "consistent," whereas others will probably use the query "consistency," or some other form of this word. In order to obtain a common (unified) form for user queries and the different word forms in the document, all word forms are/should be transformed to one common form, namely to their lexical root (or stem); in this case to "consist."

The operation of stemming introduces a partition of the words of a language into—not necessarily disjoint—equivalence classes. Every equivalent class consists of the words that have the same lexical root. We should note, however, that there are languages (e.g., Hungarian) in which some word forms can be stemmed to multiple lexical roots.

For the English language, a widely used stemming algorithm is the Porter algorithm,[2] which is based on successively truncating the characters of a word, according to grammatical rules for suffixes, etc., until the lexical root of the word is obtained. In practice, a dictionary containing the lexical roots of as many words as possible can also be used. Thus, the process of stemming may reduce to dictionary look up. However, especially in agglutinative languages (e.g., Hungarian) the number of word forms may be in the billions. Because a stemmer is a software module that may/should be used in real time (e.g., in stemming query words), the process of stemming may not exceed a certain time limit. This is an important programming problem. Stemming algorithms have been developed for several other languages as well.[3] (Stemmers and stoplists exist for English, French, Spanish,

---

[2] http://www.tartarus.org/~martin/PorterStemmer

[3] http://snowball.tartarus.org

Portuguese, Italian, Romanian, German, Dutch, Swedish, Danish, Norwegian, Russian, Hungarian, and Turkish.)

4.5 Inverted File Structure

Let $E = \{E_1,\ldots,E_j,\ldots,E_m\}$ denote a set of entities to be searched in a future retrieval system, and let

$$D = \{D_1,\ldots,D_j,\ldots,D_m\} \tag{4.3}$$

denote the documents corresponding to E. After word identification, stoplisting, and stemming, the following set of *terms* is identified:

$$T = \{t_1,\ldots,t_i,\ldots,t_n\}. \tag{4.4}$$

The set T can be used to construct an inverted file structure as follows:

Construction of Index Table

1. Sort the terms $t_1,\ldots,t_i,\ldots,t_n$ alphabetically. For this purpose, some appropriate (fast) sorting algorithm should be used (e.g., quick sorting or some other sorting algorithm depending on the number n of terms, on the available (internal or external) memory for sorting. (see, e.g., (Weiss 1995).
2. Create an index table I in which every row r_i contains exactly one term t_i together with the codes (identifiers) of documents D_j in which that term t_i occurs (**Table 4.4**).

Table 4.4. Index Table I

Terms in alphabetical order	Codes of documents in which the term occurs
t_1	$D_{11},\ldots,D_{1k}$
…	
t_i	$D_{i1},\ldots,D_{is}$
…	
t_n	$D_{n1},\ldots,D_{np}$

As every document D_j uniquely identifies its corresponding entity E_j, a structure *IF* (*inverted file*) consisting of the index table I and of the entities (*master file*) of set E can be constructed (usually on a disk; **Fig. 4.1**).

The *codes* in the index table I can also contain the disk addresses (pointers) of the corresponding entities in the master file.

The inverted file structure *IF* is used in the following way:

1. Let t denote a query term. A binary search (or other appropriate search algorithm) locates t in table I, i.e., the result of the search is the row:

$$[t \mid D_{t1},\ldots,D_{tu}]. \tag{4.5}$$

2. Using the codes $D_{t1},\ldots,D_{tu}$, we can read the corresponding entities $E_{t1},\ldots,E_{tu}$ from the master file for further processing.

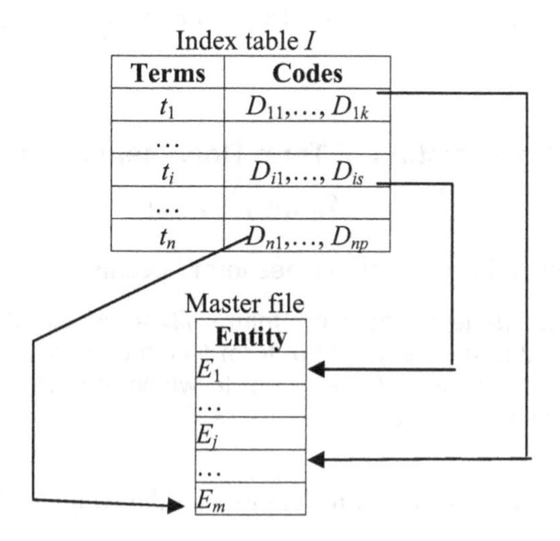

Fig. 4.1. Inverted file structure (*IF*).

Other data can also be stored in an inverted file structure, such as:

- The number of occurrences of term t_i in document D_j.
- The total number of occurrences of term t_i in all documents.
- And so on.

The inverted file structure is a logical one. Its physical implementation depends on the properties of the particular computer hardware, operating system, programming language, database management system, etc., available.

4.6 Term-Document Matrix

Just as before, let $E = \{E_1,\ldots,E_j,\ldots,E_m\}$ denote a set of entities to be searched in a future computerized retrieval system, and let

$$D = \{D_1,\ldots,D_j,\ldots,D_m\} \tag{4.6}$$

denote the documents corresponding to E. After word identification, stoplisting, and stemming, the following set of terms is constructed

$$T = \{t_1,\ldots,t_i,\ldots,t_n\}. \tag{4.7}$$

The set T can be used for the construction of *term-document matrix TD* as follows:

Construction of Term-Document Matrix *TD*

($i = 1,\ldots,n$, $j = 1,\ldots,m$)

1. Establish f_{ij}: the number of times term t_i occurs in document D_j, $\forall i, j$.
2. Construct the term-document matrix $TD = (w_{ij})_{n \times m}$, where the entry w_{ij} is referred to as the *weight* of term t_i in document D_j. The weight is a numerical measure of the extent to which the term reflects the content of the document.

There are several methods for computing the weights. Perhaps the most obvious are:

1. Binary weighting method:

$$w_{ij} = \begin{cases} 1 & \text{if } t_i \text{ occurs in } D_j \\ 0 & \text{otherwise} \end{cases}, \tag{4.8}$$

2. Frequency weighting method:

$$w_{ij} = f_{ij}. \tag{4.9}$$

There are also more advanced methods that offer a more balanced and realistic measurement of content (Belew, 2000), and these can be expressed in a unified manner as follows:

Theorem 4.1. (Dominich 2002) *The entries w_{ij} of a term-document matrix $TD = (w_{ij})_{n \times m}$ can be computed using the following generalized weighting method:*

$$w_{ij} = \frac{w'_{ij}}{[n]} = \frac{w'_{ij}}{\left[(\max_{1 \le k \le n} w'_{kj})^{\nu_1} \times \sqrt[\nu_3]{\sum_{k=1}^{n} \left(w'_{kj}\right)^{\nu_2}} \right]},$$

where

$$w'_{ij} = [l] \times [g] = \left[f_{ij}^{\lambda_1} \cdot \left(\frac{f_{ij}}{\max_{1 \le k \le n} f_{kj}} \right)^{\lambda_2} \cdot \ln^{\lambda_3}(e \cdot f_{ij}) \right] \times \left[\left(\frac{\gamma_1}{F_i} + \log^{\gamma_2} \frac{m - \gamma_3 F_i}{F_i} \right) \right],$$

and F_i denotes the number of documents in which term t_i occurs; further $\lambda_1, \lambda_2, \lambda_3, \gamma_1, \gamma_2, \gamma_3$ $\nu_1, \nu_2, \nu_3 \in \{0, 1, 2, 3, 4, \infty\}$.

Proof. It is shown that the usual special cases of each factor (normalization, local weighting, and global weighting) are obtained for certain values of the parameters.

Normalization factor $[n]$:

SUM: $\nu_1 = 0, \nu_2 = 1, \nu_3 = 1$; $[n] = \sum_{k=1}^{n} \left(w'_{kj}\right)$

COSINE: $\nu_1 = 0, \nu_2 = 2, \nu_3 = 2$; $[n] = \sqrt{\sum_{k=1}^{n} \left(w'_{kj}\right)^2}$

4th: $\nu_1 = 0, \nu_2 = 4, \nu_3 = 1$; $[n] = \sum_{k=1}^{n} \left(w'_{kj}\right)^4$

MAX: $_1 = 1, \nu_2 = 0, \nu_3 = \infty$; $[n] = \max_{1 \le k \le n} w'_{kj}$

NONE: $\nu_1 = 0, \nu_2 = 0, \nu_3 = \infty$; $[n] = 1$

Local weighting factor $[l]$:

FREQ: $\lambda_1 = 1, \lambda_2 = 0, \lambda_3 = 0$; $[l] = f_{ij}$

LOG: $\lambda_1 = 0, \lambda_2 = 0, \lambda_3 = 1$; $[l] = ln(f_{ij}) + 1$

MAXNORM: $\lambda_1 = 0, \lambda_2 = 1, \lambda_3 = 0;$ $[l] = \dfrac{f_{ij}}{\max\limits_{1 \le k \le n} f_{kj}}$

Global weighting factor [*g*]:

NONE: $\gamma_1 = 0, \gamma_2 = 0, \gamma_3 = 0;$ $[g] = 1$

INVERSE: $\gamma_1 = 0, \gamma_2 = 1, \gamma_3 = 0;$ $[g] = \log \dfrac{m}{F_i}$

SQUARED: $\gamma_1 = 0, \gamma_2 = 2, \gamma_3 = 0;$ $[g] = \log^2 \dfrac{m}{F_i}$

PROBABILISTIC: $\gamma_1 = 0, \gamma_2 = 1, \gamma_3 = 1;$ $[g] = \log \dfrac{m - F_i}{F_i}$

FREQUENCY: $\gamma_1 = 1, \gamma_2 = 0, \gamma_3 \neq \infty;$ $[g] = \dfrac{1}{F_i}$. □

The explicit forms of the weighting schemes widely used in practice are as follows:

(a) *max-tf*, max-normalized method:

$$w_{ij} = \frac{f_{ij}}{\max\limits_{1 \le k \le n} f_{kj}}. \tag{4.10}$$

(b) *norm-tf*, length-normalized method:

$$w_{ij} = \frac{f_{ij}}{\sqrt{\sum_{k=1}^{n} f_{kj}^2}}. \tag{4.11}$$

(c) *tf-idf*, term frequency inverse document frequency method:

$$w_{ij} = f_{ij} \times \left(\log \frac{m}{F_i} \right), \tag{4.12}$$

where F_i denotes the number of documents in which term t_i occurs.

(d) *norm-tf-idf*, length normalized term frequency inverse document frequency method:

$$w_{ij} = \frac{f_{ij} \times \left(\log \frac{m}{F_i}\right)}{\sqrt{\sum_{k=1}^{n} \left(f_{kj} \times \left(\log \frac{m}{F_k}\right)\right)^2}}. \tag{4.13}$$

A more recent weighting scheme that has given good results on large databases is the Okapi-BM25 formula (Cummins and O'Riordan 2006):

$$w_{ij} = \frac{f_{ij}}{f_{ij} + k\left(1 - b + b\frac{l_j}{l_{avg}}\right)} \times \log \frac{m - F_i + 0.5}{F_i + 0.5}, \tag{4.14}$$

where k and b are tuning parameters, l_j denotes the length (in arbitrary units) of document d_j, and l_{avg} denotes average document length.

4.7 General Architecture of a Retrieval System

Figure 4.2 shows the general architecture of an IR system.

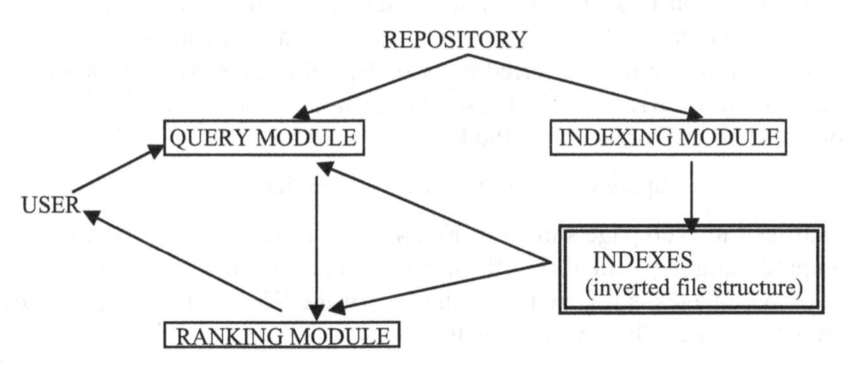

Fig. 4.2. General architecture of an IR system.

- **REPOSITORY**. The entities (documents) to be searched are stored in a central REPOSITORY (on computer disks). They are collected and entered into the REPOSITORY manually or using specialized computer programs.

- **INDEXING MODULE**. Using the documents stored in the REPOSITORY, the INDEXING MODULE creates the INDEXES in the form of inverted file structures. These structures are used by the QUERY MODULE to find documents that match the user's query.

- **QUERY MODULE**. This module reads in the user's query. The QUERY MODULE, using INDEXES, finds the documents that match the query (typically, the documents that contain the query terms). It then passes the located documents to the RANKING MODULE.

- **RANKING MODULE**. This module computes similarity scores (using INDEXES) for the documents located by the QUERY MODULE. Then, the documents are ranked (sorted descendingly) on their similarity score, and are presented to the user in this order. (This list is called a *hit list*.) For the computation of similarity scores, several methods can be used, and these are dealt with in subsequent chapters.

4.8 Elements of Web Retrieval Technology

4.8.1 World Wide Web

The World Wide Web (Web, for short) is a network of electronic documents stored on dedicated computers (*servers*) around the world. Documents can contain different types of data, such as text, image, or sound. They are stored in units referred to as *Web pages*. Each page has a unique code, called a URL (universal resource locator), which identifies its location on a server. For example, the URL

http://www.dcs.vein.hu/CIR/i2rmeta/i2rmeta.cgi

identifies the Web page shown in **Fig. 4.3**. Pages are typically written in a computer language called HTML (hypertext markup language). The number of Web pages is referred to as the *size* of the Web, which is estimated at more than 12 billion pages to date.

4.8.2 Major Characteristics of the Web

In what follows, the major characteristics of the Web that are relevant for IR are reviewed briefly.

Most Web documents are in HTML format and contain many *tags*. Tags can provide important information about the page. For example, the tag <b>, which is a bold typeface markup, usually increases the importance of the term to which it refers. In **Fig 4.3**, the tag <title> defines a title text for the page.

In traditional IR, documents are typically well structured. For example, every scientific journal defines its own characteristic structure for authors of research papers to follow. Similarly, books and newspaper articles have their typical formats and structures. Such documents are carefully written and are checked for grammar and style. On the other hand, Web pages can be less structured (there is no generally recommended or prescribed format that should be followed when writing a Web page). They are also more diverse:

- They can be written in many languages; moreover, several languages may be used within the same page.
- The grammar of the text in a page may not always be checked very carefully.
- The styles used vary to a great extent.
- The length of pages is virtually unlimited (if at all, then the limits are posed by, e.g., disk capacity, memory).

Web pages can contain a variety of data types, including:

- Text
- Image
- Sound
- Video
- Executable code

Many different formats are used, such as:

- HTML
- XML
- PDF,
- MSWord
- mp3
- avi
- mpeg
- etc.

While most documents in classical information retrieval are considered to be static, Web pages are dynamic, i.e., they can be:

- Updated frequently.
- Deleted or added.
- Dynamically generated.

Web pages can be hyperlinked, which generates a linked network of Web pages. Various factors can provide additional information about the importance of the target page, such as:

- A URL from one Web page to another page.
- Anchor text.
- The underlined, clickable text.

The size of the Web, i.e., the number of Web pages and links between them, is orders of magnitudes larger than the size of corpuses and databases used in classical IR. For example, the size of classical test databases (such as ADI, TIME, CISI, CACM, TREC databases, etc.) can be measured in the range from kilobytes to terabytes. The quantity of data stored on the Web is practically incomparable to these sizes: it is very much larger and very hard to estimate (owing to the fact that the number of Web pages can only be estimated very roughly and the size of a page can vary to a very great extent).

The number of users of, e.g., a university library system can be in the range of, say, tens of thousands, whereas the number of users of a banking intranet system may be in the range of, say, thousands. However, the number of Web users is in the range of billions, and it is increasing rapidly. Moreover, the users of the Web are more diverse than the users of, say, a university library system in terms of:

- Interest.
- Search experience.
- Languages spoken.
- And so on.

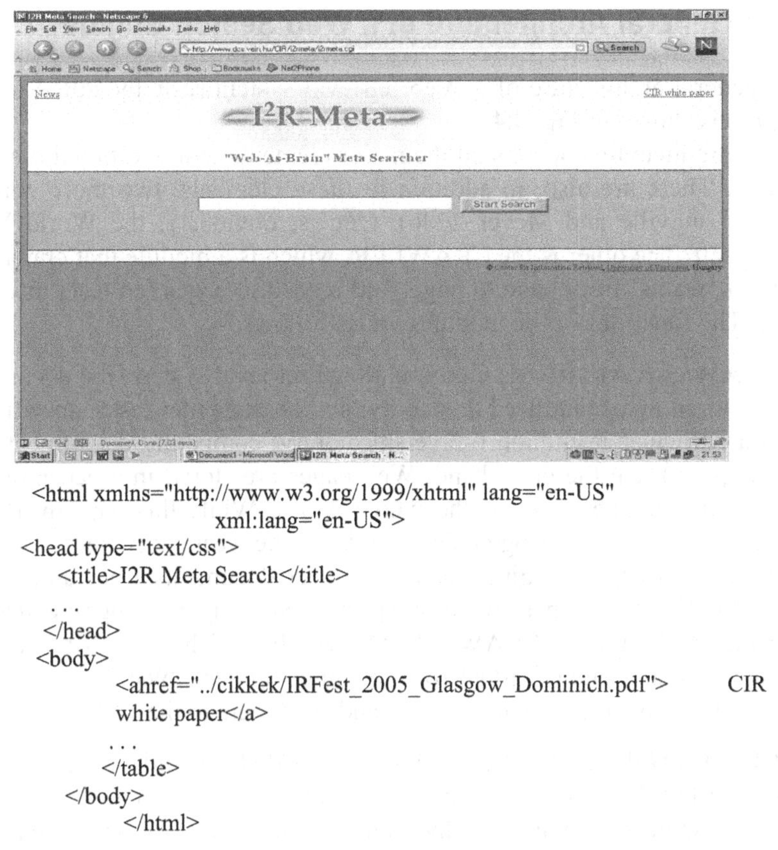


```
<html xmlns="http://www.w3.org/1999/xhtml" lang="en-US"
                    xml:lang="en-US">
<head type="text/css">
   <title>I2R Meta Search</title>
  . . .
  </head>
 <body>
        <ahref="../cikkek/IRFest_2005_Glasgow_Dominich.pdf">        CIR
        white paper</a>
        . . .
       </table>
   </body>
       </html>
```

Fig. 4.3. Example of a Web page (above: screen image, below: fragment of its HTML format).

All of the aforementioned characteristics (and others not touched upon here) represent challenges to Web retrieval. Web retrieval methods and systems should be able to:

- Address these characteristics (e.g., take into account the languages spoken by a user or his/her fields of interest).
- Cope with the dynamic nature of the Web (i.e., for instance, to observe when a new page has been added or a page deleted, or to realize that a link between two pages has disappeared, etc.).
- Scale up with size (i.e., for instance, the computational complexity, and thus physical running time, of the retrieval methods and algorithms used should be kept within polynomial limits such that running time does not exceed certain acceptable limits).

4.8.3 General Architecture of a Web Search Engine

The general architecture of a Web retrieval system (usually called *search engine*) is shown in **Fig. 4.4**.

The architecture contains all the major elements of a traditional retrieval system. There are also, in addition to these elements, two more components (Langville and Meyer 2006). One is, obviously, the World Wide Web itself. The other is the CRAWLER, which is a module that crawls the Web: it 'walks' from page to page, and reads the pages (collects information). The functions of the modules are as follows:

- **CRAWLER MODULE**. In a traditional retrieval system, the documents are stored in a centralized repository, i.e., on computer disks, specifically in a particular institution (university library, computing department in a bank, etc.). On the other hand, Web pages are stored in a decentralized way in computers around the whole world. While this has advantages (e.g., there are no geographic boundaries between documents), it also means that search engines have to collect documents from around the world. This task is performed by specialized computer programs that together make up the CRAWLER MODULE, which have to run all the time, day and night. Virtual robots, named *spiders*, 'walk' on the Web, from page to page, download them, and send them to the REPOSITORY.
- **REPOSITORY**. The Web pages downloaded by spiders are stored in the REPOSITORY (which physically means computer disks mounted on computers belonging to the company that runs the search engine). Pages are sent from the REPOSITORY to the INDEXING MODULE for further processing. Important or popular pages can be stored for a longer (even a very long) period of time.
- **INDEXING MODULE**. The Web pages from the REPOSITORY are processed by the programs of the INDEXING MODULE (HTML tags are filtered, terms are extracted, etc.). In other words, a compressed representation is obtained for pages by recognizing and extracting important information.
- **INDEXES**. This component of the search engine is logically organized as an inverted file structure. It is typically divided into several substructures. The *content structure* is an inverted structure that stores, e.g., terms, anchor text, etc., for pages. The *link structure* stores connection information between pages (i.e., which page has a link to which page). The spider may access the link structure to find addresses of uncrawled pages. The inverted structures are physically implemented in compressed ways in order to save memory.

- **QUERY MODULE**. The QUERY MODULE reads in what the user has typed into the query line and analyzes and transforms it into an appropriate format (e.g., a numeric code). The QUERY MODULE consults the INDEXES in order to find pages that match the user's query (e.g., pages containing the query terms). It then sends the matching pages to the RANKING MODULE.
- **RANKING MODULE**. The pages sent by the QUERY MODULE are ranked (sorted in descending order) according to a similarity score. The list obtained is called a *hit list*, and it is presented to the user on the computer screen in the form of a list of URLs together with a *snippet* (excerpt from the corresponding page). The user can access the entire page by clicking on its URL. The similarity score is computed based on several criteria and uses several methods. (The most important methods will be dealt with in Chapter 11.) The similarity scores are calculated based on a combination of methods from traditional information retrieval and Web-specific factors. Typical factors are: page content factors (e.g., term frequency in the page), on-page factors (e.g., the position of the term in the page, the size of characters in the term), link information (which pages link to the page of interest, and which pages it links to), and so on.

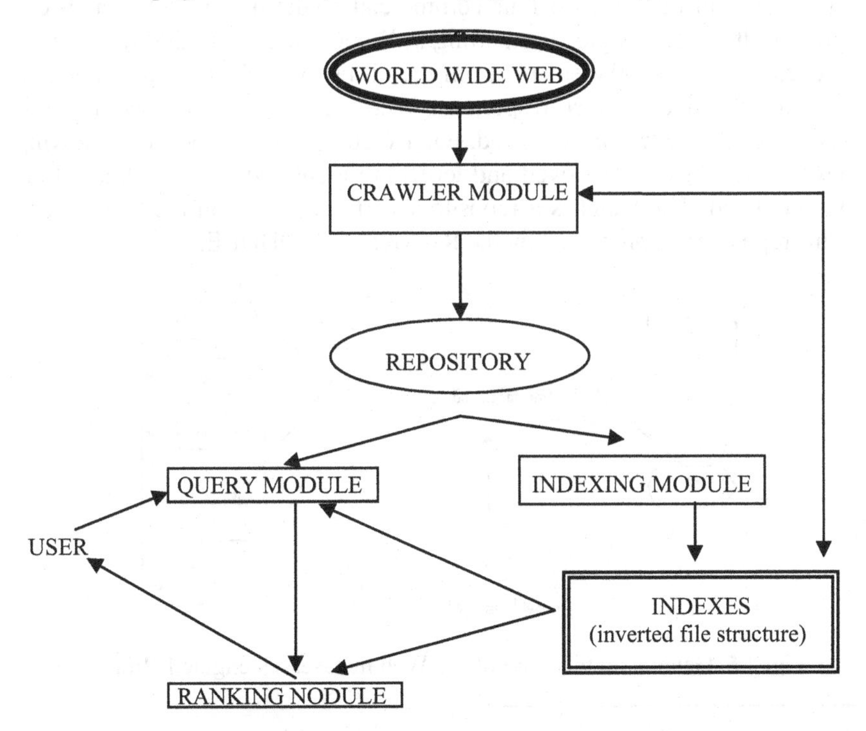

Fig. 4.4. General architecture of a Web search engine.

4.8.4 General Architecture of a Web Metasearch Engine

Web search engines are the most important retrieval systems used to find information on the Web.

Each search engine has its own ranking method, which is usually different from one used by another search engine. On the other hand, the hit list presented by a search engine can be very long in many cases (even in the millions), albeit that users typically consult at most 10–20 hits.

With the aim of returning fewer but more relevant pages (by taking advantage of different ranking methods simultaneously), *metasearch engines* can be developed. Typically, a metasearch engine reads in the user's request, sends it to several search engines, downloads some of the pages they return in response to the query, and then produces its own hit list using those pages. **Figure 4.5** shows the general architecture of the I^2RMeta metasearch engine[4] (whose interface screen is shown in **Fig. 4.2**) as an example of such an architecture (Dominich 2003).

- **INTERFACE MODULE.** It is written in PERL and works online. The communication with the Web server is performed by CGI. The query is entered as a set of terms (separated by commas); the terms are Porter-stemmed and then sent to four commercial spider-based Web search engines (Altavista, Google, Northernlight, WebCrawler as of 2003) as HTTP requests. The first 50 elements from the hit list of each Web search engine are considered, and the corresponding Web pages are downloaded in parallel (*parallel user agent*) for speed. Each Web page undergoes the following processing: tags are removed and terms are identified, stoplisted, and Porter-stemmed. The result is a repository of these pages on the server disk. This repository is processed by the RANKING MODULE.

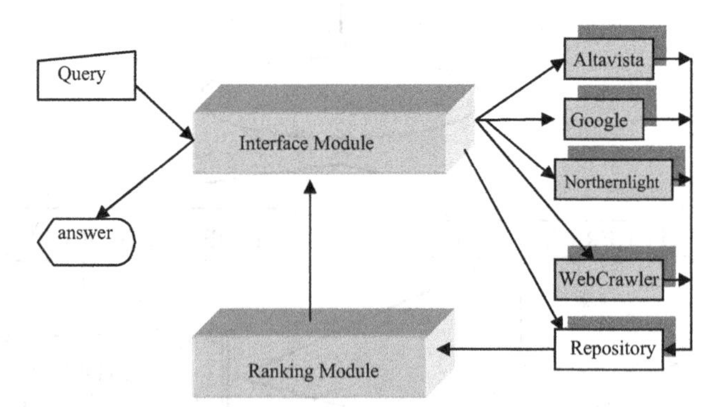

Fig. 4.5. General architecture of the Web metasearch engine I^2Rmeta.

[4] www.dcs.vein.hu/CIR

- **REPOSITORY MODULE.** It stores the data sent by the INTERFACE MODULE on the server disk, i.e., the transformed Web pages downloaded by the INTERFACE MODULE. This file is created "on the fly" during the process of answering the query.
- **RANKING MODULE.** This module is written in C and works online. Using the query and the Web pages in the repository, it creates a network based on page links as well as terms occurring in both pages and query. The hit list will contain the most important pages, i.e., the pages that are most strongly linked to each other, starting from the query. The hit list is sent to the INTERFACE MODULE, which screens it out (answer).

4.9 Measurement of Relevance Effectiveness

4.9.1 Relevance

In general, the meaning of the word *relevance* is: "A state or quality of being to the purpose; a state or quality of being related to the subject or matter at hand" [The *Cambridge English Dictionary*, Grandreams Limited, London, English Edition, 1990]. Relevance is a complex and widely studied concept in several fields, including philosophy, psychology, communication theory, artificial intelligence, library science, and so on. Yet, it is not completely understood, nor is it mathematically defined in an acceptable way.

Relevance also plays a major role in information science. Moreover, information science emerged *on its own* and *not as a part* of some other discipline because scientific communication has to deal not with *any* kind of information but with *relevant* information. The creators of the mathematical information theory, Shannon and Weaver (1949), begin their landmark book by pointing out that relevance is a central problem in communication: Is an American news program relevant to a Russian who does not speak English?

4.9.2 Measures

The *effectiveness* of an IR system (or method) means how well (or badly) it performs. Effectiveness is expressed numerically by *effectiveness measures*, which are elaborated based on different categories such as (Meadow et al. 1999):

- Relevance
- Efficiency

- Utility
- User satisfaction

Within each category, there are different specific effectiveness measures:

- Relevance: precision, recall, fallout, etc.
- Efficiency: cost of search, amount of search time, etc.
- Utility: worth of search results in some currency, etc.
- User satisfaction: user's satisfaction with precision or intermediary's understanding of request, etc.

Relevance effectiveness is the ability of a retrieval method or system to return relevant answers. The traditional (and widely used) measures are:

- *Precision*: the proportion of relevant documents out of those returned.
- *Recall*: the proportion of returned documents out of the relevant ones.
- *Fallout*: the proportion of returned documents out of the ones that are nonrelevant.

Obviously, these measures are neither unmistakable nor absolute. To quote Heine (1999): "The concept of relevance does not have a prior existence, but is rather created 'on the fly', at least in some cases." For instance, the estimation of recall requires the a priori (i.e., before retrieval) knowledge of the total number of relevant documents in the entire collection (for a given query). However paradoxical this may sound, experimental results have shown that users are more concerned with high recall than precision (Su 1994).

Attempts to balance these measures have been made and various other complementary or alternative measures have been elaborated. Cooper (1968) suggests expected search length, i.e., the number of nonrelevant documents before finding the relevant ones. Van Rijsbergen (1979) proposes a weighted combination of recall and precision:

$$1 - \frac{\alpha \times Precision \times Recall}{\beta \times Precision + Recall}. \tag{4.15}$$

Bollmann-Sdorra and Raghavan (1993) suggest another measure called R_{norm}:

$$R_{norm} = 0.5 \times (1 + R^{+} - I^{-}), \tag{4.16}$$

where R^+ denotes the number of times a relevant document occurs before a nonrelevant one in the retrieval order and I^- is the number of times a nonrelevant document occurs after a nonrelevant one.

In what follows, the following three widely accepted and used measures are defined:

- Precision
- Recall
- Fallout

The precision-recall measurement method of relevance effectiveness that is being used in laboratories is also delineated.

Let D denote a collection of documents and q a query. Further,

- $\Delta \neq 0$ denotes the total number of relevant documents to query q.
- $\kappa \neq 0$ denotes the number of retrieved documents in response to query q.
- α denotes the number of retrieved and relevant documents.

From the point of view of practice, it is reasonable to assume that the total number of documents to be searched, M, is greater than the number of those retrieved, i.e., $|D| = M > \Delta$. The usual relevance effectiveness measures are defined formally as:

1. *Recall* ρ is defined as $\rho = \dfrac{\alpha}{\Delta}$.

2. *Precision* π is defined as $\pi = \dfrac{\alpha}{\kappa}$.

3. *Fallout* φ is defined as $\varphi = \dfrac{\kappa - \alpha}{M - \Delta}$.

Figure 4.6 helps one to better understand the meaning of these measures. From the above definitions 1., 2., 3., it follows that:

- $0 \leq \rho \leq 1$.
- $0 \leq \pi \leq 1$.
- $\rho = 0 \Leftrightarrow \pi = 0$.
- $\pi = 1 \Leftrightarrow \varphi = 0$.
- $\alpha = \kappa = \Delta \Leftrightarrow (\rho = \pi = 1 \wedge \varphi = 0)$.

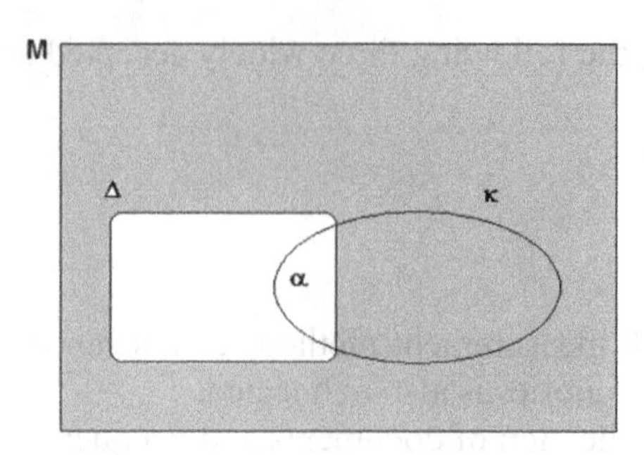

Fig. 4.6. Visual representation of quantities that define precision, recall, and fallout.

Other measures are as follows (van Rijsbergen 1979, Meadow et al 1999):

$$\pi + \rho, \tag{4.17}$$

$$\pi + \rho - 1, \tag{4.18}$$

$$\frac{\rho - \varphi}{\rho + \varphi - 2\rho\varphi}, \; 0 \le \varphi \le 1, \tag{4.19}$$

$$1 - \frac{1}{\frac{1}{2}\left(\frac{1}{\pi}\right) - \frac{1}{2}\left(\frac{1}{\rho}\right)}, \tag{4.20}$$

$$\text{F-measure: } \frac{2\rho\pi}{\rho + \pi}, \tag{4.21}$$

$$\text{Heine measure: } 1 - \frac{1}{\frac{1}{\pi} + \frac{1}{\rho} - 1}, \tag{4.22}$$

$$\text{Vickery measure: } 1 - \frac{1}{2\left(\frac{1}{\pi}\right) + 2\left(\frac{1}{\rho}\right) - 3}, \tag{4.23}$$

$$\text{Meadow measure: } 1 - \frac{\sqrt{(1-\pi)^2 + (1-\rho)^2}}{\sqrt{2}}. \tag{4.24}$$

From Salton and Lesk (1968):

$$R_{norm} = \frac{1}{M-\Delta}\sum_{i=1}^{M}\rho_i - \frac{\Delta+1}{2(M-\Delta)}, \tag{4.25}$$

where R_{norm} is normalized recall (for a given query), M is the number of documents, and ρ_i is the recall at the ith hit in the ranked hit list.

4.9.3 Precision-Recall Graph Method

The *precision-recall graph method* is used for the measurement of retrieval effectiveness under laboratory conditions, i.e., in a controlled and repeatable manner (Baeza-Yates and Ribeiro-Neto, 1999).

This measurement method employs *test databases* (*test collections*).[5] Each test collection is manufactured by specialists, and has a fixed structure:

- The documents d are given.
- The queries q are given.
- The relevance list is given, i.e., it is known exactly which document is relevant to which query.

For every query, retrieval should be performed (using the retrieval method whose relevance effectiveness is to be measured). The hit list is compared with the relevance list (corresponding to the query of interest). The following recall levels are considered standard:

0.1; 0.2; 0.3; 0.4; 0.5; 0.6; 0.7; 0.8; 0.9; 1.

(These levels can also be given as percents, e.g., 0.1 = 10%). For every query, pairs of recall and precision are computed. If the computed recall value is not standard, it is approximated. The precision values corresponding to equal recall values are averaged.

Let R_q denote the relevant documents to query q. Let us assume, for instance, that

$$R_q = \{d_2, d_4, d_6, d_5, d_9, d_1\}, \Delta = 6,$$

[5] For example, ADI, CRAN, TREC, etc.

and that the retrieval method under measurement returns the following ranked hit list (for q):

1. d_1—
2. d_8
3. d_6—
4. d_7
5. d_9—

where the "—" sign marks a relevant document (as a result of comparison with R_q).

Document d_1 is relevant, which means that one-sixth of the documents of R_q have been retrieved, and so precision is 100% at the recall level one-sixth. The third element, d_6, is also relevant. Thus, precision is two-thirds at recall level two-sixths. The fifth element of the hit list is d_9, which is also relevant. Hence, precision is three-fifths at the recall level three-sixths.

When the computed recall value r is not equal to a standard level, the following interpolation method can be used to calculate the precision value $p(r_j)$ corresponding to the standard recall value r_j:

$$p(r_j) = \max_{r_{j-1} < r \leq r_j} p(r), j = 1,\ldots,10. \tag{4.26}$$

It is known from practice that the values $p(r_j)$ are monotonically decreasing. Thus, the value $p(r_0)$ is usually determined to have $p(r_0) \geq p(r_1)$. For all queries q_i, the precision values $p_i(r_j)$ are averaged at all standard recall levels:

$$P(r_j) = \frac{1}{n}\sum_{i=1}^{n} p_i(r_j) \;, j = 0,\ldots,10, \tag{4.27}$$

where n denotes the number of queries used. **Figure 4.7** illustrates a typical precision-recall graph (for the test collection ADI).

The average of the values $P(r_j)$ is called MAP (*mean average precision*). MAP can also be computed just at the recall values 0.3, 0.6, and 0.9.

Apart from MAP, the following measures can also be used:

- *P@n* (*precision* at n): only the first n elements of every hit list are considered; typical values for n are 10, 20, 30, 100.
- *R-prec* (*R precision*): for each query q, only the first Δ_q elements of the hit list are considered (i.e., $\Delta_q = R_q$).

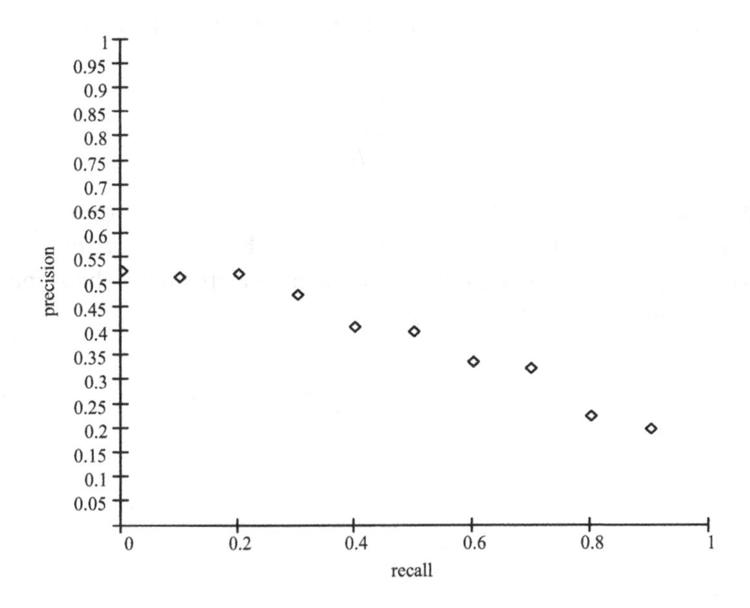

Fig. 4.7. Typical precision-recall graph (for the test collection ADI).

4.9.4 Uncertainty of Measurement

A test collection consists of three parts: documents, queries, and relevance assessments. All three parts are fixed and—usually—are provided as plain text files. Relevance assessments are produced by human experts and are provided as a table or list indicating which document is relevant to which query. Documents and queries typically are texts, shorter or longer, homogeneous or heterogeneous in content (e.g., taken from journals or newspapers). **Table 4.5** lists the names, the number of documents, and queries in the most commonly used classical test collections.

Table 4.5. Parameters of Classical Test Collections

Name	Number of documents	Number of queries
ADI	82	35
MED	1033	30
TIME	423	83
CRANFIELD	1400	225
NPL	11429	93
CACM	3204	64
CISI	1460	111

In Dominich (2001), it is shown that the following relationship holds for every query q:

$$\frac{\varphi\pi}{\rho(1-\pi)} = \frac{R}{M-R}. \tag{4.28}$$

The left-hand side of Eq. (4.28) defines a surface in three-dimensional Euclidean space called the *effectiveness surface* (**Fig. 4.8**). The effectiveness surface has the property that it has query-independent shape but a query-dependent actual position in space.

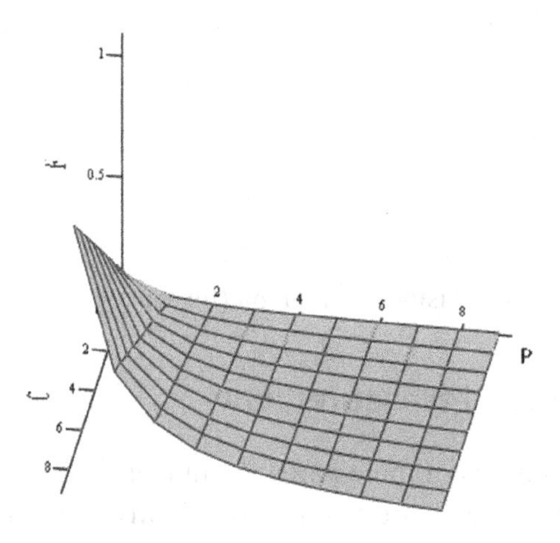

Fig. 4.8. A typical effectiveness surface. It has query-independent shape, whereas its specific position in the space depends on the query and the number of documents in the collection. C = recall (0 to 10 scale), F = fallout, P = precision (0 to 10 scale).

Let us denote the right-hand side of Eq. (4.28) by $f(R)$. From $R_1 \leq R_2$ it follows that $M - R_1 \geq M - R_2$, and thus $f(R_1) \leq f(R_2)$. In other words, $f(R)$ is monotonic with respect to R, i.e., $R_1 \leq R_2 \Leftrightarrow f(R_1) \leq f(R_2)$. If we take R_1 to correspond to the query that has the lowest number of relevant documents, $R_1 = R_{min}$ (R_{min} is the minimum number of relevant documents) and R_2 to correspond to the query that has the highest number of relevant documents, $R_2 = R_{max}$ (R_{max} is the maximum number of relevant documents), we find that the left-hand side of Eq. (4.28) is bounded for every query as follows:

$$f(R_{min}) \leq \frac{\varphi\pi}{\rho(1-\pi)} \leq f(R_{max}) \tag{4.29}$$

Table 4.6 shows the lower- and upper-bound values, $f(R_{min})$ and $f(R_{max})$ for widely used test collections.

Table 4.6. Lower- and Upper-Bound Values in Test Collections for the Effectiveness Surface

Name	**Number of documents**	**Minimum number of relevant documents**	**Maximum number of relevant documents**	**Lower-bound value**	**Upper-bound value**
	M	R_{min}	R_{max}	$f(R_{min})$	$f(R_{max})$
ADI	82	2	33	0.025	0.673
MED	1033	9	39	0.009	0.039
TIME	423	1	18	0.002	0.044
CRANFIELD	1400	2	40	0.001	0.029
NPL	11429	1	84	0.0001	0.074
CACM	3204	1	51	0.0003	0.016
CISI	1460	1	155	0.0007	0.118
Genomics 2005	4591008	2	709	0.000000435	0.000154
Disk12 TREC	741856	14	1141	0.00001887	0.00154
Disk45 TREC	528155	3	448	0.00000568	0.000848
Wt2g	247491	6	148	0.0000242	0.000598
Wt10g	1692096	1	519	0.00000059	0.000306
Terabyte	25205179	4	617	0.000000158	0.0000244

Owing to these limitations, even if a retrieval method can perform better, a test collection may not be 'able' to 'observe' it (or to 'show' it to us). Test collections only allow us to 'see' those values of effectiveness measures that are situated in the space region bounded by the lower and upper effectiveness surfaces (regardless of the retrieval method being tested). These space regions are specific for each test collection, as shown in **Fig. 4.9**:

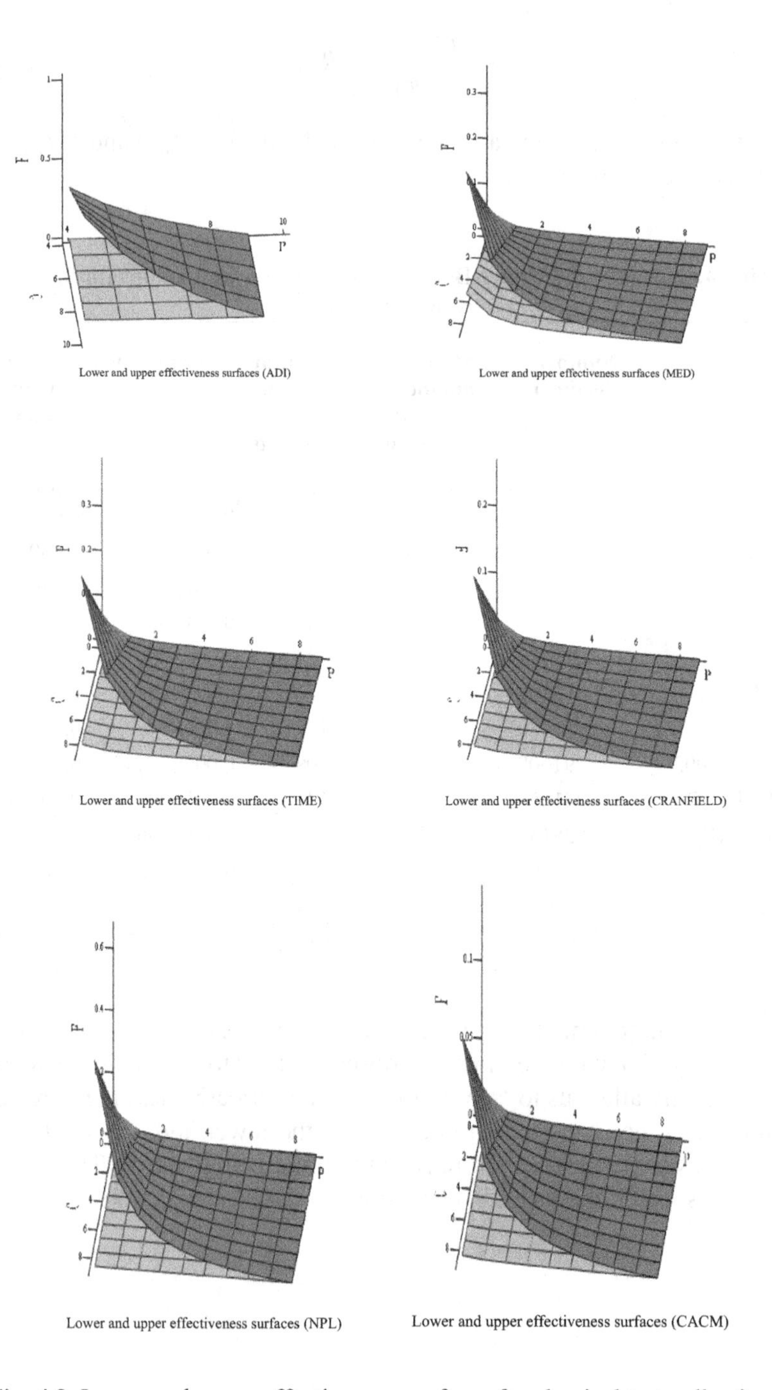

Lower and upper effectiveness surfaces (ADI)

Lower and upper effectiveness surfaces (MED)

Lower and upper effectiveness surfaces (TIME)

Lower and upper effectiveness surfaces (CRANFIELD)

Lower and upper effectiveness surfaces (NPL)

Lower and upper effectiveness surfaces (CACM)

Fig. 4.9. Lower and upper effectiveness surfaces for classical test collections.

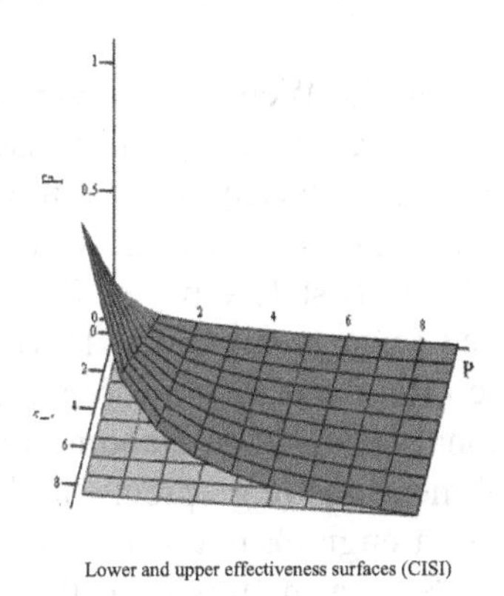

Lower and upper effectiveness surfaces (CISI)

Fig. 4.9. (Continued)

Using **Table 4.6**, we can perform calculations to obtain the following results.

ADI allows for 'seeing' both precision and recall in the range 0.4 to 0.8, but with fairly high fallout values and within a large range of approximately 0.97. MED and CRANFIELD allow for 'seeing' both precision and recall in the whole range 0 to 1 with fairly low fallout values within a range of approximately 0.378 and 0.252, respectively, dropping quickly for precision values from 0 to 0.2, and then decreasing slowly. NPL allows for 'seeing' both precision and recall in the whole range 0 to 1 with fallout values dropping quickly from the fairly high value of 0.666 to almost 0 at very low precision values, and then remaining near 0. CACM allows for showing both precision and recall in the whole range 0 to 1 with fairly low fallout values throughout, dropping quickly from 0.141 to almost 0 at very low precision values, and then remaining near 0. CISI allows for showing both precision and recall in the whole range 0 to 1 with fairly high fallout values at low to mid-precision values, dropping from 1 to almost 0, and then remaining near 0. These results may help in selecting which test collection to use for which purpose. Thus, if one wishes to measure precision and recall in an entire spectrum at low fallout values, then CACM, MED, or CRANFIELD is recommended. If, however, one wishes, for some reason, to monitor the sensitivity to fallout of a retrieval method being tested, then ADI or NPL would be recommended.

4.10 Measurement of Search Engine Effectiveness

Owing to the characteristics of the Web, the measurement of relevance effectiveness of a Web search engine is, typically, user centered (Borlund 2003). It is an experimentally established fact that most users generally examine the first two pages of a hit list. Thus, the search engine should rank the most relevant pages in the first few pages. The traditional measures cannot always be computed (e.g., recall and fallout). This means that the measurement of relevance effectiveness of search engines requires measures other than the traditional ones. When elaborating such new measures, one tries to use traditional measures (e.g., precision that can also be calculated for a hit list of a search engine), but also takes into account various characteristics of the Web. Several methods for the measurement of relevance effectiveness of a search engine have been elaborated thus far, and they can be grouped as follows:

User-Based Methods. These methods measure user satisfaction. In Nielson (1993), a method is given to measure utility and satisfaction. Su et al. (1998) involved real users to measure the effectiveness of the following search engines: Altavista, Infoseek, Lycos, OpenText. Tang and Sun (2003) co-opted Ph. D. students to measure the *20 full precision* for Google, Altavista, Excite, and Metacrawler using the following formula:

$$\frac{1}{20 \times 4} \sum_{i=1}^{20} (weight\_of\_i^{th}\_hit), \tag{4.30}$$

as well as *search length* as equal to *the number of irrelevant hits seen before getting i=2 relevant ones.*

Measurement of Precision. Chu and Rosenthal (1996) used ten queries to measure the precision of the search engines Altavista, Excite, and Lycos. A hit was relevant, irrelevant, or partially relevant (i.e., a page that was irrelevant but pointed to a relevant one).

In Gwizdka and Chignell (1999), a four-degree relevance scale (most relevant, partially relevant, hardly relevant, irrelevant) was used to propose different types of precisions:

- *Best precision* = the proportion of the most relevant hits.
- *Useful precision* = the proportion of the most relevant hits and of those which point to them.
- *Objective precision* = the proportion of hits containing the query.

They measured the effectiveness of Altavista, HotBot, and Infoseek, and found that Altavista's best precision was the highest.

Measurement of Recall. Clark and Willett (1997) proposed a method for the measurement of relative recall using the merged hit lists of several search engines. Shafi and Rather (2005) measured the recall and precision of Altavista, Google, HotBot, Scirus, and Bioweb with regard to finding scientific papers. Twenty queries were used from biotechnology, and the first ten hits from every hit list were examined. The following four-degree relevance scale was used: 3 = full paper, 2 = abstract, 1 = book, 0 = other. A relative recall was defined as follows: the relative recall of a given search engine was the proportion of the relevant hits returned by that search engine out of the total number of relevant hits returned by all search engines. They found that Scirus had the highest relative recall, whereas Bioweb had the lowest.

Measurement of Other Characteristics. Chu and Rosenthal (1996) studied several characteristics of Altavista, Excite, and Lycos: *coverage* (size of the index, update frequency of the index), search options (Boolean search, truncation, proximity search).

4.10.1 M-L-S Method

Leighton and Srivastava (1999) proposed a general method for the measurement of the extent to which a search engine is able to rank relevant hits within the first n hits of the hit list (*first n-precision*). The principles of the method are:

- Definition of relevance categories.
- Definition of groups.
- Weighting of hits.

Each hit on a hit list returned in response to a query was assigned to only one category. The hit list was divided into s_i groups having c_i weights ($i = 1,...,m$). The value of first n-precision was defined as the sum of the weights of relevant hits divided by the maximum sum. The method was applied to give a *first 20-precision* algorithm with which AltaVista, Excite, HotBot, Infoseek, and Lycos were measured (in 1997). It was found that AltaVista was the best.

Based on the above principles and taking into account that most users (85%) only assess at most the first two pages of a hit list (Silverstein et al. 1998), Dominich (2003) proposed the following measurement method, known as the *modified Leighton-Srivastava method* (*M-L-S method*) The M-L-S method measures the ability of a search engine to rank relevant hits within the first five or ten hits of the hit list.

M-L-S Method (First 5/10-Precision)

1. Select search engine to be measured.
2. Define relevance categories.
3. Define groups.
4. Define weights.
5. Give queries q_i (i = 1,...,s).
6. Compute $P5_i$ and/or $P10_i$ for q_i (i=1,...,s).
7. The first 5/10-precision of the search engine is:

$$Pk = \frac{1}{s}\sum_{i=1}^{s} Pk_i \text{, where } k = 5 \text{ or } k = 10.$$

The relevance categories are:

- 0—category (irrelevant hit).
- 1—category (relevant hit).

When measuring first 5-precision, the first five hits are assigned to one of two groups:

1. Group: the first two hits (on the grounds that they are usually on the first screen).
2. Group: the following three hits.

When measuring first 10-precision, the first ten hits are assigned to one of the following three groups:

1. Group: the first two hits.
2. Group: the next three hits.
3. Group: the rest of five hits.

Groups 1 and 2 are based on the assumption that, in practice, the most important hits are the first five (usually on the first screen).

Hits within the same group receive equal weights. The weights reflect the fact that the user is more satisfied if the relevant hits appear on the first screen. For first 5-precision, the weights are:

1. For group 1: 10.
2. For group 2: 5.

Obviously, instead of 10 and 5, other but proportional values may be used. For the first 10-precision, the weights are:

1. For group 1: 20.
2. For group 2: 17.
3. For group 3: 10.

Just as before, obviously, instead of 20, 17, and 10, other but proportional values may be used.

The definition of queries is a very important step. However, it is almost impossible to give a generally valid method for it. It is advisable to define a topic first, and the queries after that. The topic should be broad enough to be able to see how well the search engine performs at a general level. In order to avoid bias, define both general and specialized queries. As most users prefer unstructured queries, such queries should be defined. It is very important that the weights be defined prior to obtaining any hits, or else our assessments would be more subjective or biased (because, in this case, we already know how the search engine 'behaves' for certain queries).

The *P5* measure is defined as

$$P5 = \frac{r\_hit_{1.-2.hit} \times 10 + r\_hit_{3.-5.hit} \times 5}{35 - miss\_hit_{1.-5.hit} \times 5}, \tag{4.31}$$

where

- *r_hit* denotes the number of relevant hits in the respective group.
- The numerator is the weighted sum of the relevant hits within the first five hits.
- *miss_hit* denotes the number of missing hits,
- In the denominator, 35 is the weighted sum in the best case (i.e., when the first five hits are all relevant): (2 × 10) + (3 × 5) = 35. For every missing hit out of five, 5 is subtracted.

The measure *P5* is given for the case in which multiple hits are not penalized. If we want to penalize multiple hits, then a multiple hit is considered as many different hits as its multiplicity.

Example 4.2

Let us assume that in response to query "WWW" three hits are returned and that all are relevant. Thus, the numerator is (2 × 10) + (1 × 5) = 25. The first two hits belong to the first group, so their weight is 10. The third hit belongs to group 2; thus its weight is 5. The denominator is 35 – (2 × 5) = 25. So, *P5* = 25:25 = 1.

Let the query be "VLSI." Five hits are returned, out of which three are relevant: the second, the third, and the fourth. Thus, the numerator is (1 × 10) + (2 × 5) = 20, so *P5* = 20:35 = 0.571. If the first three hits were relevant, then *P5* = [(2 × 10) + (1 × 5)] : 35 = 0.714. The two values obtained for *P5* are different, which reflects the ranking difference of relevant hits.

Let us assume that for the query "Network" five hits are returned, and these are relevant, but the third and the fifth are the same (i.e., we have a double hit). In this case, we have $P5 = [(2 \times 10) + (2 \times 5)] : (35 - 1 \times 5) = 1$ (without penalty); and $P5 = [(2 \times 10) + (2 \times 5)] : 35 = 0.857$ (with penalty). It can be seen that taking penalty into account yields lower effectiveness. □

The $P10$ measure is defined in a similar manner:

$$P10 = \frac{r\_hit_{1.-2.hit} \times 20 + r\_hit_{3.-5.hit} \times 17 + r\_hit_{6.-10.hit} \times 10}{141 - miss\_hit_{1.-10.link} \times 10}. \tag{4.32}$$

The penalized version is similar to that for $P5$.

4.10.2 RP Method

We know that precision is defined as (Section 4.9.2):

$$p = \frac{r}{k}, \tag{4.33}$$

where p denotes precision, k the number of returned items, and r the relevant items out of the k returned.

A Web metasearch engine uses the hit lists of search engines to produce its own hit list. Thus, also taking into account the definition of precision, a method to compute a *relative precision* (called as the *RP method*) can be given (Dominich 2003). The idea of the RP method is that if the hits of a metasearch engine are compared to the hits of the search engines used, then a relative precision can be defined for the metasearch engine. We note that earlier Clark and Willett (1997) defined a relative recall measure in a similar manner.

Let q be a query. Let V be the number of hits returned by the metasearch engine being considered and T those hits out of these V that were ranked by at least one of the search engines used within the first m of its hits. Then, the *relative precision* $RP_{q,m}$ of the metasearch engine is calculated as follows:

$$RP_{q,m} = \frac{T}{V}. \tag{4.34}$$

The value of m can be, e.g., $m = 10$ or $m = 5$, or some other value depending on several factors (the range of the measurement, etc.). The value of relative precision should be computed for several queries, and an average should be taken.

Example 4.3

Let us assume that a metasearch engine uses four search engines. Let the query q be "Download ICQ Message Archive," and let us assume further that the metasearch engine returns five hits, i.e., $V = 5$.

By analyzing the hit lists of all the search engines, we see that the first hit of the metasearch engine is the third on the hit list of the first search engine, the second hit was the first in the second search engine, the third was the fourth in the third search engine, the fourth was the second in the fourth search engine, and the last one was the third in the second search engine.

Thus, $T = 5$, and for $m = 10$ the relative precision is $RP_{q,10} = 5{:}5 = 1$. □

RP Method

(Relative Precision of a Web Metasearch Engine)

1. Select the metasearch engine to be measured.
2. Define queries q_i, $i = 1,...,n$.
3. Define the value of m; typically $m = 5$ or $m = 10$.
4. Perform searches for every q_i using the metasearch engine

as well as the search engines used by the metasearch engine, $i = 1,...,n$.

5. Compute relative precision for q_i as follows: $RP_{q_i,m} = \frac{T_i}{V_i}$, $i = 1,...,n$.

6. Compute average: $\sum_{i=1}^{n} RP_{q_i,m}$.

The *RP* method relies heavily on the hypothesis that the hit lists of search engines contain relevant hits. In other words, the *RP* measure is only as good as the hit lists.

4.11 Exercises and Problems

The exercises below are best solved, and can only be really helpful, if there is a computing infrastructure (computer, software, test databases) at your disposal. Of course, some calculations (e.g., the computation of a weight) can be done by hand, but the actual usefulness of the retrieval technologies can only be grasped and felt in a real computer setting.

1. Take a collection of texts of your choice (e.g., papers, stories, etc.). Verify the validity of the power law using different methods: least squares, linear regression. (*Note*: The collection should be fairly large to obtain meaningful results. Thus, you should not work manually. Write adequate computer programs.)

2. Create term-document matrices using the weighting schemes given in Theorem 4.1. Observe, analyze, and compare the running times necessary to create the matrices. Discuss memory usage to store the matrices on disk and in the main memory. (Try to use matrix storage methods that allow economical storage.) Observe and discuss the relation between economical storage and the ease of using the matrices in computations.

3. Analyze in more depth the characteristics of the World Wide Web. Identify and discuss characteristics other than those presented in Section 4.8.1.

4. Using a standard test collection or a data collection of your choice, measure relevance effectiveness of a retrieval method of your choice using the precision-recall graph method. Experiment with other interpolation (averaging) formulas [other than Eqs. (4.25) and (4.26); e.g., instead of maximum use average in Eq.(4.25)].

5 Lattice-Based Retrieval Systems

Have in mind the physical methods and mechanisms used to instrument models.
(Calvin N. Mooers)

This chapter describes the application of lattices in retrieval systems (Mooers, FaIR, BR-Explorer, Rajapakse-Denham, FooCA). The use of lattices for visualization or navigation is not considered, nor are programming and implementation issues dealt with, as these fall outside our scope.

The goal of describing these systems is to present the way in which lattices have been used to represent document structures, term relationships, and term-document matrices in actual retrieval systems developed thus far.

Mathematical properties (with proofs) of the lattices applied are established. The principal finding is that they are not modular.

Further, a method is given to transform a term-document matrix into a Galois (concept) lattice.

The chapter ends with exercises and problems that are designed to enhance understanding of the properties of the lattices applied.

5.1 Mooers' Model

Mooers (1959) seems to have been the first individual to offer a detailed and comprehensive treatment of the application of the lattice concept in IR. His model has the merit of being able to capture relationships between terms (i.e., to formally express the fact that words are not necessarily independent of each other, which is a widely accepted hypothesis in many retrieval methods today). The model focuses on what Mooers calls symbols (known today as terms) that together form a query and on the relationship between the query and a subset of documents that can be selected from a collection of documents.

5.1.1 Lattice of Documents

The document subsets can be formed from, e.g., a library collection (books, articles, etc.). If one denotes a document subset by A and the entire library collection by $L = \{D_1,\ldots,D_n\}$, then the document subsets $A \subseteq D$ form a Boolean algebra $(\wp(L), \cap, \cup, \backslash)$ with respect to set intersection $\cap$, set union $\cup$, and set complement $\backslash$, where $\wp(L)$ denotes the powerset of L, i.e., the set of all subsets of L. In other words, the structure $(\wp(L), \cap, \cup, \backslash)$ is a complemented and distributive lattice.

5.1.2 Lattice of Unstructured Queries

A query Q is conceived as consisting of one or several terms (in the latter case constructed from one-term queries). For example, the one-term query $Q = A$ is modeled as the following lattice: $\{\mathbf{0}, A\}$ (**Fig. 5.1**):

Fig. 5.1. One-term query $Q = A$ represented as a lattice $\{\mathbf{0}, A\}$, $\mathbf{0} \leq A$.

A new lattice P can be obtained by taking the product $\times$ of one-term lattices (**Fig. 5.2**):

1. Let A_i denote (or correspond to) the one-element lattice $\{\mathbf{0}, A_i\}$, $i = 1,\ldots,n$.
2. The product lattice $P = \times_{i=1}^{n} \{\mathbf{0}, A_i\}$ is given by the lattice $P=(\wp(T), \subseteq)$, where $T = \{A_1,\ldots,A_n\}$.

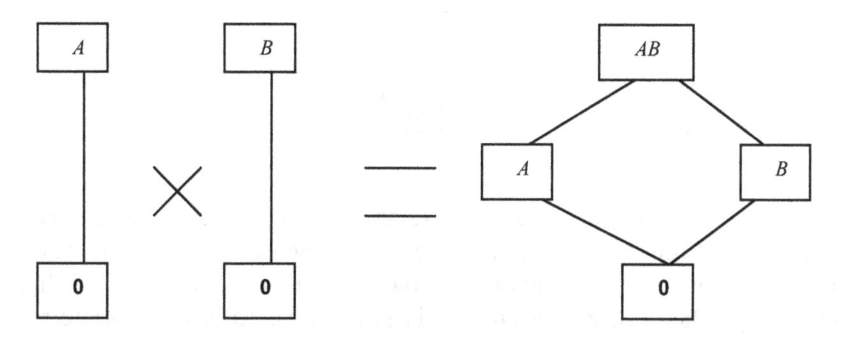

Fig. 5.2. Product P of two one-term lattices: $\{\mathbf{0}, A\}$ and $\{\mathbf{0}, B\}$.

The lattice P contains all the possible queries that can be formed using the given terms. It can be seen that the lattice P is a Boolean algebra and thus a complemented and distributive lattice.

Retrieval is formally viewed as follows. Given a query Q, i.e., an element of lattice P, there may be other elements in P preceded by Q, or elements that precede Q. Retrieval is some procedure that locates those elements of $\wp(T)$ that precede and are preceded by Q. However, the retrieval procedure is not described.

5.1.3 Lattice of Term Hierarchies

Mooers treats, very briefly, the case of term hierarchies. For example, the term "clothing" is broader in sense than the term "shoe." Thus, the following lattice (reflecting a hierarchy) can be constructed:

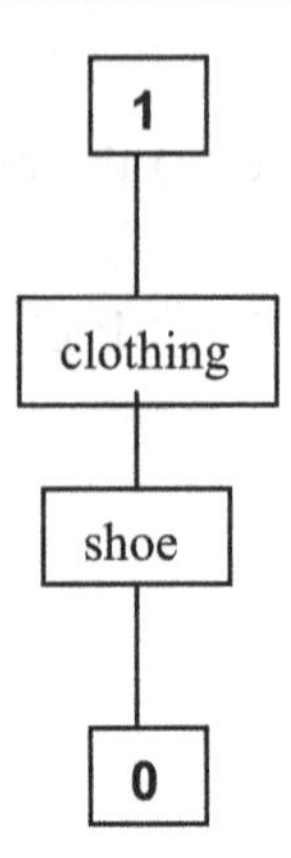

Taking into account hierarchical relations between terms should have an impact upon retrieval. In such lattices, some elements may be preceded by more than one element; this is referred to as a "system with weak hierarchy" (e.g., the U.S. Patent Office classification). After appropriate reductions, another lattice can be obtained from this one that does not allow for any element to be preceded by more than one element. Such a reduced lattice is referred to as a "system with strong hierarchy" (e.g., the Dewey Decimal classification).

5.1.4 Lattice of Boolean Queries and Documents

Mooers also outlines the case of Boolean (i.e., structured) queries (which he calls "characters with logic"). He starts by emphasizing that, "symbolic logic is a stylized view of things, and the symbolism or method which is found useful in that discipline need not necessarily be the most appropriate symbolism for information retrieval."

Given the terms $A_1,\ldots,A_n$, a Boolean query is conceived as being an element of a lattice $(L, \subseteq)$ obtained as follows:

1. The atoms are $A_1,\ldots,A_n$.
2. The elements greater than the atoms, and immediately above them, are given by the Cartesian product $\{A_1, \neg A_1\} \times \{A_2, \neg A_2\} \times \ldots \times \{A_n, \neg A_n\}$, where $\neg$ denotes negation.
3. The elements of point (2) are "topped" by the maximal element of the lattice.

If the query is a conjunction of terms (e.g., $A_1 \wedge A_2$), all terms are equally relevant to the subject matter of the document. When the query is a disjunction of terms (e.g., $A_1 \vee A_2$), then—based on the interpretation of $\vee$ in mathematical logic—either one, or either two, and so on either all terms may be relevant to the subject matter of the document. "This is ridiculous," —Mooers says. He continues by asking: "For example, how good is a retrieval system that treats the query

$$\text{red} \vee \text{square}$$

as a logical expression?" Thus, he notes:

- Disjunction should not be a permissible operation in queries for retrieval; the only permissible operations should be conjunction and negation.
- On the other hand, a document should be represented as a lattice using only negation and disjunction of terms.

For example, if two terms are used, say A and B, then all possible documents (i.e., the 'space' whose element a document may be) are given by the lattice shown in **Fig. 5.3**.

The document lattice can be obtained as a product of two-element lattices. **Figure 5.4** shows the two 2-element lattices whose product lattice is shown in **Fig. 5.3**.

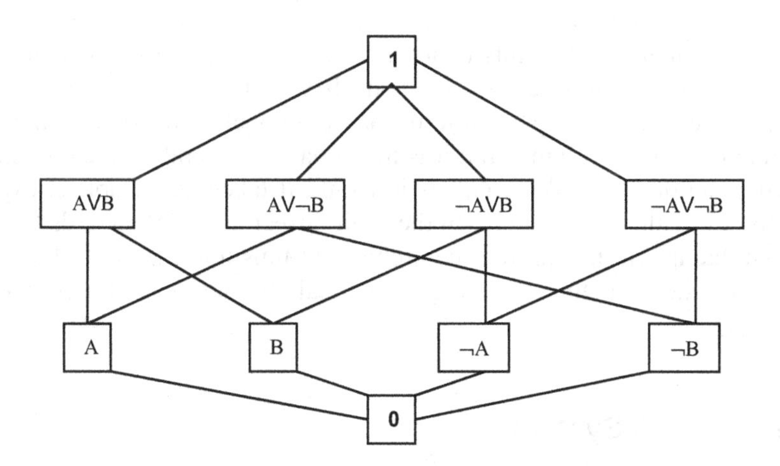

Fig. 5.3. Document lattice for Boolean queries using two terms A and B.

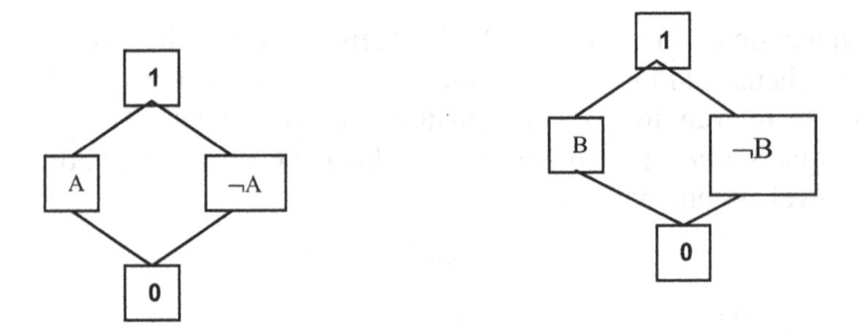

Fig. 5.4. The two 2-element lattices whose product is the lattice in Fig. 5.3.

The query lattice is similarly generated, but instead of disjunction we have conjunction. Mooers does not treat retrieval for such representations. At the same time, he makes a number of general and interesting observations:

- If a term A has not been used as an index term for a document, then the query "$\neg A$" should not retrieve that document merely because it does not contain A. In other words, in retrieval, absence is not necessarily synonymous with negation.
- It may be important to know the frequency with which terms are used. Thus, one can attach frequencies as "scalars" to lattice elements.
- In Boolean logic, the operations commute, e.g., $A \vee B$ is the same as $B \vee A$. If the information in a document is structured as a lattice, the ideas are commutative. But this is not always the case. The words that make up a term may form a sequence, but they do not always commute. For example, "street lamp" as a term may be modeled formally as the conjunction "street $\wedge$ lamp" but it is not equal, in general, to the commuted conjunction "lamp $\wedge$ street," as it should if taken as a Boolean expression of mathematical logic. In the term "street lamp," the words "street" and "lamp" do not form a hierarchy (and thus not a lattice); they form some other structure (e.g., a grammatical structure called a compound word).

5.2 The FaIR System

For a long time after Mooers's paper was published, the application of lattices in information retrieval was seen more as a theoretical possibility or curiosity. There were no further developments until Priss (2000) proposed a practical retrieval system, called FaIR, based on lattices, and thus showed that such a retrieval system could be built effectively.

The FaIR system uses domain knowledge (in the form of a thesaurus) to generate term lattices. The thesaurus consists of a set T of terms that is partitioned into classes (called facets). The facets are lattices. Every node in such a lattice represents a term (word or phrase), and the lattice expresses thesaurus relationships such as "broader than," "narrower than," etc. Every such lattice, i.e., facet, is conceptually complete (their terms express one concept).

Documents are represented by (or assigned to) as many terms as needed, but at most one term from one facet. Documents that contain only one of the terms of the facet are mapped to that concept. Documents that contain several terms of the facet are mapped to the join of the concepts. **Figure 5.5** illustrates an example of a facet lattice.

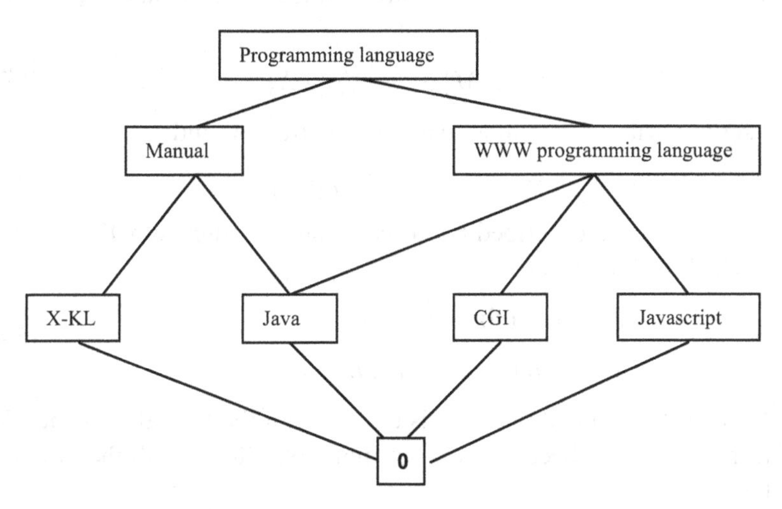

Fig. 5.5. Facet lattice "programming language." A document containing both CGI and Java is assigned to the node "WWW programming language," which is the join of these terms.

The elements of the facet lattice may be formally conceived as containing (i.e., being equal to the union of) the elements that are below it (equivalently, down to atoms). Every document is mapped to a single concept over all facets.

A query Q is a Boolean expression of terms. For example, the query Q = "Java" retrieves the documents containing exactly and only the term "Java" (in the exclusive search) and the documents that also contain the more general term "WWW programming language" (in the inclusive search).

5.3 Galois (Concept) Lattice-Based Models

5.3.1 Galois (Concept) Lattice

Concepts are basic units of language and thought. The *extension* G of a concept consists of all objects that belong to it. The *intension* M of a concept consists of all attributes (properties) that apply to the elements of its extension. A formal context K is defined as a binary relation, i.e., as the set of relationships between objects and attributes to denote which object has a given property (Wolff 1993, Kim and Compton 2004, Wille 2005):

$$K = (G, M, I), I \subseteq G \times M. \tag{5.1}$$

The following *derivation operations* are defined for arbitrary $X \subseteq G$ and $Y \subseteq M$:

$$X \rightarrow X^I = \{m \in M \mid g\,I\,m, \forall g \in X\}, \tag{5.2}$$

i.e., the set of attributes common to all objects from X, and

$$Y \rightarrow Y^I = \{g \in G \mid g\,I\,m, \forall m \in Y\}, \tag{5.3}$$

i.e., the set of objects described by at least one attribute from Y. A *formal concept* (A, B) is defined as

$$\begin{gathered}(A, B) \text{ is a formal concept} \Leftrightarrow \\ (A \subseteq G, B \subseteq M, A = B^I, B = A^I).\end{gathered} \tag{5.4}$$

The set A is the *extent* and the set B is the *intent* of the formal concept. The set of formal concepts becomes a poset [notation: $\Re(K)$] with the ordering relation

$$\begin{gathered}(A_1, B_1) \leq (A_2, B_2) \Leftrightarrow \\ A_1 \subseteq A_2 (\Leftrightarrow B_2 \subseteq B_1).\end{gathered} \tag{5.5}$$

The poset $\Re(K)$ can be turned into a complete lattice, denoted by $\underline{\Re}(K)$, with the following definitions of infimum and supremum:

$$\begin{gathered}\text{infimum: } \wedge_j (A_j, B_j) = \left(\bigcap_j A_j, \left(\bigcup_j B_j\right)^{II}\right) \\ \text{supremum: } \vee_j (A_j, B_j) = \left(\left(\bigcup_j A_j\right)^{II}, \bigcap_j B_j\right).\end{gathered} \tag{5.6}$$

Concept lattices are useful for the representation of conceptual structure of data. There are efficient procedures for constructing formal concepts and the concept lattice from a given formal context (Kim and Compton 2004, Wille 2005).

5.3.2 Term-Document Matrix and Concept Lattice

Cheung and Vogel (2005) view the term-document matrix $TD_{n,m}$ (in its Boolean form, i.e., adjacency matrix, **Table 5.1**) as a formal context that is transformed into a concept lattice (**Fig. 5.6**).

Table 5.1. Term-Document Matrix

	D_1	D_2	D_3	D_4
T_1	1	1	0	0
T_2	1	0	1	0
T_3	0	1	0	1
T_4	0	0	1	1

Thus, term T_1 occurs in documents D_1 and D_2, term T_4 occurs in documents D_3 and D_4, and so on. The term-document matrix can be transformed into a concept lattice using the following method:

Generation of a Concept Lattice from the Term-Document Matrix

1. The least element, **0** (as well as the greatest element, **1**) of the concept lattice is introduced artificially.
2. The lattice is built in a bottom-up fashion (i.e., from **0** to **1**).
3. Every term T_i corresponds to an atom.
4. For every column j (j = 1,...,m), if $TD_{i,j} = TD_{k,j} = 1$, then document D_j is the meet (superconcept) of terms T_i and T_k.

This method can be applied even when the TD matrix is not Boolean, i.e., when it contains (nonbinary) weights. In this case, the condition in point 4 is rewritten as $TD_{i,j}$, $TD_{k,j} \neq 0$. (The superconcepts, as given by the TD matrix, may not be unique. However, in the concept lattice only one is allowed.)

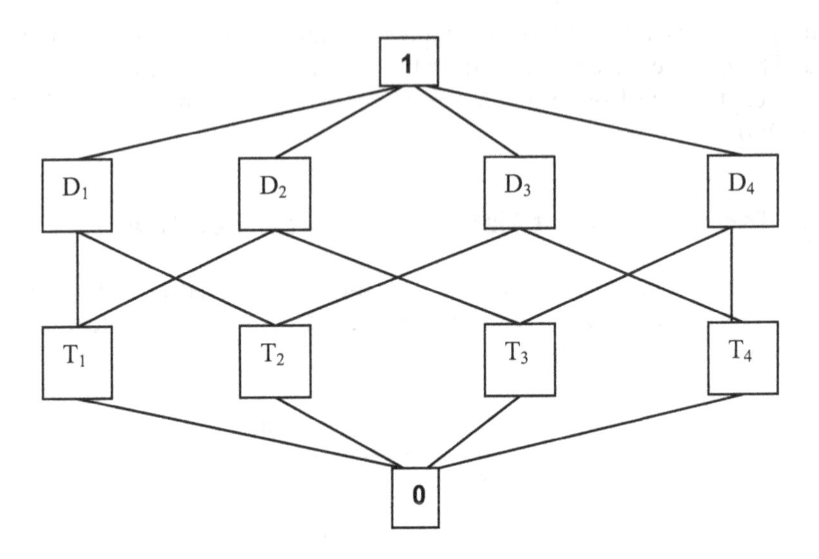

Fig. 5.6. Concept lattice obtained from the term-document matrix of Table 5.1.

A method to obtain a lattice from a term-document matrix is given in (Godin et al. 1989).

Each element of the lattice is a couple (d, t) such that

- d is the set of documents described by at least the terms in t.
- t is the set of terms common to all the documents in d.

Thus, t is the set of terms appearing in a conjunctive query retrieving exactly the documents in d. The set of all such couples is a lattice with the following partial order defined from the corresponding order on the term sets:

$$c_1 = (d_1, t_l) < c_2 = (d_2, t_2) \Leftrightarrow t_1 \subset t_2. \quad (5.7)$$

Equivalently,

$$t_1 \subset t_2 \Leftrightarrow d_2 \subset d_1. \quad (5.8)$$

The partial order is used to generate the Hasse diagram of the lattice. There is an edge (c_l, c_2) if $c_l < c_2$, and there is no other element c_3 such that $c_l < c_3 < c_2$.

5.3.3 BR-Explorer System

Messai et al. (2006) propose a retrieval system called BR-Explorer based on concept lattices. The term-document Boolean matrix (conceived as a formal context) is first transformed into a concept lattice L. A query Q is conceived as being a set of terms (attributes). In order to answer query Q, this is inserted into the concept lattice L first (e.g., by building L from scratch, or using some more efficient method as suggested by Messai et al.).

Relevance is defined as follows. The document d is relevant to query Q if they share at least one attribute. Messai et al. give a retrieval method (i.e., traversal of the concept lattice) that retrieves documents already ranked by their relevance degree.

5.3.4 Rajapakse-Denham System

Rajapakse and Denham (2006) proposed another application of lattices to IR. Documents and queries are represented as individual lattices. Concepts extracted from documents are used to construct a lattice. A document or query is conceived as a structure of objects, attributes, and their relationships from which a lattice is generated. The atoms are the elements consisting of objects that have identical attributes. On the next level, the elements are the objects that share most of their attributes, and so on. The smallest element of the lattice, at the bottom, is an artificial empty element, whereas the largest element, on the top, is the union of all the objects.

Retrieval is defined as follows. The relevance of a document to a query is determined on the basis of their common concepts. This is achieved by comparing nodes of the query lattice with the nodes of the document lattice. A partial match between the query lattice and the document lattice is defined as being the meet between their corresponding objects and attributes. When any of these two meets are empty, a keyword match is applied.

Rajapakse and Denham showed that their model worked by building an experimental system whose relevance effectiveness was measured on the Cranfield test collection. Moreover, the system was enhanced with a (personalized) learning strategy. In a relevance feedback process, all the terms of the query that are not present in a relevant document are added to that document. Weighting strategies were also used to refine the model.

5.3.5 The FooCA System

The FooCA system applies formal concepts to enhance Web retrieval (Koester 2006a,b, 2005). The system runs under Linux and is written in PERL. FooCA lets the user enter a query, which it sends to Google (other Web search engines can also be used).

The hit list returned by the search engine is used to construct a formal context and formal concepts as follows. The snippets (i.e., the short excerpts returned by the search engine) are used to extract terms. If there is no snippet, then the page address is used instead. The URLs of pages are viewed as objects, whereas its terms are viewed as attributes.

The hit list returned by the search engine is presented to the user as a table in which the rows correspond to objects (in the ranked order given by the search engine) and the columns to attributes. The table can be navigated using the mouse, and the corresponding row is highlighted. When the user clicks on a row, he/she is taken to the corresponding page.

The table can also be used for query refinement. If the user clicks on an attribute, then he/she can launch another search using the clicked attribute as the query or he/she can include or exclude that term into/from the original query.

FooCA can be used to visualize the hierarchy of formal concepts, which helps the user to assess the hits better and thus to know which hit to view first.

5.3.6 Query Refinement, Thesaurus Representation

Carpineto and Romano (2005) offer an excellent overview of other uses of concept lattices in retrieval. One such application of concept lattices is query refinement. Given a Boolean query (i.e., a Boolean expression of terms), the matching documents are found first. Then, the set of common terms in the retrieved documents is determined and used to build a concept lattice. The query can be refined by choosing the most general term (concept) that contains all the query terms.

Another use of lattices in retrieval is the representation of a thesaurus as a concept lattice by taking into account the ordering suggested by the thesaurus. The concept lattice of a document collection may be used as an underlying clustering structure. The query is merged into this lattice. Each document is ranked according to the shortest path between the query and the document concept.

Concept lattices can also be used to bound the search space or for navigational purposes. A possible application relates to Web searching. A set of pages returned by a search engine is parsed, and a concept lattice is built using the pages as objects and the terms as attributes. The user is presented with this lattice to initiate the next search interaction.

5.4 Properties of the Lattices Applied

As seen in the retrieval systems that have been described, lattices are used as mathematical models of the entities involved: objects and their relationships (document sets, structure of a document, within document-term relationships, queries, term relationships in general, and concepts). These lattices have different meanings (or roles) such as query, document, or concept. According to their role, they are subjected to appropriate processing. Retrieval is defined as a matching between a document and a query lattice, or between lattices (a query lattice and a document lattice). Different specific matching algorithms were proposed.

The lattices used are complex (albeit that there are attempts to find methods to reduce their complexity), and their construction is not an easy task.

Godin et al. (1998) showed that the number H of nodes in a concept lattice has linear complexity with the number n of documents:

$$H = \mathrm{O}(n \cdot 2^k), \tag{5.9}$$

where O denotes "big-Oh" (upper bound) from computational complexity, and k denotes the maximum number of terms/document. Experimental results showed that the ratio H/n was fairly constant and much lower than 2^k.

Cheung and Vogel (2005) applied SVD (singular value decomposition) to reduce the size of the term-document matrix, whereby the corresponding lattice became considerably smaller than the one corresponding to the original matrix.

From a purely formal point of view, the application of lattices in retrieval systems can be characterized as follows. Let

$$T = \{t_1, t_2, \ldots, t_n\} \tag{5.10}$$

denote a set of elements (e.g., terms, documents, etc.). Several types of lattices are defined over T, such as:

- Atomic, complete.
- Boolean algebra.
- Complete, atomic, complemented, nonmodular (hence not distributive).

- Complete, atomic, nonmodular (hence not distributive), not complemented.

We show now that, in Mooers's model, the following property holds:

Theorem 5.1. *The lattices of Boolean documents and queries are*

1. *Atomic and complete.*
2. *Complemented.*
3. *Not modular.*

Proof (using **Fig. 5.3** as a model). Point (1) is straightforward. To prove point (2), note that the complement A^C of any atom A is equal to any expression of which A is not a member. The complement A^C of any nonatom A is equal to its negated counterpart. For point (3), we give a counterexample. By definition, a lattice L is modular if

$$(\forall A, B, C \in L \text{ for which } A \leq C) \Rightarrow$$
$$A \vee (B \wedge C) = (A \vee B) \wedge C.$$

For example, $A \subseteq C = \{A, B\}$, i.e., $A \leq (A \vee B)$ (recall that $\vee = \cup$ and $\wedge = \cap$). By taking $B = \{\neg A, \neg B\}$, we now have

$$A \cup (\{\neg A, \neg B\} \cap \{A, B\}) = A \cup \mathbf{0} = A,$$

which is not equal to

$$(A \cup \{\neg A, \neg B\}) \cap \{A, B\} = \mathbf{1} \cap \{A, B\} = \{A, B\}. \square$$

Further, we can prove that the facet lattices used in the FaIR system have the following property:

Theorem 5.2. *The facet lattice is*

1. *Atomic and complete.*
2. *Not modular.*

Proof. Point (1) is straightforward. For point (2), we give a counterexample (using **Fig. 5.5**). For A = 'X-KL', B = 'CGI' and C = 'Manual', we have $A \leq C$, and

$$A \cup (B \cap C) = A \cup \mathbf{0} = \text{“X-KL”},$$

which is not equal to

$$(A \cup B) \cap C = \text{“Programming language”} \cap C = \text{“Manual.”} \square$$

Before proceeding with the analysis of the properties of lattices used in retrieval systems, we introduce further concepts related to lattices.

Definition 5.1. Two lattices $(L_1, \wedge_1, \vee_1)$ and $(L_2, \wedge_2, \vee_2)$ are *isomorphic* if there exists a bijective function $f: L_1 \rightarrow L_2$ such that

$$f(a \wedge_1 b) = f(a) \wedge_2 f(b),$$
$$f(a \vee_1 b) = f(a) \vee_2 f(b), \forall\, a, b \in L_1. \square \tag{5.11}$$

Definition 5.2. The structure $(L_1, \wedge, \vee)$, where $L_1 \subseteq L$ and $L_1 \neq \varnothing$, is a *sublattice* of the lattice $(L, \wedge, \vee)$ if

$$a \wedge b \in L_1,\ a \vee b \in L_1, \forall a, b \in L_1. \square \tag{5.12}$$

Definition 5.3. A lattice L_1 can be *embedded* into a lattice L_2 if there exists a sublattice of L_2 isomorphic to L_1. □

There is an important relationship between the pentagon lattice and nonmodularity, namely:

Theorem 5.3. (Burris and Sankappanavar, 2000) *A lattice L is not modular if and only if the pentagon lattice can be embedded into L.*

Proof. It is clear that if the pentagon lattice can be embedded into a lattice L, then L is not modular (because the pentagon lattice itself is not modular; see Section 3.6). In order to prove the reverse, let us assume that L is not modular. This means that for $A, B, C \in L$ such that $A \leq C$ we have $A \vee (B \wedge C) < (A \vee B) \wedge C$. Let $D = A \vee (B \wedge C)$, then

$$\begin{aligned} B \vee D = B \vee (\, A \vee (B \wedge C)\,) = \\ B \vee (\, (B \wedge C) \vee A\,) = \\ (\, B \vee (B \wedge C)\,) \vee A = \\ B \vee A. \end{aligned}$$

Now let $E = (A \vee B) \wedge C$; then

$$\begin{aligned} B \wedge E = \\ B \wedge (\, (A \vee B) \wedge C\,) = \\ (\, B \wedge (A \vee B)\,) \wedge C = \\ B \wedge C. \end{aligned}$$

$D < E$ by assumption. Also $B \wedge C < A \vee (B \wedge C) = D$. Thus $B \wedge C < D < E$, and so

$$\begin{aligned} B \wedge C < \\ B \wedge D < \\ B \wedge E = B \wedge C. \end{aligned}$$

Hence, $B \wedge D = B \wedge E = B \wedge C$. Likewise, $B \vee E = B \vee D = B \vee A$. **Figure 5.7** shows the copy of the pentagon lattice in L.

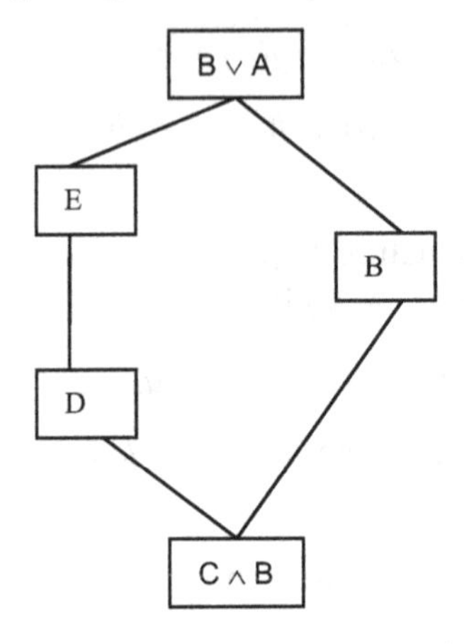

Fig. 5.7. Copy of the pentagon lattice as an embedded lattice.

□

Wille (2005) gives many examples of concept lattices, which model a wide range of different real situations:

- Geography (bodies of water).
- Sociology (economic concepts of young persons).
- Geometry (types of triangles, inversion of a circle).
- Medicine (examination of anorectic patient, functional rooms in a hospital, Ph-level of children with diabetes).
- Tourism (leisure activities).
- Information science (information and knowledge processing).
- Urbanism (town and traffic).
- Music (musical attributes).
- Biology (animals).

The concept lattice of the Ph-level of children with diabetes is shown in **Fig 5.8**. (This concept lattice was generated using the data from 111 children and 22 attributes in collaboration with medical experts.) It can be seen that this concept lattice is nothing other than the pentagon lattice, and hence it is not modular.

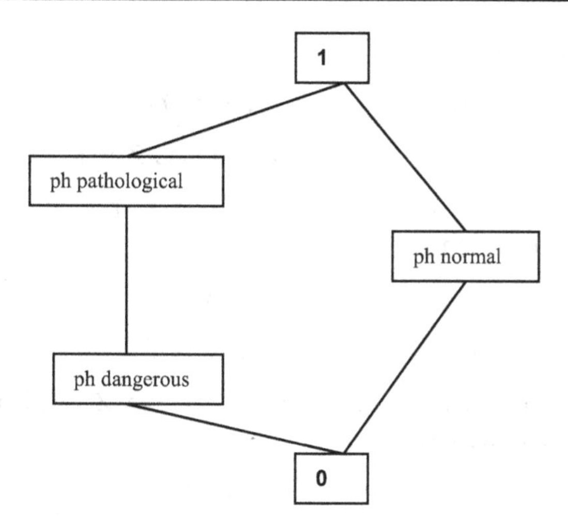

Fig. 5.8. Concept lattice of Ph-level of children with diabetes.

If one examines the other lattices given by Wille, it turns out that other concept lattices (e.g. those of economic concepts of young persons, of types of triangles, and of animals) also have sublattices isomorphic with the pentagon lattice, which means that they are not modular. Indeed, we can show that in general:

Theorem 5.4. *Galois (concept) lattices are not modular.*

Proof. A lattice L is, by definition, modular if

$$(\forall A, B, C \in L \text{ for which } A \leq C) \Rightarrow$$
$$A \vee (B \wedge C) = (A \vee B) \wedge C.$$

For Galois lattices, the definition of modularity becomes

$$(\forall (X_1, X_2), (Y_1, Y_2), (Z_1, Z_2) \in L \text{ such that } (X_1, X_2) \leq (Z_1, Z_2)) \Rightarrow$$
$$(X_1, X_2) \vee ((Y_1, Y_2) \wedge (Z_1, Z_2)) =$$
$$((X_1, X_2) \vee (Y_1, Y_2)) \wedge (Z_1, Z_2).$$

Using the definitions of join $\vee$ and meet $\wedge$ in concept lattices, and the hypothesis $(X_1, X_2) \leq (Z_1, Z_2)$, we rewrite the modularity condition as follows:

$$((X_1 \cup (Y_1 \cap Z_1))^{II}, X_2 \cap Y_2^{II}) =$$
$$((X_1 \cup Y_1)^{II} \cap Z_1, ((X_2 \cap Y_2) \cup Z_2)^{II}),$$

which is equivalent to

$$(((X_1 \cup Y_1) \cap Z_1)^{II}, X_2 \cap Y_2) =$$
$$((X_1 \cup Y_1)^{II} \cap Z_1, ((Z_2 \cap (Y_2 \cup Z_2)^{II}).$$

Let us take now $(X_1, X_2) \leq (Y_1, Y_2)$, i.e., $X_1 \subseteq Y_1, Y_2 \subseteq X_2$. Then, the previous condition is rewritten as

$$((Y_1 \cap Z_1)^{II}, Y_2) =$$
$$(Y_1^{II} \cap Z_1, ((Z_2 \cap (Y_2 \cup Z_2)^{II}).$$

The equality holds when $(Y_1 \cap Z_1)^{II} = Y_1^{II} \cap Z_1$, which is true because $Y_1^{II} = Y_1$, and so we have:

$$(Y_1 \cap Z_1)^{II} = Y_1 \cap Z_1;$$

and when

$$Y_2 = (Z_2 \cap (Y_2 \cup Z_2)^{II} = Z_2 \cap (Y_2 \cup Z_2).$$

But if $Y_2 \subset Z_2$, then $Z_2 \cap (Y_2 \cup Z_2) = Z_2 \cap Z_2 = Z_2$, which is different from Y_2. □

However, in nonmodular lattices certain pairs of elements may satisfy the modularity condition.

Definition 5.4. An ordered pair (A, B) of elements—i.e., in this order A is the first element of the pair and B is the second element of the pair—of a lattice L is referred to as a *modular pair* (notation: $A\mathrm{M}B$) if

$$\begin{aligned} &(\forall\, C \in L \text{ such that } C \leq B) \Rightarrow \\ &C \vee (A \wedge B) = (C \vee A) \wedge B. \end{aligned} \qquad (5.13)$$
□

If A and B are not modular pairs, then this is denoted by $A\not\mathrm{M}B$. Albeit that Galois (concept) lattices are not, in general, modular, in the concept lattices in Wille (2005), which all have a sublattice isomorphic to the pentagon lattice, there are modular pairs that correspond to the following modular pair in the pentagon lattice:

$$\begin{aligned} &\mathbf{0} \leq x \Rightarrow \\ &\mathbf{0} \vee (x \wedge y) = \mathbf{0} \vee x = x = \\ &(\mathbf{0} \vee x) \wedge y = x \wedge y \\ &= x. \end{aligned} \qquad (5.14)$$

This means that y and x are a modular pair: $y\mathrm{M}x$. A Galois lattice has a sublattice isomorphic with the pentagon lattice. Hence, it also has a modular pair of elements. We should note, however, that $x\not{M}y$. We have $x \leq y$, but $x \vee (z \wedge y) = x$, which is not equal to $(x \vee z) \wedge y = y$. This means that x and y are not a modular pair: $x\not{M}y$.

It is known that the logical propositions of mathematical logic form a complemented and distributive lattice $(\{T, F\}, \vee, \wedge, \neg)$. As concept lattices are not modular (Theorem 5.4), they are not distributive either. Hence, the main difference between concept lattices and the lattice of propositions relates to the presence/absence of distributivity. In logic, distributivity is a property that connects conjunction and disjunction and is the expression of compatibility between any two propositions P and Q in the sense that

$$(P \wedge Q) \vee (P \wedge \neg Q) = P \vee (Q \wedge \neg Q) = P. \tag{5.15}$$

The fact that concept lattices are not distributive means that, in general, there are objects and/or properties that are not compatible, i.e., about which we cannot always reason in the sense of mathematical logic. For example, in the concept lattice of **Fig. 5.6**, we have

$$(D_1 \wedge D_2) \vee (D_1 \wedge \neg D_2^{C}) = T_1, \tag{5.16}$$

which means that reasoning with documents D_1 and D_2 does not result in some other document, but rather in a term (which is a different type of entity). In other words, in concept lattices, reasoning may lead to an object having a different nature or quality than the nature of objects on which reasoning has operated (a situation unimaginable in mathematical logic).

5.5 Exercises and Problems

1. Using a document collection of your choice, construct the corresponding concept lattice (using the term-document matrix).
2. Show that facet lattices are not distributive.
3. Are concept lattices uniquely complemented?
4. Is the Boolean algebra of documents uniquely complemented?
5. Are concept lattices orthomodular?

6 Boolean Retrieval

To give expression to the fundamental laws of those operations of the mind by which reasoning is performed in the symbolic language of Calculus...but did not try to treat the mysterious depths of actual thought.
(George Boole)

The Boolean retrieval method is a very important one as it is widely used in database systems (e.g., Oracle, SQL) and World Wide Web search engines. In principle, it is a simple method, but all the more important for that.

This chapter describes the Boolean retrieval method (both formally and using an example) and the application of lattices in Boolean retrieval. An effective method is presented to answer Boolean queries in relational databases.

The chapter ends with exercises and problems that are designed to promote a deeper understanding of the theory and applications of Boolean retrieval.

6.1 Boolean Retrieval Method

The *Boolean retrieval method*, which is used by virtually all commercial database and retrieval systems today, is based on mathematical logic and set theory. Both the documents to be searched and the user's query are conceived as sets of terms. Retrieval is based on whether or not the documents contain the query terms. There is a finite set

$$T = \{t_1, t_2, \ldots, t_j, \ldots, t_m\} \tag{6.1}$$

of elements called *terms* (e.g., words or expressions—which may be stemmed—describing or characterizing documents such as, e.g., keywords given for a journal article), and a finite set

$$D = \{D_1, \ldots, D_i, \ldots, D_n\},\ D_i \in \wp(T) \tag{6.2}$$

of elements called *documents*.

Traditionally, a "real" document can be a journal article (or its abstract or title), or a newspaper article, etc. These documents are formally conceived, for retrieval purposes, as being represented by sets of terms. Practically, the original document and its representation are two different entities. In principle, however, from a formal mathematical point of view, it is not a restriction to refer to document representations as documents, for two reasons:

- There is a correspondence between the original document and its representation.
- For retrieval purposes, the document representation is used rather than the original document.

Given a Boolean expression—in a normal form—Q called *query* is express as

$$Q = \bigwedge_k \left(\bigvee_j \theta_j\right),\ \theta_j \in \{t_j, \neg t_j\}. \tag{6.3}$$

Equivalently, Q can also be given in a disjunctive normal form. As any Boolean expression can be transformed into an equivalent normal form [i.e., the original Boolean expression and its normal form have the same logical value for the same truth values of the variables (Kneale and Kneale 1962)], any query Q, which may be any arbitrary Boolean expression of terms, can be transformed into an equivalent normal form as shown by Eq. (6.3). Viewing Q as a normal form allows for a compact formal description of the Boolean retrieval method:

1. Sets S_j of documents are obtained that match Q, i.e., $S_j = \{D_i | \theta_j \in D_i\}$, where $\neg t_j \in D_i$ means $t_j \notin D_i$.
2. The documents that are retrieved in response to Q are those that are the result of set operations corresponding to the logical operators in Q (i.e., set union corresponds to disjunction, and set intersection corresponds to conjunction): $\bigcap_k (\bigcup_j S_j)$.

Example 6.1

Let the set of original documents be $O = \{O_1, O_2, O_3\}$, where

O_1 = Bayes's principle: The principle according to which, when estimating a parameter, one should initially assume that each possible value has equal probability (uniform prior distribution).
O_2 = Bayesian decision theory: A mathematical theory of decision-making that presumes utility and probability functions, and according to which the act to be chosen is the Bayes act, i.e., the one with highest subjective expected utility.
O_3 = Bayesian epistemology: A philosophical theory that holds that the epistemic status of a proposition (i.e., how well proven or well established it is) is best measured by a probability and that the proper way to revise this probability is given by Bayesian conditionalization or similar procedures. A Bayesian epistemologist would use probability to define concepts such as epistemic status, support, or explanatory power and explore the relationships among them.

Let the set T of terms be:

$$T = \{t_1 = \text{Bayes's principle}, t_2 = \text{probability}, t_3 = \text{decision-making}\}.$$

Then, the set D of documents is as follows: $D = \{D_1, D_2, D_3\}$, where

$$D_1 = \{\text{Bayes's principle, probability}\},$$
$$D_2 = \{\text{probability, decision-making}\},$$
$$D_3 = \{\text{probability}\}.$$

Let the query Q be:

$$Q = \text{probability} \land \text{decision-making}.$$

Step 1: The following sets S_1 and S_2 of documents D_i are obtained:

$$S_1 = \{D_i | \text{probability} \in D_i\} = \{D_1, D_2, D_3\},$$
$$S_2 = \{D_i | \text{decision-making} \in D_i\} = \{D_2\}.$$

Step 2: The following documents D_i are retrieved in response to Q:

$$\{D_i | D_i \in S_1 \cap S_2\} = \{D_1, D_2, D_3\} \cap \{D_2\} = \{D_2\}.$$

This means that the original document O_2 (corresponding to D_2) is the answer to Q. □

Obviously, if there is more than one document with the same representation, every such document is retrieved. Such documents are, in Boolean retrieval, indistinguishable (or, in other words, equivalent).

Example 6.2

Let us consider the following document: O_4 = Bayes's principle is used to construct an equation that is of basic importance in probability theory. With the set T of index terms of Example 6.1, O_4 has the same representation, D_4, as O_1, i.e., $D_4 = D_1$ = {Bayes's principle, probability}, and thus O_1 and O_4 are indistinguishable. □

6.2 Technology of Boolean Retrieval

From a formal mathematical point of view, the Boolean retrieval method is fairly simple. However, it is all the more important because every Web search engine offers Boolean search capability, as does any database management system. The basic technology for Boolean retrieval is described in Chapter 4 (inverted file structure). Step 1 usually means a binary search in the inverted file structure, which results in the sets S_j. Another possibility (e.g., when computer memory is large enough) is the use of a binary term-document matrix:

Boolean Retrieval Using Binary Term-Document Matrix

1. Construct a term-document matrix $TD = (w_{ji})_{m \times n}$, where $w_{ji} = 1$ if term t_j occurs in document D_i, and $w_{ji} = 0$ otherwise.
2. Formulate a Boolean query Q = $\underset{k}{\wedge}(\underset{j}{\vee} t_j)$.
3. The sets S_j of documents that match Q are: $S_j = \{D_i | t_j \in D_i\} = \{D_i | w_{ji} = 1, i = 1,\ldots,n\}$, i.e., as given by the jth row of matrix TD: $row_{TD}(j)$.
4. The documents retrieved in response to Q are the result of logical operations on the rows selected in Step 3, i.e., $\underset{k}{\wedge}(\underset{j}{\vee} row_{TD}(j))$.

6.3 Lattice-Based Boolean Retrieval

In principle, the lattice characterization of Boolean retrieval is simple. As we have already seen, documents are represented as sets of terms. Thus, they are elements of the document lattice $L_D = \wp(T)$ encountered in Mooers's model (Section 5.1).

In database operations, retrieving data from a database is an important topic in both theory (data model) and practice (running time). In relational databases, which are based on entity-relationship models (Ullman 1980), data are grouped into entity sets. Retrieval of data (in response to a query) is performed by accessing instances (i.e., entity sets) of the entities involved.

The notion of Boolean algebra (i.e., complemented and distributive lattices) can be applied to design a retrieval method that is more efficient (in terms of the number of accesses to the database and thus in terms of running time) than traditional methods (Yang and Chen 1996).

In what follows, the method is described together with an example.

Step 1. Let *ENT* denote an entity type as well the corresponding entity set (which may seem strange, but in our context it will not lead to any confusion). In other words, *ENT* denotes a table in which

- The columns correspond to attributes

$$A_1,\ldots,A_i,\ldots,A_n$$

  (i.e., properties, terms),

- The rows correspond to entity instances (i.e., documents) that contain specific values of the attributes (e.g., weights).

Table 6.1 shows an example for the entity set *ENT*. The attributes of *ENT* are as follows: A_1 = Name, A_2 = Age, A_3 = Job, A_4 = Salary.

Table 6.1. Entity set *ENT*

Entity instance	Name	Age	Job	Salary
e_1	B	68	Secretary	21000
e_2	C	46	Professor	52000
e_3	L	24	Professor	38000
e_4	M	52	Secretary	35000
e_5	W	43	Professor	42000
e_6	D	25	Professor	36000
e_7	O	21	Secretary	18000
e_8	P	25	Secretary	30000

Step 2. The attributes of *ENT* are partitioned. $P_i = \{A_{i1},...,A_{ik}\}$ denotes a partition of attribute A_i. For example, let the partition of attribute A_3 = Job be P_3 = {Secretary, Professor}. (*Note:* The attribute "Professor" can be further partitioned into Professor = {Assistant, Associate, Full}. Depending on the needs of the application of interest, any attribute may be broken down into further partitions.)

The partition of A_i can be represented as a graph G_i in which A_i corresponds to the root of the graph, while the other vertices and the edges reflect the dependence relations in the partition. This graph is a lattice and is called *descendance graph* or *d-graph*. **Figure 6.1** shows the *d*-graph of P_3.

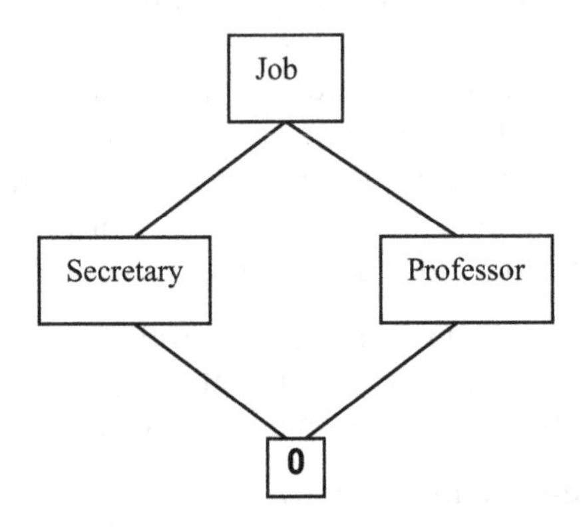

Fig. 6.1. The *d*-graph (partitioning lattice) of attribute "Job."

Let $D_i = \{v_{i1},...,v_{ij},...,v_{ik}\}$ denote the vertices of *d*-graph G_i. It can be seen that the *d*-graph is the Hasse diagram of the lattice $(D_i, \leq_i)$, where $v_{ij} \leq_i v_{is}$ if v_{ij} is a descendant of v_{is}, $\forall j,s$. This lattice is referred to as the *partitioning lattice* L_i of A_i (**Fig. 6.1**).

The attribute "Age" can be partitioned as follows. If Age ≤ 40, then he/she is Young, and Notyoung otherwise. The attributes "Name" and "Salary" form their own partitions. **Figure 6.2** shows the four partition lattices L_1, L_2, L_3, L_4 thus obtained.

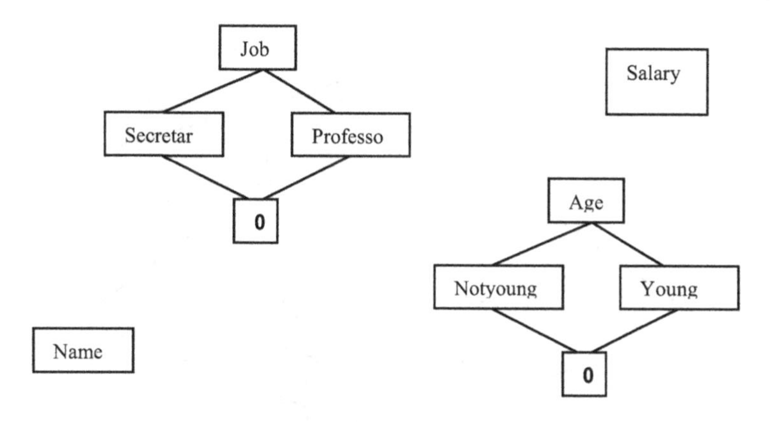

Fig. 6.2. The partition lattices for the entity *ENT*.

Step 3. Build the product lattice $L = \times_i L_i$. In our example,

$$
\begin{aligned}
L &= L_1 \times L_2 \times L_3 \times L_4 \\
&= \{a_1, a_2, a_3, a_4, a_5, a_6, a_7, a_8, a_9\} \\
&= \{ \quad \text{(Name, Age, Job, Salary)}, \\
&\qquad \text{(Name, Age, Secretary, Salary)}, \\
&\qquad \text{(Name, Age, Professor, Salary)}, \\
&\qquad \text{(Name, Notyoung, Job, Salary)}, \\
&\qquad \text{(Name, Notyoung, Secretary, Salary)}, \\
&\qquad \text{(Name, Notyoung, Professor, Salary)}, \\
&\qquad \text{(Name, Young, Job, Salary)}, \\
&\qquad \text{(Name, Young, Secretary, Salary)}, \\
&\qquad \text{(Name, Young, Professor, Salary)} \}.
\end{aligned}
$$

Figure 6.3 shows the Hasse diagram of the product lattice L for entity *ENT*.

Step 4. The entity set is divided into as many groups $n_1,...,n_k$ as the number of atoms in the product lattice L, based on the meaning of the atoms. For the entity set *ENT*, the number of atoms is $k = 4$, and the atoms are a_5, a_6, a_8, and a_9. For the entity *ENT*, the groups are:

$n_1 = e_1, e_4$ (i.e., the two entity instances in which Secretary is Notyoung),

$$n_2 = e_7, e_8,$$
$$n_3 = e_2, e_5,$$
$$n_4 = e_3, e_6.$$

Let $S = \{n_1,...,n_k\}$ denote the set of groups. Then, $(\wp(S), \subseteq)$ is a Boolean algebra.

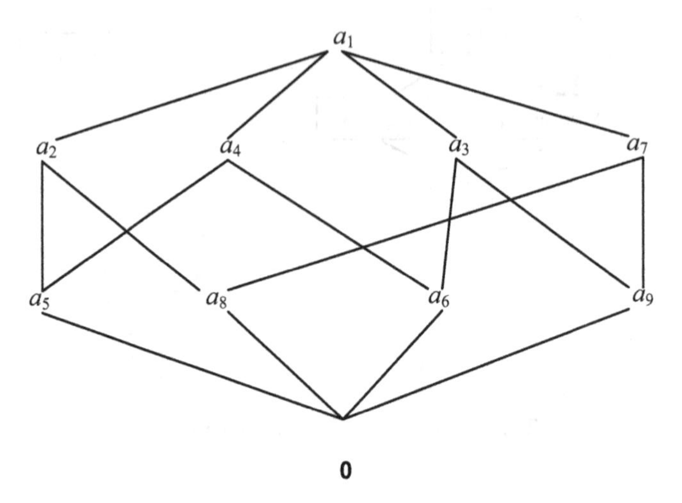

Fig. 6.3. Product lattice L for the entity *ENT*.

Step 5. The Boolean lattice $(\wp(S), \subseteq)$ can be used to answer queries as follows. In practice, we do not have to construct the entire lattice. By using the operations in a Boolean algebra, we only have to know what the atoms are, and then we can use the usual set operations $\cup$, $\cap$, and \ to retrieve data. This retrieval method is shown for the following query (as an example):

> What are the names of the employees who are secretaries or who are young?

Using the notations:

$$\text{A} = \{\text{Name, Secretary}\},\ \text{B} = \{\text{Young, Job}\},$$

the query as well as the result is:

$$\text{A} \cup \text{B} = \{n_1, n_2\} \cup \{n_2, n_4\} = \{n_1, n_2, n_4\} = \{e_1, e_4, e_7, e_8, e_3, e_6\}.$$

6.4 Exercises and Problems

1. Given the following information need: "I am interested in a route planner to plan a European journey by car." Formulate Boolean queries expressing this information need and experiment by supplying them to several Web search engines.

2. Given the following information need: "I am interested in the opening hours of chair museums in the United Kingdom, but I do not want any hits on cars having type Seat or on chairs of organizations." Formulate Boolean queries expressing this information need and perform searching by supplying them to several Web search engines.

3. Draw and discuss a parallel between the Boolean retrieval method and the functioning of the selection operator $\sigma_E(R)$ encountered in relation algebra (database theory), where E denotes a Boolean expression (i.e., the selection criteria), and R denotes a relation (i.e., a table from which the rows satisfying E are selected).

4. Use an appropriate software tool (e.g., database management system, programming language) to implement a small retrieval system using the Boolean retrieval method for a small text collection of your choice.

5. Let D denote a set of car parts (e.g., pedal, speedometer, seat, weight sensor, etc.) and T denote a set of symbols (or names) of electronic components (e.g., LED, chip of some kind, etc.) used in car parts. T indexes D. Develop a retrieval system using the Boolean retrieval method to find components for car parts.

6. Let D denote a set of medical images, namely CT (computer tomography) images of the human brain as follows: $D = \{d_1,\ldots,d_i,\ldots,d_n\}$, where d_i denotes a package of image slices, $i = 1,\ldots,n$. Each d_i is associated with a medical report (which is a piece of text) written by a neurologist when he/she examines the CT images of a patient. Develop a retrieval system for the use of physicians using the Boolean retrieval method to find the images of a specific patient having a specific diagnosis.

7 Lattices of Subspaces and Projectors

Nature does not demand a numeric description of us.
(Imre Fényes)

This chapter presents the notions and results (metric space, complete space, linear space, subspace, linear operator, Banach space, Hilbert space, Euclidean space, projection theorem, projector, lattice of subspaces) that are applied in Chapters 8 and 9 (on vector space retrieval and algebra-based retrieval methods).

Every notion is illustrated with detailed intuitive or mathematical examples to promote better understanding of their meaning.

The Gram-Schmidt procedure for defining an orthonormal basis of a subspace of a linear space is described, the lattice of closed subspaces of a Hilbert space is defined, and the way in which this lattice expresses the underlying geometry of the space is shown.

The chapter ends with exercises and problems in IR.

7.1 Metric Space

Let X denote a set. A function

$$\delta: X \times X \to \mathbb{R}_+, \tag{7.1}$$

where $\mathbb{R}_+$ denotes the set of positive real numbers, and $X \times X$ denotes the Cartesian product of set X with itself, is called a *pseudometric* if the following properties hold:

1. $x = y \Rightarrow \delta(x, y) = 0$.
2. $\delta(x, y) = \delta(y, x), \forall x, y \in X$. (symmetry)
3. $\delta(x, z) \leq \delta(x, y) + \delta(y, z), \forall x, y, z \in X$ (triangle inequality).

If, in addition to properties (1)–(3), the function δ obeys also the property:

4. $\delta(x, y) = 0 \Rightarrow x = y$,

then δ is called a *metric*. Set X with a (pseudo-)metric δ is called a *(pseudo-)metric space*, expressed as (X, δ).

The notion of metric space is a very important one (used in several scientific disciplines), and it may not always be an easy or trivial one. The concept of metric space embodies two major characteristics:

- A "distance" ("closeness") can be measured between any two elements of the space.
- The "distance" obeys (1)–(4) above, which are special rules (e.g., symmetry means that the same "distance" is measured from an element x to an element y as from y to x).

Example 7.1

(a) From an intuitive point of view, the concept of metric can be well illustrated by the notion of the usual physical distance in the three-dimensional physical space in which we live our everyday lives. When, e.g., the plan of the base of a future house is drawn on the ground, the ground is conceived as a two-dimensional space (i.e., a plane) in which we use the well-known Euclidean distance as a metric to measure physical distances and to make the drawing (plan) of the future base. In this case, the ground and the "meter" together constitute a metric space (**Fig. 7.1**). Using the meter, we can measure the physical distance between any two points on the ground. Indeed, in a metric space, the distance between any of its elements is defined (and, hence, can be measured).

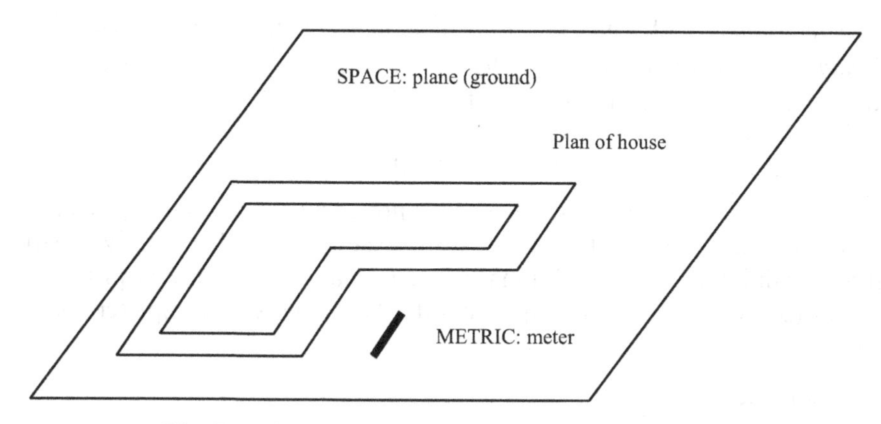

Fig. 7.1. Plane and meter: an example of metric space.

(b) To better illustrate another important characteristic of the notion of metric space as a structure, let us consider the following set:

X = {Grandparents, Mother, Father, Children, Hobbies, Friends, Memories, Love, Preferences, House, Garden, Professions, Car, Money}.

Can X be organized as a metric space? This example is very relevant to IR. The elements of X may be conceived as being documents. A metric should be a measure of how "close" (or "distant") they are to each other. Function δ defined as $\delta(a, b) = 1$ if $a \neq b$ and $\delta(a, b) = 0$ if $a = b$ is a metric. Thus, the structure (X, δ) is, from a mathematical point of view, a metric space. However, the space thus obtained would hardly be appropriate for retrieval purposes: more suitable measures have to be defined (as they indeed have been defined over time). □

7.2 Complete Metric Space

Let (X, δ) denote a metric space. A sequence $x_1,\ldots,x_n,\ldots \in X$ is said to be *convergent* if it has a *limit*, denoted by, say, L; i.e.,

$$\forall \varepsilon \in \mathbb{R}\ \exists n_\varepsilon \in \mathbb{N} \text{ such that } \delta(x_n, L) < \varepsilon,\ \forall n > n_\varepsilon, \tag{7.2}$$

where $\mathbb{N}$ denotes the set of natural numbers and $\mathbb{R}$ the set of real numbers. In words, condition (7.2) means that after some index n_ε all the remaining terms of the sequence "rush" to the value L (i.e., the limit), which they "approach" as closely as desired.

A metric space (X, δ) is said to be *complete* (relative to metric δ) if the *Cauchy principle of convergence* holds in space (X, δ), i.e., sequence $x_1,\ldots, x_n,\ldots \in X$ is convergent if and only if

$$\forall \varepsilon \in \mathbb{R}\ \exists n_\varepsilon \in \mathbb{N} \text{ such that } \delta(x_m, x_n) < \varepsilon,\ \forall m, n > n_\varepsilon. \tag{7.3}$$

The completeness of a metric space is not a trivial or intuitive property. It means that convergent sequences are exactly those that satisfy [apart from condition (7.2)] an additional property [namely the Cauchy principle (7.3)]. Example 7.2 should help to clarify the property of completeness of a metric space.

Example 7.2

Let us consider the real line with the usual Euclidean distance as a metric on it as a metric space. Then, the convergence of any sequence of numbers is equivalent to the Cauchy principle. For example, let us imagine that we walk, with decreasing steps, straight toward a cricket hole starting from, say, 2 meters away. We make each step at half the distance remaining to the hole. Then, for a stick of any length (i.e., ε), after some number of steps (i.e., n_ε), the distance between any two further successive steps (i.e., m and n) will be shorter than the stick (**Fig. 7.2**).

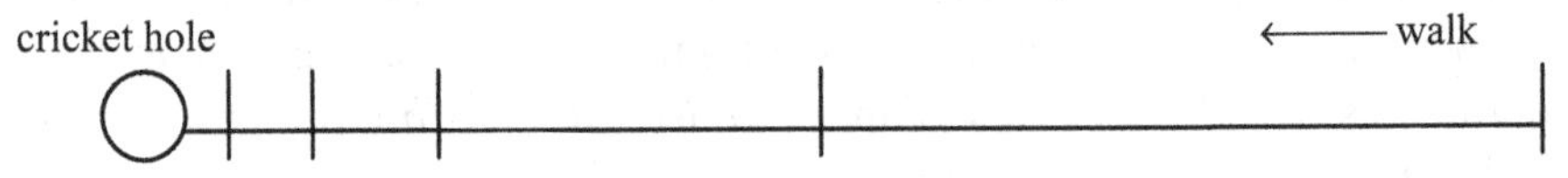

Fig. 7.2. Illustration of the Cauchy principle of convergence.

The condition formulated in the Cauchy principle of convergence is far from being intuitive. As an example, try to imagine a similar walk, but not in a plane, as in the previous walk to a cricket hole, but rather on a "creased" surface (**Fig. 7.3**).

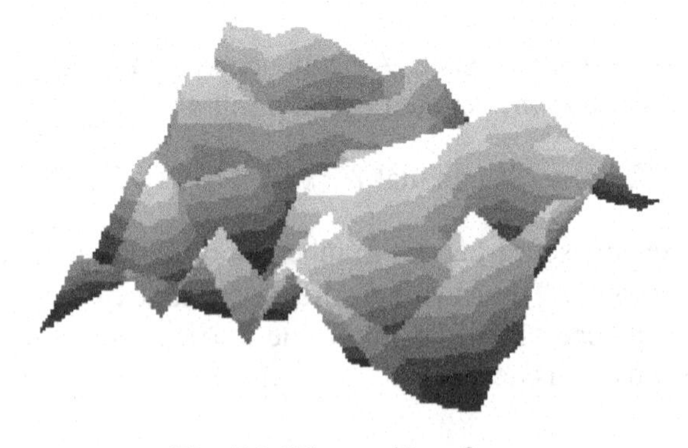

Fig. 7.3. "Creased" surface.

The notion of a complete metric space is not trivial either. A convergent sequence always satisfies the Cauchy principle of convergence, but the reverse is not always true: not every sequence that satisfies the Cauchy principle is convergent (i.e., has a limit) in the space. For example, in the metric space ($\mathbb{Q}$, δ), where $\mathbb{Q}$ denotes the set of rational numbers and $\delta = |a - b|$, $\forall a, b \in \mathbb{Q}$, the sequence

$$x_n = \left(1 + \frac{1}{n}\right)^n$$

does satisfy the Cauchy principle, but it is not convergent in ($\mathbb{Q}$, δ); its limit is an irrational number, namely e (and $e \notin \mathbb{Q}$).

7.3 Linear Space

A *linear space* (or *vector space*) over a field F (for our purposes it is sufficient to assume that F is equal to the set $\mathbb{R}$ of real numbers or to the set $\mathbb{C}$ of complex numbers, both endowed with the usual operations of addition (i.e., +) and multiplication (i.e., $\times$) of real or complex numbers, respectively) is the structure (L, $\oplus$, $\otimes$, F), where $\oplus$ and $\otimes$ denote two binary operations:

$$\oplus\colon \mathsf{L} \times \mathsf{L} \to \mathsf{L} \text{ and } \otimes\colon \mathsf{F} \times \mathsf{L} \to \mathsf{L}, \tag{7.4}$$

if the following properties hold:

- $a \oplus b = b \oplus a$, $\forall a, b \in \mathsf{L}$ (commutativity).
- $\exists e \in \mathsf{L}$ such that $a \oplus e = a$, $\forall a \in \mathsf{L}$ (e is referred to as the *null* vector).
- $\forall a \in \mathsf{L}\ \exists a' \in \mathsf{L}$ such that $a \oplus a' = e$ (a' is called the *inverse* of a);
- $a \oplus (b \oplus c) = (a \oplus b) \oplus c$, $\forall a, b, c \in \mathsf{L}$ (associativity).

Further, for $\forall r, p \in \mathsf{F}$, $\forall a, b \in \mathsf{L}$ we have:

- $(r + p) \otimes a = (r \otimes a) \oplus (p \otimes a)$.
- $r \otimes (a \oplus b) = (r \otimes a) \oplus (r \otimes b)$.
- $(r \times p) \otimes a = r \otimes (p \otimes a)$.
- $1 \otimes a = a$.

A linear space is denoted in short by L. The elements of a linear space L are traditionally called *vectors* and are usually denoted by bold letters, e.g., **v**, while the elements of F are called *scalars*.

Example 7.3

(a) Let us assume that we are standing on side A of a river and that we are pulling a boat from side B toward us (**Fig. 7.4**). Pulling the boat means that a force **P** is acting on the boat. The water is also acting on the boat by some force **R**. Then, actually, a resultant force **P** + **R** is acting on the boat. The sum **P** + **R** is also a vector, i.e., a quantity of the same type as **P** and **R**. Forces and operations with them form a linear space.

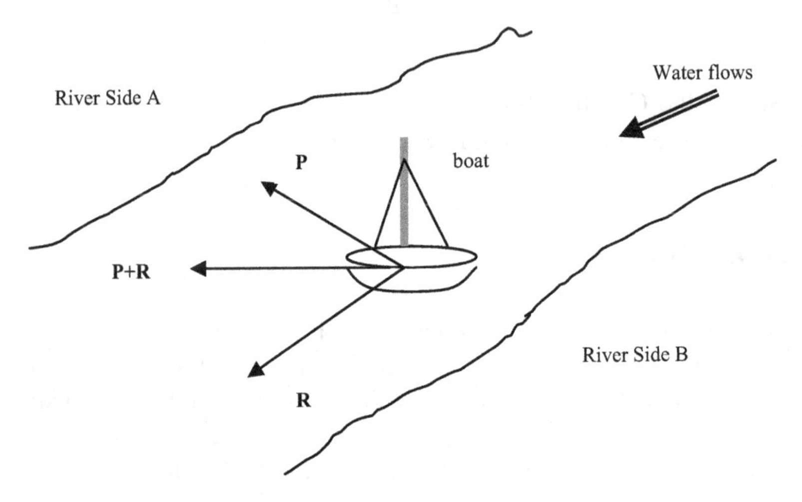

Fig. 7.4. Space of forces.

(b) As another example, let us consider a room and the objects within it. The position of any object in the room can be given relative to where we stand by specifying its distance (say, in meters) to our left or right, above or below us, in front of or behind us (**Fig. 7.5**). The position of objects in the room is a quantity with magnitude and direction, i.e., a vector.

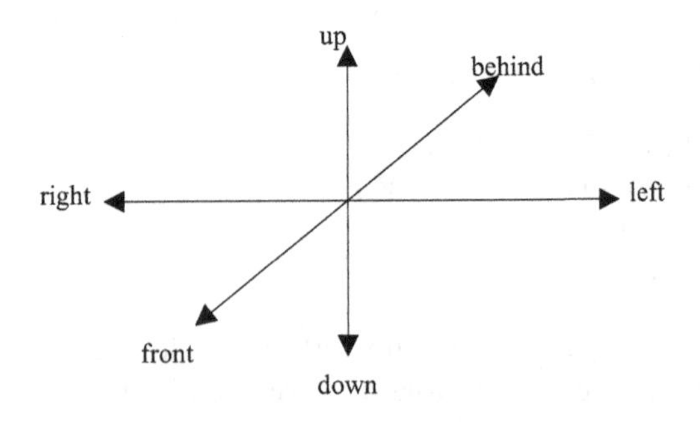

Fig. 7.5. Space of position vectors.

The expression $(r_1 \otimes \mathbf{v}_1) \oplus \ldots \oplus (r_m \otimes \mathbf{v}_m)$ is called a *linear combination* of vectors $\mathbf{v}_1,\ldots,\mathbf{v}_m$ $(r_1,\ldots,r_m \in \mathsf{F})$. When the linear combination is equal to e if and only if $r_1 = \ldots = r_m = 0$, then vectors $\mathbf{v}_1,\ldots,\mathbf{v}_m$ are said to be *linearly independent*, and *linearly dependent* otherwise. A set of linearly independent vectors forms an *algebraic basis* (basis for short) of L if any vector of the space can be written as a linear combination of them. Every linear space has at least one basis. Each basis contains the same number of vectors, and this number is referred to as the *dimension* of the space.

If $\mathbf{b}_1,\ldots,\mathbf{b}_n \in \mathsf{L}_n$ denote basis vectors of an n-dimensional linear space L_n, then every vector $\mathbf{v} \in \mathsf{L}_n$ can be written as a linear combination of basis vectors:

$$\mathbf{v} = (p_1 \otimes \mathbf{b}_1) \oplus \ldots \oplus (p_n \otimes \mathbf{b}_n), \tag{7.5}$$

where the scalars $p_1,\ldots,p_n \in \mathsf{F}$ are called the *coordinates* of vector $\mathbf{v}$; expressed as $\mathbf{v} = (p_1,\ldots,p_n) = [p_1,\ldots,p_n]^\mathsf{T}$, where $^\mathsf{T}$ denotes the transpose, i.e.,

$$[p_1,\ldots,p_n]^\mathsf{T} = \begin{bmatrix} p_1 \\ \ldots \\ p_n \end{bmatrix}.$$

7.4 Subspace of Linear Space

A subset $A \subseteq \mathsf{L}$, $A \neq \varnothing$, of space L is a *subspace* of L if A is itself a linear space. Equivalently:

- $a \oplus b \in A, \forall a, b \in A.$
- $r \otimes a \in A, \forall r \in \mathsf{F}, \forall a \in A.$

Example 7.4

The ground (see Example 7.1) may be viewed as a linear space L of position vectors. The line A in the plane is a subspace of L.

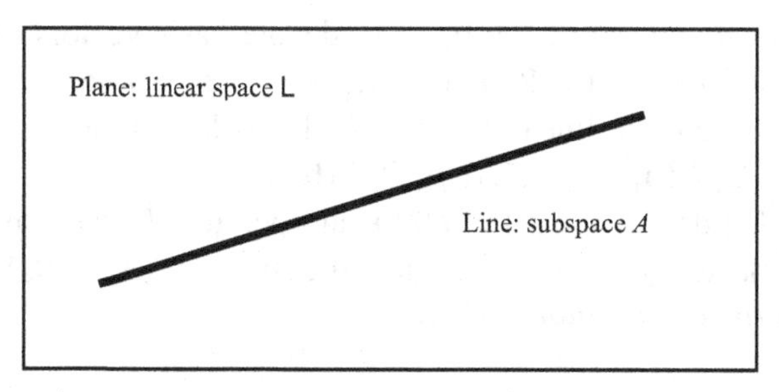

Fig. 7.6. Line A is a subspace of the plane as a linear space L of position vectors.

A subset A of L is *closed* if and only if the limit of any convergent sequence $x_1, x_2, \ldots \in A$ belongs to A. A subset A of the linear space (L, ℝ) is *convex* if

$$a, b \in A \Rightarrow r \otimes a \oplus (1 - r) \otimes b \in A, \forall r \in [0; 1].$$

It is easy to see that every subspace of a linear space is convex.

The *direct sum* ✚ of two subspaces A_1 and A_2 of linear space L is defined as (**Fig. 7.7**):

$$A_1 + A_2 = \{\mathbf{x} \oplus \mathbf{y} \mid \mathbf{x} \in A_1, \mathbf{y} \in A_2\}.$$

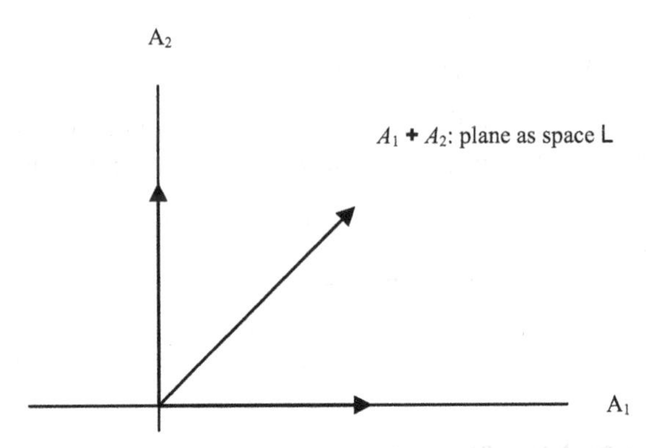

Fig. 7.7. Direct sum of lines A_1 and A_2 as subspaces of the plane.

7.5 Linear Operator

Let (L_1, $\oplus_1$, $\otimes_1$, F) and (L_2, $\oplus_2$, $\otimes_2$, F) denote two linear spaces. A function U: $L_1 \rightarrow L_2$ with the properties

$$U(\mathbf{v} \oplus_1 \mathbf{w}) = U(\mathbf{v}) \oplus_2 U(\mathbf{w}), \tag{7.6}$$

$$U(a \otimes_1 \mathbf{v}) = a \otimes_2 U(\mathbf{v}),$$

is called a *linear operator*. Let $\mathbf{b}_1,\ldots,\mathbf{b}_n$ denote basis vectors of space (L_1, $\oplus_1$, $\otimes_1$, F), and let $\mathbf{v} = (p_1 \otimes_1 \mathbf{b}_1) \oplus_1 \ldots \oplus_1 (p_n \otimes_1 \mathbf{b}_n)$ denote an arbitrary vector of L_1. Then, the linear operator U: $L_1 \rightarrow L_2$ is uniquely determined by $U(\mathbf{v}) = (p_1 \otimes_2 U(\mathbf{b}_1)) \oplus_2 \ldots \oplus_2 ((p_n \otimes_2 U(\mathbf{b}_n))$.

Let $\Lambda(L_1, L_2)$ denote the set of all linear operators U from space (L_1, $\oplus_1$, $\otimes_1$, F) into space (L_2, $\oplus_2$, $\otimes_2$, **F**). Then, the structure (Λ, +, ×, F) is the *linear space of linear operators*, where

$$(U_1 + U_2)(\mathbf{v}) = U_1(\mathbf{v}) \oplus_2 U_2(\mathbf{v}),$$
$$(a \times U)(\mathbf{v}) = a \otimes_2 U(\mathbf{v}). \quad (7.7)$$

The name linear operator may be misleading, as it may suggest that some trivial operation (e.g., the drawing of a straight line) has been generalized to a needlessly complicated abstract formulation. That this is by far not the case is illustrated in Example 7.5.

Example 7.5

Let us consider the well-known Euclidean plane. Let P denote a point in this plane, and let its position vector be $\mathbf{r} = (b, a)$. Let us consider now an operation U that "mirrors" every vector by the horizontal axis (**Fig. 7.8**). This means that U is defined as follows:

$$U(b, a) = (b, -a). \quad (7.8)$$

Operator U is given by the following matrix:

$$U = \begin{pmatrix} 1 & 0 \\ 0 & -1 \end{pmatrix}, \ U(b,a) = \begin{pmatrix} 1 & 0 \\ 0 & -1 \end{pmatrix} \times \begin{pmatrix} b \\ a \end{pmatrix} = \begin{pmatrix} b \\ -a \end{pmatrix}. \quad (7.9)$$

Operator U thus defined is a linear operator on the plane. It mirrors any point P by the horizontal axis into the point P'. □

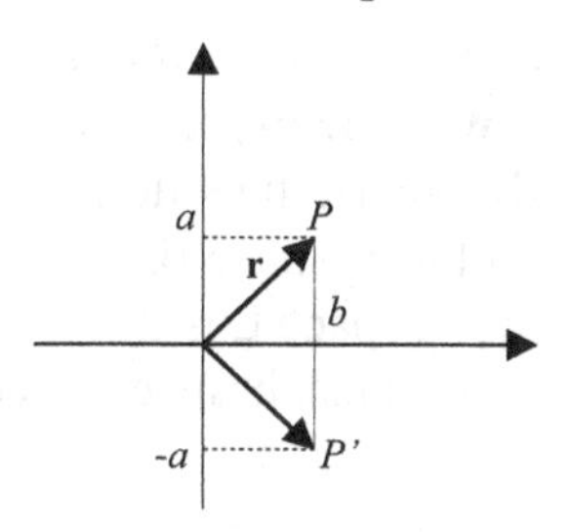

Fig. 7.8. Mirroring of vectors in a plane by the horizontal axis: linear operator.

7.6 Banach Space

Let L be a linear space. A function $\nu : \mathsf{L} \to \mathbb{R}_+$ is called a *pseudonorm* if the following properties hold:

- If $\mathbf{v}$ denotes the null vector of L, then $\nu(\mathbf{v}) = 0$.
- $\nu(r \otimes \mathbf{v}) = |r| \times \nu(\mathbf{v}), \forall r \in \mathsf{F}, \forall \mathbf{v} \in \mathsf{L}$.
- $\nu(\mathbf{v} \oplus \mathbf{w}) \leq \nu(\mathbf{v}) + \nu(\mathbf{w}), \forall \mathbf{v}, \mathbf{w} \in \mathsf{L}$.

If, in addition, function ν obeys also the following property:

- If $\nu(\mathbf{v}) = 0$, then $\mathbf{v}$ is the null vector of L,

then function ν is called a *norm*. Usually, $\nu(\mathbf{v})$ is denoted by $||\mathbf{v}||$. A linear space L with a norm is called a *normed* (linear) *space*.

The notion of norm is not trivial. It endows the space with a special character. Example 7.6 illustrates this.

Example 7.6

- The absolute value $|r|$ of any real number $r \in \mathbb{R}$ is a norm in the set of real numbers $\mathbb{R}$.
- It may happen that the same space can be endowed with more than one norm. For example, the space $C[a, b]$ of real and continuous functions defined on $[a, b]$ can be organized as a normed space with the norm $\nu(f(x)) = \max_{x \in [a, b]} | f(x) |$, and as another normed space with the norm

$$\nu(f(x)) = \sqrt{\int_a^b | f(x) |^2 \, dx} \, . \, \square$$

A normed linear space $(\mathsf{L}, ||.||)$ defines a metric space $(\mathsf{L}, \mathbb{R})$ with the metric $\delta(\mathbf{v}, \mathbf{w}) = ||\mathbf{v} \oplus (-1) \otimes \mathbf{w}||$. The expression $||\mathbf{v} \oplus (-1) \otimes \mathbf{w}||$ is usually written as $||\mathbf{v} - \mathbf{w}||$, and the metric thus defined is called a *metric induced by the norm*. If a normed linear space $(\mathsf{L}, ||.||)$ is complete relative to the metric induced by the norm $||.||$, space L is called a *Banach space*.

As can be seen, the notion of Banach space encompasses two highly nontrivial properties:

- That of being normed.
- That of being complete.

In a Banach space, one can measure distances as defined by the norm, and all convergent sequences are exactly those that satisfy the Cauchy principle of convergence. The fact that not every normed space is a Banach space, i.e., that this is not trivial, is illustrated in Example 7.7.

Example 7.7

- The set $\mathbb{R}$ of real numbers is a Banach space relative to the metric induced by the norm given by the absolute value.

- The space $C[a, b]$ of real and continuous functions defined on $[a, b]$ is a Banach space with the norm $\nu(f(x)) = \max_{x \in [a,b]} | f(x) |$.
- The space $C[a, b]$ of real and continuous functions defined on $[a, b]$ is not a Banach space with the norm

$$\nu(f(x)) = \sqrt{\int_a^b | f(x) |^2 \, dx} . \square$$

7.7 Hilbert Space

Let (L, ⊕, ⊗, F) be a linear space. A mapping π: L × L → F satisfying the properties is called a *scalar* (or *inner* or *dot*) *product*:

- $\pi(\mathbf{v} \oplus \mathbf{w}, \mathbf{u}) = \pi(\mathbf{v}, \mathbf{u}) + \pi(\mathbf{w}, \mathbf{u})$, $\forall \mathbf{v}, \mathbf{w}, \mathbf{u} \in$ L.
- $\pi(r \otimes \mathbf{v}, \mathbf{w}) = r \times \pi(\mathbf{v}, \mathbf{w})$, $\forall r \in$ F, $\forall \mathbf{v}, \mathbf{w} \in$ L.
- $\pi(\mathbf{v}, \mathbf{w}) = (\pi(\mathbf{w}, \mathbf{v}))^*$, $\forall \mathbf{v}, \mathbf{w} \in$ L, where $(\pi(\mathbf{w}, \mathbf{v}))^*$ denotes conjugate if L = $\mathbb{C}$.
- $\pi(\mathbf{v}, \mathbf{v}) \geq 0$, $\forall \mathbf{v} \in$ L.
- $\pi(\mathbf{v}, \mathbf{v}) = 0$ if and only if $\mathbf{v} = \mathbf{0}$ ($\mathbf{0}$ is the null vector of space L).

Instead of $\pi(\mathbf{x},\mathbf{y})$ the following shorter notations may also be used: $(\mathbf{x}, \mathbf{y})$, $\mathbf{x} \cdot \mathbf{y}$, $\langle\mathbf{x} \mid \mathbf{y}\rangle$, $\langle\mathbf{x}, \mathbf{y}\rangle$, $\mathbf{xy}$. We use the notation: $\langle\mathbf{x}, \mathbf{y}\rangle$.

A Banach space (L, ⊕, ⊗, $\mathbb{C}$) in which the norm is defined using the scalar product $||\mathbf{v}|| = \langle\mathbf{v}, \mathbf{v}\rangle^{1/2}$ is called an *abstract Hilbert space* (or *Hilbert space* for short). As can be seen, the notion of Hilbert space is special in that the norm is defined using a very special function: the scalar product. Every Hilbert space is a Banach space, but the reverse is not necessarily true (the scalar product is not an 'ingredient' of a Banach space).

Let $U \in \Lambda$(L, L) be a linear operator. Then, U is self-adjoint if

$$\langle U(\mathbf{v}), \mathbf{w}\rangle = \langle\mathbf{v}, U(\mathbf{w})\rangle, \forall \mathbf{v}, \mathbf{w} \in \mathsf{L}.$$

Example 7.8

- Let $\mathbf{r} = (r_1, r_2)$ and $\mathbf{p} = (p_1, p_2)$ denote two position vectors in a plane. Then, $\pi(\mathbf{r}, \mathbf{p}) = r_1 p_1 + r_2 p_2$ is a scalar product.

- Originally, the Hilbert space was the set of sequences $x_1,\ldots,x_i,\ldots \in \mathbb{R}$ for which the series $\sum_{i=1}^{\infty} x_i^2$ was convergent, endowed with the metric

$$\delta(x,y) = \sqrt{\sum_{i=1}^{\infty}(x_i - y_i)^2}\ . \square$$

7.8 Euclidean Space

The Euclidean space, which we denote by E_n (**Figure 7.9** shows the three-dimensional space E_3), is a special Hilbert space (L, ⊕, ⊗, F) defined as:

- The set L is equal to the set of n-tuples $(v_1,\ldots,v_n) \in \mathbb{R}^n$ of real numbers, i.e., L = $\mathbb{R}^n$.
- F = $\mathbb{R}$.
- The operation ⊕ is defined as follows: ⊕ = +; $\mathbf{v} = (v_1,\ldots,v_n)$, $\mathbf{w} = (w_1,\ldots,w_n)$, $\mathbf{v} + \mathbf{w} = (v_1 + w_1,\ldots,v_n + w_n)$.
- The operation ⊗ is defined as ⊗ = ×; $r \times \mathbf{v} = (r \times v_1,\ldots, r \times v_n)$.
- The norm is defined as the *Euclidean length* of a vector, i.e.,

$$||\mathbf{v}|| = \sqrt{\sum_{i=1}^{n} v_i^2}\ .$$

- The scalar product is defined as $\langle\mathbf{v}, \mathbf{w}\rangle = v_1 \times w_1 + \ldots + v_n \times w_n = ||\mathbf{v}|| \cdot ||\mathbf{w}|| \cdot \cos\varphi$ (where φ is a measure of the angle between vectors $\mathbf{v}$ and $\mathbf{w}$).

It follows that the Euclidean distance between two vectors $\mathbf{v} = [v_1,\ldots,v_n]^{\mathrm{T}}$ and $\mathbf{w} = [w_1,\ldots, w_n]^{\mathrm{T}}$ is defined as

$$||\mathbf{v} + (-1)\times\mathbf{w}|| = \sqrt{\sum_{i=1}^{n}(v_i - w_i)^2}\ , \tag{7.10}$$

and a measure of the angle φ between vectors $\mathbf{v}, \mathbf{w} \neq \mathbf{0} \in \mathbb{R}^n$ is a real number φ such that

$$\cos\varphi = \frac{\langle\mathbf{v}, \mathbf{w}\rangle}{||\,\mathbf{v}\,|| \times ||\,\mathbf{w}\,||}\ . \tag{7.11}$$

Two vectors **v** and **w** are orthogonal to each other if $\cos\varphi = 0$. Any n-dimensional Euclidean space E_n has an orthonormal (i.e., orthogonal and unit lengths) basis (there may also be other bases that need not be orthogonal or have unit lengths). A common orthonormal basis is

$$\mathbf{e}_1 = [1,0,0,\ldots,0]^{\mathsf{T}},\ \mathbf{e}_2 = [0,1,0,\ldots,0]^{\mathsf{T}},\ \ldots,\ \mathbf{e}_n = [0,0,0,\ldots,1]^{\mathsf{T}},$$

where $<\mathbf{e}_i, \mathbf{e}_j> = \delta_{ij}$ (δ_{ij} is the *Kronecker delta* symbol, i.e., $\delta_{ij} = 1$ if $i = j$, and $\delta_{ij} = 0$ if $i \neq j$).

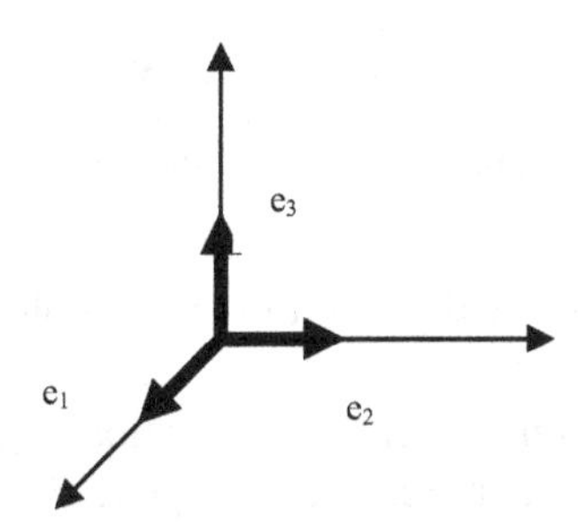

Fig. 7.9. Visualization of the three-dimensional orthonormal Euclidean space E_3. (This space is used to model, e.g., the usual physical space we live in)

7.9 Projection Theorem

Two elements **u** and **v** of the Hilbert space L ($\mathbf{u}, \mathbf{v} \in \mathsf{L}$) are said to be *orthogonal* if $<\mathbf{u}, \mathbf{v}> = 0$, expressed as $\mathbf{u} \perp \mathbf{v}$. Two subsets A and B of L are said to be orthogonal, written as $A \perp B$, if

$$A \perp B \Leftrightarrow (\mathbf{u} \perp \mathbf{v},\ \forall \mathbf{u} \in A,\ \forall \mathbf{v} \in B). \tag{7.12}$$

When subset A consists of one element, $A = \{\mathbf{u}\}$, then the notations $A \perp B$ and $\mathbf{u} \perp B$ are considered to be equivalent to each other. For a subset $A \subset \mathsf{L}$ of the Hilbert space L, the set $A^{\perp} = \{\mathbf{u} \in \mathsf{L} \mid \mathbf{u} \perp A\}$ is referred to as the *orthogonal complement* of A.

The following result, known as the *projection theorem*, is a well-known and very important result in functional analysis and quantum mechanics, and is no less important in IR (as will be seen in Chapter 8):

Theorem 7.1. *Let* L *be a Hilbert space, and A a closed subspace of* L*, $A \subset$* L*. Then, any element $\boldsymbol{u} \in$* L *can be represented as*

$$\boldsymbol{u} = \boldsymbol{v} \oplus \boldsymbol{w}$$

in a unique way, $\boldsymbol{v} \in A$, $\boldsymbol{w} \in A^{\perp}$.

Proof. As the case when L $= E_n$ (i.e., the n-dimensional Euclidean space) is important in IR, it is the one for which we give the proof.

Let $\mathbf{w}_1,\ldots,\mathbf{w}_k$ be a basis of subspace A. Let M denote the matrix formed by these basis vectors, i.e.,

$$M = [\mathbf{w}_1 \ \ldots \ \mathbf{w}_k].$$

The basis vectors $\mathbf{w}_1,\ldots,\mathbf{w}_k$ span the subspace A. It is known from the theory of matrices that:

- Subspace A is equal to the column space col(M) of matrix M, i.e., A = col(M).
- The orthogonal complement $A^{\perp}$ of subspace A is $A^{\perp}$ = col$(M)^{\perp}$ = null(M^{T}), where M^{T} denotes the transpose of matrix M, while null(M^{T}) denotes its null space, i.e., the space of vectors $\mathbf{a} \in E_n$ for which $M^{\mathsf{T}}\mathbf{a} = \mathbf{0}$.

Let $\mathbf{u} \in E_n$ be a vector of the n-dimensional Euclidean space. Then, the simultaneous system of linear equations

$$(M^{\mathsf{T}}M)\mathbf{x} = M^{\mathsf{T}}\mathbf{u}$$

has a unique solution because—by assumption—the rank of the $k \times k$ matrix $M^{\mathsf{T}}M$ is equal to k, i.e., rank$(M^{\mathsf{T}}M) = k$. From $(M^{\mathsf{T}}M)\mathbf{x} = M^{\mathsf{T}}\mathbf{u}$ we obtain

$$M^{\mathsf{T}}(\mathbf{u} - M\mathbf{x}) = \mathbf{0},$$

which means that vector $\mathbf{w} = \mathbf{u} - M\mathbf{x}$ belongs to the orthogonal complement of A, i.e., $\mathbf{w} = \mathbf{u} - M\mathbf{x} \in A^{\perp}$. Vector $M\mathbf{x}$ belongs to subspace A, i.e., $M\mathbf{x} \in A$, and can be denoted by $\mathbf{v}$: $M\mathbf{x} = \mathbf{v}$. Thus, we have that

$$\mathbf{w} = \mathbf{u} - M\mathbf{x} = \mathbf{u} - \mathbf{v}, \ \ \mathbf{u} = \mathbf{v} + \mathbf{w}. \ \Box$$

Example 7.9

For example, any vector $\mathbf{u} \in E_3$ in the three-dimensional space can be uniquely written as

$$\mathbf{u} = \mathbf{v} + \mathbf{w}, \text{ where } \mathbf{v} \in E_2 \text{ and } \mathbf{w} \in E_2^{\perp}.$$

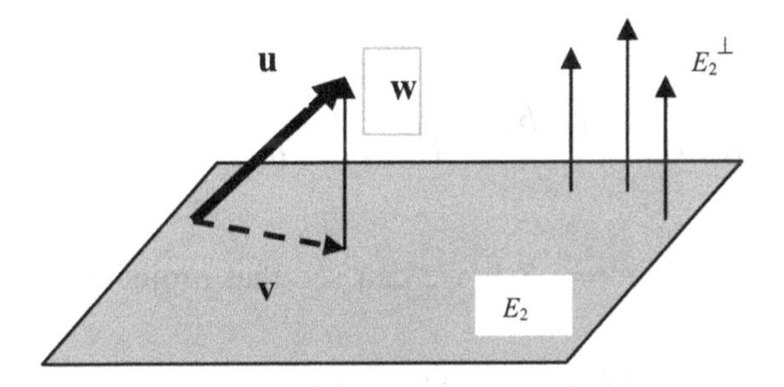

If $\mathbf{u} \in E_2$, then $\mathbf{u} = \mathbf{v} + \mathbf{w} = \mathbf{u} + \mathbf{0}$. If $\mathbf{u} \in E_2^{\perp}$, then $\mathbf{u} = \mathbf{v} + \mathbf{w} = \mathbf{0} + \mathbf{u}$. If $\mathbf{u} \notin E_2$ and $\mathbf{u} \notin E_2^{\perp}$, then vector $\mathbf{u}$ makes an angle α with the plane E_2. Let O denote the starting point of vector $\mathbf{u}$. Further, let Q denote the intersection point between the plane and the perpendicular from the endpoint P of vector $\mathbf{u}$ onto the plane. Then, the directed line segments $\overrightarrow{OQ}$ and $\overrightarrow{QP}$ are just the vectors $\mathbf{v}$ and $\mathbf{w}$:

$$\mathbf{v} = \overrightarrow{OQ}, \mathbf{w} = \overrightarrow{QP}.$$

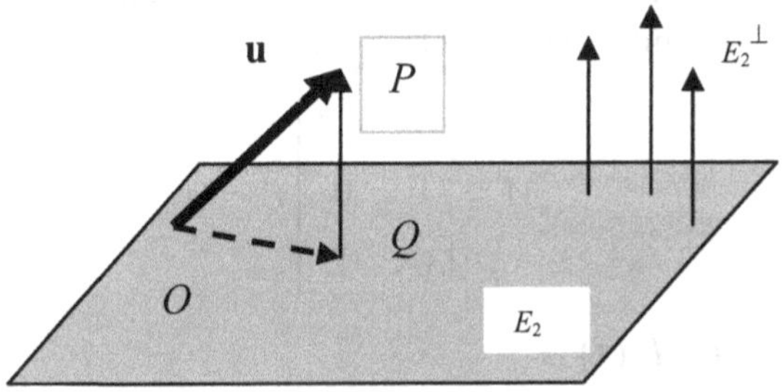

u
P
$E_2^{\perp}$
Q
O
E_2

7.10 Projector

Vector $\mathbf{v}$ in Theorem 7.1 is called the *projection* of vector $\mathbf{u}$ onto subspace A; notation: $\mathbf{v} = [A]\mathbf{u}$. An operation P defined as $P_A(\mathbf{x}) = [A]\mathbf{x}$, i.e., giving the projection of vector $\mathbf{x}$ of a Hilbert space onto subspace A, is called a *projector* (Example 7.10). Projectors are self-adjoint linear operators with the property $P^2 = P$.

Example 7.10

The following operator P projects any vector (a, b) of the plane onto the horizontal axis, the result being the vector $(a, 0)$:

$$P = \begin{pmatrix} 1 & 0 \\ 0 & 0 \end{pmatrix},\ P(a,b) = \begin{pmatrix} 1 & 0 \\ 0 & 0 \end{pmatrix} \times \begin{pmatrix} a \\ b \end{pmatrix} = \begin{pmatrix} a \\ 0 \end{pmatrix}. \square$$

From the proof of Theorem 7.1, we can see that projector $P_A(\mathbf{x})$ is given by:

$$P_A(\mathbf{x}) = M(M^T M)^{-1} M^T \mathbf{x}, \tag{7.13}$$

where the matrix

$$M(M^T M)^{-1} M^T \tag{7.14}$$

is the matrix of projector $P_A(\mathbf{x})$.

Example 7.11

Plane S is defined by the equation $x - 4y + 2z = 0$ in E_3. The matrix of projector $P_S(\mathbf{x})$, $\mathbf{x} \in E_3$, onto plane S can be obtained as follows. First, a basis for S should be given. Let $y = 1$ and $z = 0$; then $x = 4$, and for $y = 0$ and $z = 1$, we obtain $x = -2$. Thus, matrix M is

$$M = \begin{bmatrix} 4 & -2 \\ 1 & 0 \\ 0 & 1 \end{bmatrix}.$$

Matrix $\mathcal{M}$ of projector $P_S(\mathbf{x})$ is

$$\mathcal{M} = M(M^T M)^{-1} M^T = \begin{bmatrix} 0.952 & 0.19 & -0.095 \\ 0.19 & 0.238 & 0.381 \\ -0.095 & 0.381 & 0.81 \end{bmatrix}.$$

For example, the projection of vector $\mathbf{x} = (1\ 3\ 7)$ onto plane S is the vector $\mathcal{M}\mathbf{x} = (0.857\ 3.571\ 6.714)$. □

Apart from its use in IR and quantum mechanics, the projection theorem 7.1 has many applications in other areas as well. As a simple example, let us consider the simultaneous system of linear equations $A\mathbf{x} = \mathbf{b}$ (in matrix form). If we can solve it, then an exact solution is obtained. But when we cannot solve it exactly, then we seek a vector $\mathbf{x}$ that minimizes $\|\mathbf{b} - A\mathbf{x}\|$. Since $\text{col}(A) = \{\mathbf{w} \mid \exists \mathbf{y}\ \mathbf{w} = A\mathbf{y}\}$, the minimizing vector $\mathbf{x}$ will be its projection onto $\text{col}(A)$.

7.11 Basis of Subspace

We have seen that in order to find a projector, a basis of the subspace in question should be given. The problem of finding such a basis can be solved as follows. Let $S = \{\mathbf{v}_1,\ldots,\mathbf{v}_s\} \subset E_n$ be (not necessarily linearly independent) vectors in the n-dimensional Euclidean space. The set $Sp(S) = \{\mathbf{w} \mid \alpha_1\mathbf{v}_1 +\ldots+ \alpha_n\mathbf{v}_s\}$ of all linear combinations of the vectors from S is a subspace of E_n and is referred to as the subspace *spanned* by S.

An orthonormal basis for the subspace $Sp(S)$ can be obtained using the Gram-Schmidt procedure.

Gram-Schmidt Procedure

Let A denote the matrix of the vectors of S, $A = [\mathbf{v}_1 \ldots \mathbf{v}_s]$, i.e.,

$$A = \begin{bmatrix} a_{11} & \ldots & a_{1s} \\ . & \ldots & . \\ a_{n1} & \ldots & a_{ns} \end{bmatrix}.$$

The first basis vector, $\mathbf{e}_1$, is given by

$$\mathbf{e}_1 = \frac{\mathbf{v}_1}{\|\mathbf{v}_1\|}.$$

The second basis vector, $\mathbf{e}_2$, is given by

$$\mathbf{e}_2 = \frac{\mathbf{z}_2}{\|\mathbf{z}_2\|},$$

where $\mathbf{z}_2 = \mathbf{v}_2 - \lambda_{21}\mathbf{e}_1$, with λ_{21} such that $<\mathbf{z}_2, \mathbf{e}_1> = 0$, $\lambda_{21} = <\mathbf{v}_2, \mathbf{e}_1>$.

If the basis vectors $\mathbf{e}_1,\ldots,\mathbf{e}_m$ have already been obtained, the next basis vector, $\mathbf{e}_{m+1}$, is given by

$$\mathbf{e}_{m+1} = \frac{\mathbf{z}_{m+1}}{\|\mathbf{z}_{m+1}\|},$$

where

$$\mathbf{z}_{m+1} = \mathbf{v}_{m+1} - \sum_{k=1}^{m} \lambda_{m+1,k}\mathbf{e}_k,$$

with the coefficients $\lambda_{m+1,k}$ ($k = 1,\ldots,m$) so chosen as to have $<\mathbf{z}_{m+1}, \mathbf{e}_j> = 0$, $j = 1,\ldots,m$, i.e., $\lambda_{m+1,k} = <\mathbf{v}_{m+1}, \mathbf{e}_k>$.

7.12 Lattice of Subspaces

The structure ($\Re$(L), $\wedge$, $\vee$, C) is an atomic, complete, orthomodular lattice when:

- $\Re$(L) is the set of all closed subspaces of the Hilbert space L.
- $A \wedge B = A \cap B$ for every subspace A, B of L. The intersection is also a subspace of L; it is the largest subspace contained in both A and B.
- $A \vee B = A + B$; $A \vee B$ is the smallest subspace containing both A and B.
- C = $\perp$ (orthogonal complement).

(A proof can be given using the axioms defining a lattice, the definitions of $A \cap B$ and $A + B$, and the orthomodular law.) Lattice $\Re$(L) is modular if and only if Hilbert space L is finite-dimensional (e.g., L = E_n). The ordering relation $\leq$ is defined by the concept of subspace.

As there is a one-to-one correspondence between subspaces and projectors (Theorem 7.1), one may say that projectors form a poset, just as subspaces do. We say that $P_1 \leq P_2$ if for the corresponding subspaces we have $M_1 \subseteq M_2$. Further, it can be easily seen that $P_1 \leq P_2$ if $P_1P_2 = P_1$.

The main difference between lattice $\Re$(L) and a Boolean algebra—e.g., the powerset (L, $\subseteq$) a as Boolean algebra or the Boolean algebra of logical propositions ($\{T, F\}$, $\wedge$, $\vee$, $\neg$)—is related to the distributive law. This is an important difference between the two lattices in that it reflects a primary difference in their structures (and thus induced properties). The difference can be well illustrated using the following example. Let L denote the Hilbert space E_2, and let M and N be two one-dimensional subspaces of L (**Fig. 7.10**).

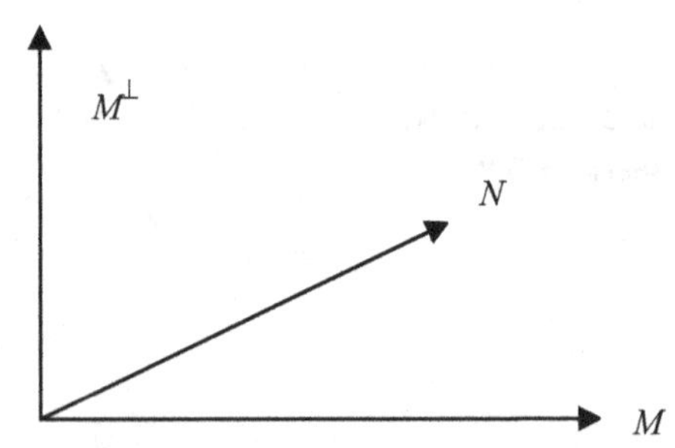

Fig. 7.10. Three one-dimensional subspaces of the two-dimensional Hilbert space E_2 that do not satisfy the distributive law.

We have:

$$N \wedge (M \vee N^{\perp}) = N \wedge \mathsf{L} = N, \tag{7.15}$$

but

$$N \wedge M = N \wedge N^{\perp} = \{0\}. \tag{7.16}$$

In any powerset lattice ($\wp(X)$, $\subseteq$), which is a Boolean algebra, the distributive law makes it possible to write the following relationship for any sets A and B of X:

$$A = A \cap X = A \cap (B \cup \mathsf{C}_X B) = (A \cap B) \cup (A \cap \mathsf{C}_X B). \tag{7.17}$$

In a subspace lattice $\Re(\mathsf{L})$, the operation corresponding to set complement C is orthocomplementation $^{\perp}$. Based on Eq. (7.17), we may say that two subspaces are *compatible* if

$$(M \wedge N) \vee (M \wedge N^{\perp}) = M. \tag{7.18}$$

The orthomodularity condition can be interpreted as an underlying property of the geometry of space, namely:

$$\begin{aligned} &N, M \in \Re(\mathsf{L}),\ M \leq N \Rightarrow \\ &N = M \vee (N^{\mathsf{C}} \wedge M) = \\ &M + (N^{\perp} \cap M) = \\ &M + (N - M), \end{aligned} \tag{7.19}$$

i.e., N is the direct sum of M and $N - M$.

As a consequence of Theorem 7.1, every subspace A_i of a Hilbert space L can be uniquely assigned a projector P_i. Thus, one may say that projectors also form a lattice.

7.13 Exercises and Problems

1. Show that $\max\limits_{1 \leq i \leq 2} |x_i - y_i|$ is a metric over E_2 (i.e., in the usual plane).

2. Let $X = D$ denote a set of documents and T a set of terms. Can you define a metric between the elements of D?

3. Let $X = D$ denote a set of documents and T a set of terms. Every document d is a set of terms, i.e., $d \in \wp(T)$. Is the function

$$\delta: D \times D \to \mathbb{N},\ \delta(d_i, d_j) = |d_i \cap d_j|$$

a metric? (*Note:* $|d_i \cap d_j|$ denotes the number of terms d_i and d_j have in common.)

4. Let $X = D$ denote a set of documents and T a set of terms. Every document d is a set of terms, i.e., $d \in \wp(T)$. Further, let w_{ij} denote the weight of term t_i in document d_j ($i = 1,\ldots,n,\ j = 1,\ldots,m$). Do the documents of D [represented as $d_j = (w_{1j},\ldots,w_{nj})$] form a linear space $(D, +, \times, \mathbb{R})$?

5. Is $\cos\varphi$ a metric on the space E_n?

6. Define projectors in E_3.

7. Show that lattice $\Re(E_n)$ of subspaces of E_n is modular.

8. Let $D = \{d_1,\ldots,d_m\}$ denote a collection of documents and $T = \{t_1,\ldots,t_n\}$ a set of terms. Let $W = (w_{ij})_{n \times m}$ be a term-document weights matrix ($n > m$).
   - Write down the subspace $Sp(W)$.
   - Determine a basis for subspace $Sp(W)$.
   - Calculate the matrix of projector $P_{Sp(W)}(\mathbf{x})$.
   - Given a query vector $\mathbf{q}$, compute the projection of the query onto the subspace of documents, i.e., $P_{Sp(W)}(\mathbf{q})$.

9. Let $D = \{d_1,\ldots,d_m\}$ denote a collection of documents and $T = \{t_1,\ldots,t_n\}$ a set of terms. Let $W = (w_{ij})_{n \times m}$ be a term-document weights matrix, and $\mathbf{q}$ a query vector. Matrix W may be conceived as the matrix of an operator O between the linear space E_n of documents and a linear space E_m of similarities (i.e., $\mathbf{s} = (s_1 \ldots s_j \ldots s_m)$, where s_j is the degree of similarity between query $\mathbf{q}$ and document d_j): $O: E_n \to E_m$. Show that the operator O is a linear operator. Is it a projector? Why?

7.14 Bibliography

Folland, G. B.: *Real Analysis: Modern Techniques and Their Applications* (Wiley, New York, 1984)

Halmos, P. R.: *Introduction to Hilbert Spaces* (Chelsea, New York, 1957)

Jauch, J. M.: *Foundations of Quantum Mechanics* (Addison-Wesley, Reading, MA, 1968)
Patterson, E. M., and Rutherford, D. E.: *Elementary Abstract Algebra* (Oliver and Boyd, Edinburgh/London, 1965)
Riesz, F., and Sz.-Nagy, B.: *Functional Analysis* (F. Ungar, New York, 1955)
Rudin, W.: *Real and Complex Analysis* (McGraw Hill, New York, 1966)

8 Vector Space Retrieval

It is not the vectors which matter, but the lattice of subspaces.
(John von Neumann)

This chapter begins with the original as well as a more formal description of vector space retrieval (VSR). An example is also given to the reader help exactly understand what the method means and how it operates.

Then, the widely used similarity measures are presented in both a compact and parameterized form (having in mind a computer programmer who prefers writing a compact code for all cases) and in their usual forms.

This is followed by a description of the use of the notion of a projector in Hilbert space for the calculation of meaning and for the expression of compatibility of relevance assessments.

The second part of the chapter is concerned with the application of lattices in VSR. It is shown that retrieving documents means projection. After introducing the concept of equivalent queries, we prove that nonequivalent queries form a nondistributive lattice (called a query lattice). We also show that VSR may be viewed as a nonsubmodular lattice-lattice mapping from the query lattice to the Boolean algebra of documents.

A parallel is drawn between the lattice-based view of quantum mechanics and the lattice-based view of IR introduced in this chapter, and that is discussed.

Thus, this chapter may help provide a deeper insight into the very mechanism or logic of VSR

The chapter ends with exercises and problems that are designed to enhance understanding of the application of the vector space method in practice.

8.1 Introduction

Given an entity described by a piece of text (traditionally called a document), if the words are ranked in decreasing order with respect to their number of occurrences (also called frequencies), then the product of the rank of any word and its number of occurrences is approximately constant (Zipf 1949). If it is assumed, naturally enough, that the most obvious place where appropriate content identifiers might be found is the document itself, then the number of occurrences of a term can give a meaningful indication of its content (Luhn 1966). Given m documents and n terms, each document can be assigned a sequence (of length n) of weights that represent the degrees to which terms pertain to (characterize) that document. If all of these sequences are put together, an $n \times m$ matrix, called a term-document matrix, of weights is obtained, where the columns correspond to documents and the rows to terms (see also Chapter 4).

Let us consider a—textual—query expressing an information need to which an answer is to be found by searching the documents. Salton (1966) proposed that both documents and queries should use the same conceptual space, and some years afterward Salton et al. (1975a) combined this idea with the term-document matrix. More than a decade later, Salton and Buckley (1988) reused this framework and gave a mathematical description that has since become known as the vector space model (VSM) or VSR:

- Both document and query weights are conceived as being vectors in the linear space (of terms).
- The degree of similarity between documents and queries is based on the scalar product of the space.
- If the scalar product is zero, then no documents are retrieved.
- Only those documents are retrieved for which the scalar product is different from zero.

Vector space retrieval has proved useful in many practical applications over time. Its retrieval effectiveness was tested under laboratory conditions almost at its inception by Salton et al. (1975a) using three test collections (CRAN, MED, TIME). They found that the mean average precision was 0.48 for CRAN, 0.57 for MED, and 0.66 for TIME. The experimental results have been confirmed time and again ever since. For example, Wong et al. (1985) found that the mean average precision was 0.25 for ADI and 0.35 for CRAN. We repeated the measurements and found the mean average precision to be: 0.33 for ADI, 0.44 for MED, 0.52 for TIME, and 0.18

for CRAN. (*Note:* Newer retrieval methods developed in the meantime gave better results on these test collections.)

There seem to be just two works in which the notion of lattice is applied in the VSR method, and both relate to the lattice of subspaces of a Hilbert space.

Widdows and Peters (2003) apply quantum logic operations to retrieval. The operations (conjunction, disjunction, negation) defining the lattice of subspaces of a Hilbert space are used to define nonclassical logical operations NOT, OR, and AND on word vectors. These are then used to find senses of words and to resolve the word sense disambiguation problem.

Van Rijsbergen (2004), applying von Neumann's ideas in quantum logic, shows how one can treat important retrieval methods (coordination level matching, relevance feedback, dynamic clustering, ostensive retrieval) within the framework of Hilbert spaces by interpreting the inner product as probability. He uses the lattice of subspaces of a Hilbert space to give an algebraic form to logical conditionals utilized in retrieval, arguing that disjunction does not commute ("observing relevance followed by topicality is not the same as observing topicality followed by relevance"), and so is not classical.

In this chapter, we deal with VSR and present a lattice theoretical approach that differs from the above-mentioned applications of lattices to VSR. Namely, we show that VSR may be conceived as a nonsubmodular lattice-lattice mapping between a nondistributive lattice and a Boolean algebra. This view of the subject should help the reader to gain a deeper insight into its formal mechanism and its underlying abstract structures.

8.2 Lattices in Vector Space Retrieval

8.2.1 Vector Space Retrieval

Salton and Buckley (1988) gave the following mathematical description, which is now known as the vector space model (VSM) of information retrieval:

> In the late 1950's, Luhn first suggested that automatic text retrieval systems could be designed based on a comparison of content identifiers attached both to the stored texts and to the users' queries. The documents would be represented by term vectors of the form $D = (t_i, t_j,\ldots,t_p)$, where each t_k identifies a content term assigned to some sample document D. Analogously, a typical query vector might be formulated as

$Q = (q_a, q_b,\ldots,q_r)$. A more formal representation of the term vectors is obtained by including in each term vector all possible content terms allowed in the system and adding term weight assignments to provide distinctions among terms. Thus, if w_{dk} (or w_{qk}) represents the weight of term t_k in document D (or query Q), and t terms in all are available for content representation, the term vectors for document and query can be written as $D = (t_0,w_{d0};\ t_1,w_{d1};\ldots;\ t_t,w_{dt})$ and $Q = (q_0,w_{q0};\ q_1,w_{q1};\ldots;\ q_t,w_{qt})$. Given the vector representations, a query-document similarity value may be obtained by comparing the corresponding vectors, using for example the conventional vector product formula similarity $(Q, D) = \Sigma w_{qk}w_{dk}$. When the term weights are restricted to 0 and 1 as previously suggested, the vector product measures the number of terms that are jointly assigned to query Q and document D. In practice it has proven useful to provide a greater degree of discrimination among terms assigned for content representation than is possible with weights of 0 and 1 alone. The weights could be allowed to vary continuously between 0 and 1, the higher weight assignment near 1 being used for the most important terms, whereas lower weights near 0 would characterize the less important terms. A typical term weight using a vector length normalization factor is $w_{dk}/(\Sigma_{vector}(w_{di})^2)^{1/2}$ for documents. When a length normalized term-weighting system is used with the vector similarity function, one obtains the well-known cosine similarity formula.

More formally, a "definition" for VSM can be given as follows:

1. Both document $D = (t_0,w_{d0};\ t_1,w_{d1};\ldots;\ t_t,w_{dt})$ and query $Q = (q_0,w_{q0};\ q_1,w_{q1};\ldots;\ q_t,w_{qt})$ are elements of the Euclidean space E_n, $n = t + 1$. In other words, we may say that the formal framework of VSM is E_n.

2. Each term t_i corresponds to a basis vector $\mathbf{e}_i$ of space E_n.

3. The degree of relevance r of a document D represented by vector $\mathbf{w}$ relative to query Q represented by vector $\mathbf{q}$ is based on the dot product $<\mathbf{w}, \mathbf{q}>$.

4. If $<\mathbf{w}, \mathbf{q}> = 0$, then the document is not relevant, and hence it is not retrieved. If $<\mathbf{w}, \mathbf{q}> \neq 0$, then the document is considered to be relevant and is retrieved.

5. Let $<\mathbf{w}_i, \mathbf{q}> \neq 0$, $i = 1,\ldots,m$, correspond to documents D_i. Then, the documents $D_1,\ldots,D_m$ are used to construct the hit list: $D_1,\ldots,D_m$ are sorted descendingly on their relevance degrees and are displayed to the user in this order.

Figure 8.1 shows a visual example, in E_3, for the mathematical formulation of VSM.

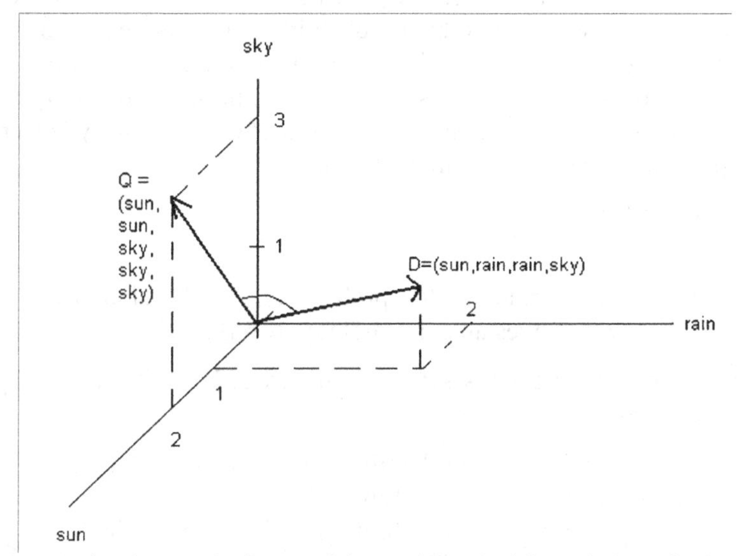

Fig. 8.1. A visual example, in E_3, of the mathematical formulation of VSM. The set T of terms is T = {sun, rain, sky}. A document D = (sun, rain, rain, sky) and a query Q = (sun, sun, sky, sky, sky) are represented as vectors. The corresponding vectors of weights (using a frequency weighting scheme) are: $\mathbf{w} = (1, 2, 1)$, $\mathbf{q} = (2, 0, 3)$. The scalar product (equivalently the cosine of the angle) between these two vectors is a measure of similarity between document and query.

One should note that from a mathematical point of view, the formal framework (i.e., linear space) adopted for VSR is, as can be seen, a precisely defined, sophisticated, and delicate mathematical structure (see Chapter 7).

Example 8.1

Let the set of original documents (to be searched) be $D = \{D_1, D_2, D_3\}$, where

D_1 = Bayes's principle: The principle that, in estimating a parameter, one should initially assume that each possible value has equal probability (uniform prior distribution).

D_2 = Bayesian decision theory: A mathematical theory of decision-making that presumes utility and probability functions, and according to which the act to be chosen is the Bayes act, i.e., the one with highest subjective expected utility. If one had unlimited time and calculating power with which to make every decision, this procedure would be the best way to make any decision.

D_3 = Bayesian epistemology: A philosophical theory that holds that the epistemic status of a proposition (i.e., how well proven or well established it is) is best measured by a probability, and that the proper way to revise this probability is given by Bayesian conditionalization or a similar procedure. A Bayesian epistemologist would use probability to define concepts such as epistemic status, support or explanatory power and to explore the relationships among them.

Let the set T of terms be:

$$T = \{t_1 = \text{Bayes's principle}, t_2 = \text{probability}, t_3 = \text{decision making}, t_4 = \text{Bayesian epistemology}, t_5 = \text{Bayes}\}.$$

Conceiving the documents as sets of terms (together with their frequencies), we can represent them as:

D_1 = {(Bayes's principle, 1); (probability, 1); (decision-making, 0); (Bayesian epistemology, 0); (Bayes, 1)}.

D_2 = {(Bayes's principle, 0); (probability, 1); (decision-making, 1); (Bayesian epistemology, 0); (Bayes, 2)}.

D_3 = {(Bayes's principle, 0); (probability, 3); (decision-making, 0); (Bayesian epistemology, 2); (Bayes, 3)}.

Here (Bayes's principle, 1) means that term t_1 = "Bayes's principle" occurs once in document D_1, etc. As there are five terms, documents are represented in the form of weight vectors $\mathbf{w}_1$, $\mathbf{w}_2$, and $\mathbf{w}_3$ in the five-dimensional Euclidean space E_5:

$$\mathbf{w}_1 = (1, 1, 0, 0, 1), \mathbf{w}_2 = (0, 1, 1, 0, 2), \mathbf{w}_3 = (0, 3, 0, 2, 3).$$

Let the query Q be

$$Q = \{(\text{probability}, 1); (\text{decision-making}, 1); (\text{Bayes}, 1)\}.$$

Thus, the query vector is $\mathbf{q} = (0, 1, 1, 0, 1)$. The similarity between query and documents may be given by the values of the inner product: $\langle \mathbf{w}_1, \mathbf{q} \rangle = 2$, $\langle \mathbf{w}_2, \mathbf{q} \rangle = 4$, $\langle \mathbf{w}_3, \mathbf{q} \rangle = 6$. □

8.2.2 Technology of Vector Space Retrieval

In what follows, an automatic method is presented that consists of the following steps (see Chapter 4 for details on technological aspects):

1. Given a set D of documents.
2. Identify terms.
3. Exclude stopwords.
4. Apply stemming to remaining words.
5. Compute for each document D_j and term t_i a weight w_{ij}.
6. A query Q_k coming from a user is also conceived as being a document; a weight vector $\mathbf{v}_k$ can be computed for it as well, in a similar way.
7. Retrieval is defined as follows:
   Document D_j is retrieved in response to query Q_k if the document and the query are "similar enough," i.e., a similarity measure s_{jk} between the document (identified by $\mathbf{v}_j$) and the query (identified by $\mathbf{v}_k$) is over some threshold K.

There are a number of similarity measures used in VSR (Meadow et al. 1999), which can be expressed in a unified form as follows:

Theorem 8.1. (Dominich 2002) *The similarity measures used in VSR can be expressed in a compact form as*

$$\rho = \frac{<\mathbf{w}_j, \mathbf{w}_Q>}{(\|\mathbf{w}_j\| \cdot \|\mathbf{w}_Q\|)^a \cdot (2^{c-b}(|\mathbf{w}_j| + |\mathbf{w}_Q|) - c \cdot <\mathbf{w}_j, \mathbf{w}_Q>)^b \cdot (\min(|\mathbf{w}_j|, |\mathbf{w}_Q|))^d},$$

where

- w_j *denotes the document vector of document* D_j.
- w_Q *denotes the query vector of query* Q.
- $|.|$ *denotes the sum of coordinates, i.e.,* $|w| = |w_{1j} \ldots w_{nj}| = w_{1j} + \ldots + w_{nj}$.
- $\|.\|$ *denotes the Euclidean norm.*
- $a, b, c, d \in \{0, 1\}$.

Proof. It is shown that the usual similarity measures are obtained for certain values of the parameters, as follows:

Dot product measure. If $a = 0, b = 0, c = c, d = 0$, then

$$\rho = \langle \mathbf{w}_j, \mathbf{w}_Q \rangle.$$

Cosine measure. If $a = 1, b = 0, c = c, d = 0$, then

$$\rho = \frac{\langle \mathbf{w}_j, \mathbf{w}_Q \rangle}{||\mathbf{w}_j|| \cdot ||\mathbf{w}_Q||}$$

Dice coefficient measure. For $a = 0, b = 1, c = 0, d = 0$, we have

$$\rho = \frac{2\langle \mathbf{w}_j, \mathbf{w}_Q \rangle}{|\mathbf{w}_j| + |\mathbf{w}_Q|}$$

Jaccard coefficient measure. For $a = 0, b = 1, c = 1, d = 0$, we have

$$\rho = \frac{\langle \mathbf{w}_j, \mathbf{w}_Q \rangle}{|\mathbf{w}_j| + |\mathbf{w}_Q| - \langle \mathbf{w}_j, \mathbf{w}_Q \rangle}$$

Overlap coefficient measure. If $a = 0, b = 0, c = 0, d = 1$, then

$$\rho = \frac{\langle \mathbf{w}_j, \mathbf{w}_Q \rangle}{\min(|\mathbf{w}_j|, |\mathbf{w}_Q|)}$$

□

Table 8.1 summarizes the similarity measures depending on the values of the parameters *a*, *b*, *c*, and *d*.

Table 8.1. The Similarity Measures Used in VSR Depending on the Values of the Parameters *a*, *b*, *c* and *d*

a	*b*	*c*	*d*	Similarity measure
0	0	*c*	0	Dot product
1	0	*c*	0	Cosine
0	1	0	0	Dice coefficient
0	1	1	0	Jaccard coefficient
0	0	0	1	Overlap coefficient

Theorem 8.1 is useful, first of all, for software developers in that the general form can be programmed as a subroutine and can then be called by appropriately particularizing the values of the parameters.

8.3 Calculation of Meaning Using the Hilbert Lattice

The operations of the Hilbert lattice $\mathfrak{H}(L)$ can be used to define (or interpret) numerically meaning of words or documents in general (Widdows and Peters 2003, Widdows 2003, 2004).

8.3.1 Queries with Negation

We have seen that the numerical expression of irrelevance in VSR is the fact that the scalar product is equal to zero: a query Q is irrelevant to document D if the query vector **q** is orthogonal to the document vector **d**, i.e., $<\mathbf{q}, \mathbf{d}> = 0$. Now let **q** be the vector $\mathbf{q} = \mathbf{a} \neg\mathbf{b}$ (e.g., the query vector **q** corresponds to the query Q = "information NOT retrieval"). In vector notation,

$$<\mathbf{q}, \mathbf{a}> \neq 0 \text{ and } <\mathbf{q}, \mathbf{b}> = 0, \tag{8.1}$$

where **a** is the vector corresponding to "information" and **b** is the vector corresponding to "retrieval." In other words, the query vector is the projection of vector **a** onto subspace $\{\mathbf{b}\}^{\perp}$.

A simple calculation shows that

$$< \mathbf{a} - \frac{<\mathbf{a}, \mathbf{b}>}{||\mathbf{b}||}\mathbf{b}, \mathbf{b} > =$$

$$<\mathbf{a}, \mathbf{b}> - \frac{<\mathbf{a}, \mathbf{b}>\, ||\mathbf{b}||}{||\mathbf{b}||} = 0, \tag{8.2}$$

which means that $\mathbf{q} \perp \mathbf{b}$. This means that query **q** is a single vector:

$$\mathbf{q} = \mathbf{a} \neg \mathbf{b} = \mathbf{a} - \frac{\langle \mathbf{a}, \mathbf{b} \rangle}{\|\mathbf{b}\|} \mathbf{b}. \quad (8.3)$$

Widdows and Peters conducted experiments to test the above negation method. They used *New York Times* data consisting of 173 million words from news articles written between July 1994 and December 1996. News articles typically prefer some meanings of ambiguous words over others. For example, the word "suit" was used mainly in a legal rather than a clothing context. They tested the effectiveness of the above negation method to find less common meanings by removing words belonging to the predominant meanings. The results showed that the method was effective in removing the "legal" meaning from the word "suit" and the "sporting" meaning from the word "play," leaving "clothing" and "performance," respectively. Moreover, removing a particular word also removes concepts related to the negated word. Using the single-word query "suit" returned the following list:

suit, lawsuit, suits, plaintiff, sued, damages, appeal.

Using the negated query "suit NOT lawsuit" returned:

pants, shirt, jacket, silk, dress, trousers,
sweater, wearing, satin, plaid, lace.

8.3.2 Queries with Disjunction

From a theoretical point of view, answering a disjunctive query may proceed along the following lines.

Let $\mathbf{q} = \mathbf{a} \vee \mathbf{b}$ be a Boolean query. Traditionally, $\mathbf{q}$ triggers the retrieval of all documents containing either or both terms whose vectors are $\mathbf{a}$ and $\mathbf{b}$, respectively. In the Hilbert lattice $\mathfrak{R}(L)$, the $\vee$ operation is the join, which means the direct sum of the vectors involved. Thus, the query $\mathbf{q} = \mathbf{a} \vee \mathbf{b}$ may be taken to represent the spanned subspace $Q = \{p\mathbf{a} + r\mathbf{b}\}, p, r \in \mathbb{R}$. The similarity $s(\mathbf{d}, Q)$ between a document vector $\mathbf{d}$ and subspace Q may be defined as

$$s(\mathbf{d}, Q) = \langle \mathbf{d}, P_Q(\mathbf{d}) \rangle, \quad (8.4)$$

which is the scalar product of the document vector **d** and its projection onto subspace Q (i.e., the component of **d** that lies in Q). In order to compute the projection $P_Q(\mathbf{d})$, an orthonormal basis $\{\mathbf{b}_i\}$ for Q should first be constructed (e.g., using the Gram-Schmidt procedure, Section 7.11). Thus, we have

$$s(\mathbf{d}, Q) = <\mathbf{d}, P_Q(\mathbf{d})> = \\ <\mathbf{d}, \sum_i <\mathbf{d}, \mathbf{b}_i>\mathbf{b}_i> = \\ \sum_i <\mathbf{d}, \mathbf{b}_i>. \tag{8.5}$$

8.4 Compatibility of Relevance Assessments

Van Rijsbergen (2004) elaborates the basics for a formal expression of relevance assessments in information retrieval.

Let Q denote a single-term query, and let us assume the use of Boolean retrieval to answer Q. Then, a document D is retrieved if it contains Q (i.e., D 'is about' Q), and it is not retrieved if it does not contain Q (i.e., D is not about Q). Once D has been retrieved, the user can decide whether it is relevant or not. The traditional working assumption is that relevance assessments are independent of one another and are binary (i.e., a document is either relevant or not). This can be expressed formally as follows:

$$(D \wedge R) \vee (D \wedge \neg R), \tag{8.6}$$

i.e., document D is either relevant (R) or nonrelevant ($\neg R$). This can be rewritten as

$$(D \wedge R) \vee (D \wedge \neg R) = D \wedge (R \vee \neg R), \tag{8.7}$$

which is the well-known distributive law in the Boolean lattice ({True, False}, $\vee$, $\wedge$, $\neg$) of mathematical logic, and is an expression of compatibility between D and R.

However, experiments confirmed the opposite: assessing the relevance of retrieved documents D in response to Q in between two retrievals for P is characterized by certain cognitive activity that affects the assessment of relevance. (Thus D, which was found relevant in the first retrieval, may be found irrelevant when assessed again after retrieval for some other query P.) Hence, relevance assessments do not seem to be totally independent of each other. In order to model this situation, which is justified by practice, van Rijsbergen proposed using the Hilbert lattice.

As any projector in a Hilbert space has two eigenvalues (0 and 1, meaning, e.g., irrelevance and relevance), it may be interpreted as (or assigned) a logical proposition. Using the Hilbert lattice $\mathfrak{R}(L)$ of projectors, we can express the compatibility condition as

$$(D \wedge R) \vee (D \wedge R^{\perp}),\ \ D, R \in \mathfrak{R}(L). \tag{8.8}$$

As the Hilbert lattice is not distributive, we have

$$(D \wedge R) \vee (D \wedge R^{\perp}) \neq D \wedge (R \vee R^{\perp}), \tag{8.9}$$

which means that compatibility does not hold (which is in accordance with experimental results).

8.5 Vector Space Retrieval: Lattice-Lattice Mapping

We first show the following:

Theorem 8.2. *The orthogonal complement $A^{\perp}$ of any subset $A \subset E_n$ of space E_n is a closed subspace of E_n.*

Proof. Let us show first that $A^{\perp}$ is a subspace of E_n. Let $\mathbf{a}$, $\mathbf{b} \in A^{\perp}$ denote two arbitrary vectors of $A^{\perp}$. We have to show that $\mathbf{a} + \mathbf{b} \in A^{\perp}$ and $r\mathbf{a} \in A^{\perp}$, $\forall\, r \in \mathbb{R}$. For $\mathbf{a} = (a_1,\ldots,a_n)$ and $\mathbf{b} = (b_1,\ldots,b_n)$ in $A^{\perp}$, it follows that $\mathbf{a} \perp \mathbf{x}$, $\mathbf{b} \perp \mathbf{x}$, $\forall \mathbf{x} = (x_1,\ldots,x_n) \in A$, i.e.,

$$a_1x_1 + \ldots + a_nx_n = 0,\ \ b_1x_1 + \ldots + b_nx_n = 0.$$

From this, we obtain:

$$a_1x_1 + \ldots + a_nx_n + b_1x_1 + \ldots + b_nx_n =$$

$$(a_1 + b_1)x_1 + \ldots + (a_n + b_n)x_n = 0,$$

which means that $\mathbf{a} + \mathbf{b} \perp \mathbf{x}$ and thus $\mathbf{a} + \mathbf{b} \in A^{\perp}$. In a similar manner, we have

$$a_1x_1 + \ldots + a_nx_n = 0 \Rightarrow$$

$$ra_1x_1 + \ldots + ra_nx_n = 0 \Rightarrow$$

$$r\mathbf{a} \in A^{\perp}.$$

We now show that $A^{\perp}$ is closed, i.e., the limit of every convergent sequence belongs to $A^{\perp}$. Let $\mathbf{y}_1,\ldots,\mathbf{y}_n,\ldots \in A^{\perp}$, $\mathbf{y}_n \neq \mathbf{0}$, denote a nontrivial convergent sequence: $\lim \mathbf{y}_n = \mathbf{y}$. This means that $\|\mathbf{y}_n - \mathbf{y}\| \rightarrow 0$. We demonstrate that $\mathbf{y} \in A^{\perp}$. Let us assume the opposite, i.e., $\mathbf{y} \notin A^{\perp}$. Then, $\mathbf{y} \in A$ and so $\mathbf{y}_n \perp \mathbf{y}$ for every n, and hence $<\mathbf{y}_n, \mathbf{y}> = 0$. Thus,

$$||\mathbf{y}_n - \mathbf{y}|| \to 0 \Leftrightarrow$$

$$\sum_i (y_{ni} - y_i)^2 = \sum_i y_{ni}^2 - 2\sum_i y_{ni} y_i + \sum_i y_i^2 = \sum_i y_{ni}^2 + \sum_i y_i^2 \to 0,$$

where $\mathbf{y}_n = (y_{n1},\ldots,y_{ni},\ldots)$, $\mathbf{y}_n = (y_1,\ldots,y_i,\ldots)$, which is only possible when $\mathbf{y}_n = \mathbf{y} = \mathbf{0}$. As this contradicts the assumption $\mathbf{y}_n \neq \mathbf{0}$, we have $\mathbf{y} \in A^{\perp}$. □

Since the Euclidean space E_n is a Hilbert space, by using Theorem 7.1 and Lemma 8.2, we get the following:

Theorem 8.3. (Dominich and Kiezer 2007) *Given a Euclidean space E_n whose vectors (of weights) $\boldsymbol{d} \in E_n$ identify documents, the set $\mathfrak{R}_Q$ of documents retrieved in response to a query Q (represented as a vector $\boldsymbol{q} \in E_n$) is*

$$\mathfrak{R}_Q = \{D \mid \boldsymbol{d} = P_A(\boldsymbol{d}) + \boldsymbol{q},\ A = \{\boldsymbol{q}\}^{\perp}\}.$$

Proof. The retrieval of documents D represented as vectors $\mathbf{w}$ in response to query Q represented as a vector $\mathbf{q}$ means constructing the set

$$\mathfrak{R}_Q = \{D \mid <\mathbf{q}, \mathbf{w}> \neq 0\}.$$

The orthogonal complement $A = \{\mathbf{q}\}^{\perp}$ (i.e., the set of documents that do not share common terms with the query) corresponding to query Q is given by the documents D whose vectors $\mathbf{w}$ are perpendicular to $\mathbf{q}$, i.e.,

$$A = \{\mathbf{q}\}^{\perp} =$$

$$\{\mathbf{w} \mid \mathbf{w} \perp \{\mathbf{q}\}\} =$$

$$\{\mathbf{w} \mid <\mathbf{w}, \mathbf{q}> = 0\}.$$

Set A is a closed linear subspace of space E_n (Lemma 8.2). It follows that any element $\mathbf{d} \in E_n$ of space E_n can be uniquely written (Theorem 7.1) as

$$\mathbf{d} = \mathbf{w} + \mathbf{q},$$

where $\mathbf{w} \in A$ and $\mathbf{q} \in A^{\perp} = \{\mathbf{q}\}$.

The projector P_A for the elements $\mathbf{d} \in E_n$ of space E_n onto the set A is defined as $P_A(\mathbf{d}) = \mathbf{w}$, $\mathbf{w} \in A$. Thus,

$$\mathfrak{R}_Q =$$

$$\{D \mid \mathbf{d} = P_A(\mathbf{d}) + \mathbf{q},\ A = \{\mathbf{q}\}^{\perp}\}.\ \square$$

Theorem 8.3 makes it possible to say that obtaining the set of retrieved documents means projection. **Figure 8.2** shows a visual example for Theorem 8.3.

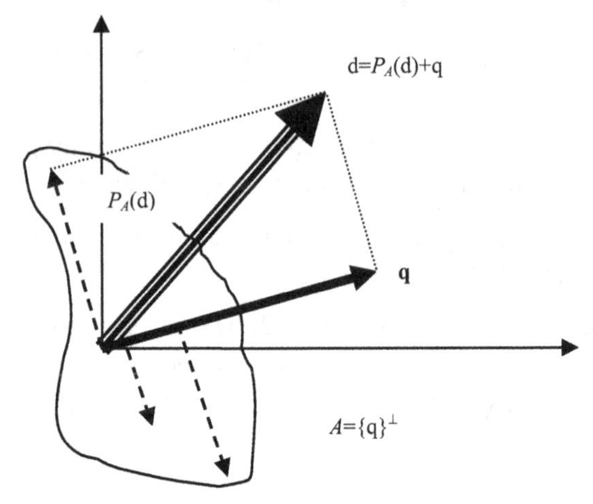

Fig. 8.2. Visual example for Theorem 8.3 in plane. The dashed-line vectors form the orthogonal complement (i.e., perpendicular) vectors $A = \{\mathbf{q}\}^\perp$ of query vector **q**. Any document vector **d** (triple-line vector) can be written as the sum of query vector **q** and its projection $P_A(d)$ onto subspace A.

From the point of view of practice, queries may be equal to or different from each other. The following properties hold:

- The complement $A = \{\mathbf{q}\}^\perp$ of any query vector **q** is a closed subspace of space E_n.
- Hence, A is a member of the subspace lattice, i.e., $A \in \Re(E_n)$.
- But not every subspace of E_n [in other words, not every element of lattice $\Re(E_n)$] is the complement of a query vector (e.g., the subspace $\{\mathbf{a}, \mathbf{b}\}^\perp$, $\mathbf{a}, \mathbf{b} \in E_n$, is not the complement of any query because a query is represented by a single vector and not by two vectors).

Thus, one may ask whether the queries (equivalently, their complements or the corresponding sublattice) form some structure or not. An answer is given by the following:

Theorem 8.4. *The set of different and nonperpendicular queries is an atomistic, complemented, modular, nondistributive lattice.*

Proof. Let **q** and **q′** denote two different and nonperpendicular query vectors. Let $\mathbf{v} \in A = \{\mathbf{q}\}^\perp$, $\mathbf{v} \neq \mathbf{0}$; hence $\langle \mathbf{v}, \mathbf{q} \rangle = 0$. We show that vector **v** cannot belong to $A' = \{\mathbf{q}'\}^{\perp}$ at the same time (i.e., **v** cannot be perpendicular to both **q** and **q′** at the same time). By Theorem 8.3, we have

$$\mathbf{q}' = P_A(\mathbf{q}') + \mathbf{q}.$$

Then,

$$<\mathbf{v}, \mathbf{q}'> = <\mathbf{v}, P_A(\mathbf{q}') + \mathbf{q}> =$$
$$<\mathbf{v}, P_A(\mathbf{q}')> + <\mathbf{v}, \mathbf{q}> =$$
$$= <\mathbf{v}, P_A(\mathbf{q}')> + 0 = <\mathbf{v}, P_A(\mathbf{q}')> \neq 0.$$

Hence, **v** cannot be perpendicular to **q′**. This means that the intersection $A \cap A'$ is empty: $A \cap A' = \varnothing$.

Thus, the following lattice can be constructed:

- $\mathbf{0} = \varnothing$.
- $\mathbf{1} = E_n$.
- The other elements of this lattice are the sets A as atoms (**Fig. 8.3**).

As this lattice is a sublattice of the—modular—lattice $\Re(E_n)$ of subspaces, it is modular as well, which means that it is orthomodular at the same time. However, it is not distributive:

$$A \vee (A' \wedge A'') = A \vee \varnothing = A \neq$$
$$(A \vee A') \wedge (A \vee A'') = E_n \wedge E_n = E_n.$$

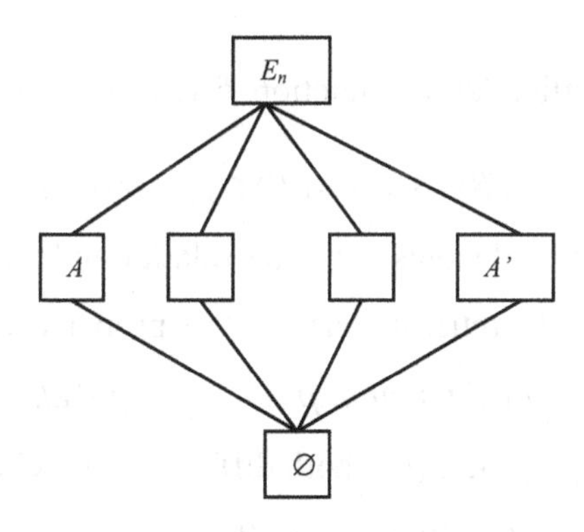

Fig. 8.3. Visualization of the query lattice.

□

Theorem 8.4 entitles us to define a query lattice as follows:

Definition 8.1. The lattice of Theorem 8.4 is called a *query lattice* and is denoted by $L(Q)$. □

Theorem 8.4 also tells us that query lattice $L(Q)$ is not a Boolean algebra. We have seen (Theorem 8.3) that obtaining the set $\Re(Q) = \{D_1,\ldots,D_i,\ldots,D_m\}$ of retrieved documents in response to a query Q means performing projection. Since the set $\{D_1,\ldots,D_i,\ldots,D_m\}$ is a subset of the set D of all documents, i.e., $\Re(Q) \in \wp(D)$, and the structure $(\wp(D), \cap, \cup, \subseteq)$ is a lattice, one may view retrieval as a lattice-lattice mapping. Thus, we introduce the following definition (which represents a lattice theoretical formulation of the VSR method):

Definition 8.2. The VSR method is a *mapping* ρ from the lattice (non-Boolean algebra) of queries, $L(Q)$, to the lattice (Boolean algebra) of documents, $\wp(D)$, based on projector P of the associated Hilbert space:

$$\rho: L(Q) \to \wp(D),$$

$$\rho(A) = \begin{cases} D & if \quad A = E_n \\ \varnothing & if \quad A = \varnothing \\ D' \subseteq D & otherwise \end{cases} \qquad \Box$$

Submodular set functions are an important class of functions (Recski 1989). They arise in many optimization applications (e.g., supply chain management), and they have a role similar to that played by convex/concave functions in continuous optimization. As an analogue of the submodular set function, we introduce a submodular law for lattice-lattice functions as follows:

Definition 8.3. A lattice-lattice function f: $(L_1, \wedge_1, \vee_1) \to (L_2, \wedge_2, \vee_2)$ is *submodular* if

$$f(A \vee_1 B) \vee_2 f(A \wedge_1 B) \le f(A) \vee_2 f(B),$$

$\forall A$ and $B \in L_1$, where $\le$ denotes the order relation in lattice L_2. □

We now show that the retrieval has just this property, namely:

Theorem 8.5. *The retrieval function ρ is not submodular.*

Proof. If we take Q, $Q' \in L(Q)$, both different from $\varnothing$ and E_n, we have

$$f(A \vee_1 B) \vee_2 f(A \wedge_1 B) =$$

$$\rho(Q \vee Q') \cup \rho(Q \wedge Q') =$$

$$\rho(E_n) \cup \rho(\varnothing) = D,$$

but

$$f(A) \vee_2 f(B) =$$

$$\rho(Q) \cup \rho(Q') \subseteq D. \square$$

In words, Theorem 8.5 is telling us that retrieval in response to the orthogonal complements of two distinct queries may not yield the entire set of documents, albeit that these complements span the whole space (i.e., they generate any conceivable query). This result is a lattice theoretical formulation of a situation that can be easily imagined if one takes, e.g., binary weights. For any two binary queries, $\mathbf{q}$ and $\mathbf{q}'$, their orthogonal complements (i.e., vectors with which their scalar product is zero; in other words, all combinations of nonquery terms) may not necessarily retrieve every document.

8.6 Discussion

8.6.1 Query Lattice and Free Will

Theorem 8.4 may also be interesting from a philosophical point of view as it shows that queries, in general, form a structure having a certain order, probably independently of users. This is not at all intuitive, as one would think that queries do not form any structure, for they originate randomly from users. Users do indeed have the freedom to formulate any query they want to. However, queries 'organize' themselves (through their complements) into a very particular structure, namely into a special kind of lattice. This is only seemingly a paradox: it originates from the fact that in the VSR model queries are members, by definition (or by assumption), of a very special space having a very sophisticated structure, i.e., linear space. Users do have the freedom to formulate their information needs, but these can only materialize in the form of queries within the properties and possibilities of the linear space.

8.6.2 Vector Space Retrieval?

As we saw in Theorem 8.3, the vector-based retrieval mechanism may be interpreted as a self-adjoint linear operator (projection) in linear space. We should draw attention to one special aspect of this view that is concerned with the space itself. It is assumed that the space is a linear space. In other words, it is assumed that documents and queries, which belong to this space, are vectors. It is well known that a vector constitutes a very special type of quantity in that:

- It is a "compound" quantity; usually this is expressed by saying that it has a direction and a magnitude (as opposed to, e.g., a real number expressing the temperature of a body),
- Vectors allow for operations to be performed on them (e.g., they can be added and the result is another vector of the same space); thus they form a particular and well-defined structure (called linear space),
- Vectors do not depend on the basis of the space they belong to (the coordinates of a vector are, in general, different in different bases; however, its magnitude and direction are unchanged), and so the scalar product is invariant with respect to the change of the basis of the space.

The question of whether the linear space is an adequate formal framework for retrieval is not a trivial one. The linear space may only be an adequate framework IF the documents and queries ARE vectors.

But are they vectors?

Wong and Raghavan (1984) showed that this case is not realistic; hence it is not realistic to assume, in general, that documents and queries ARE vectors. This may, of course, be *assumed*, but the linear space should perhaps be interpreted rather as a metaphor than as a realistic formal framework. This issue will be dealt with in detail in Chapter 9.

8.6.3 Vector Space Retrieval and Quantum Mechanics

As the Hilbert space and the Hilbert lattice formalism are also considered to be of basic importance in 'orthodox' quantum mechanics, we may not end this chapter without touching upon the connection between VSR and quantum mechanics, a connection brought about by this common formalism. Van Rijsbergen (2004) wrote extensively on this topic, mainly from the point of view of what this connection might bring into IR. Here, we wish to comment on this connection from a different angle based on Birkhoff and von Neumann (1936), Jauch (1968), Piziak (1978), Rédei (1996), and Grinbaum (2005).

The propositional calculus of quantum mechanics is usually (but maybe misleadingly) referred to as quantum logic. This calculus is expressed in terms of "yes-no" experiments, also called propositions. These are empirically verifiable propositions carrying one bit of information (i.e., they allow only two outcomes: either "yes" or "no"). Quantum logic is concerned with the study of the formal structure of such propositions.

It is commonly agreed that the empirically testable propositions of Newtonian physics form a Boolean algebra (i.e., they have the same structure as the propositions of mathematical logic or the subsets of a set in set theory). In other words, in Newtonian physics, the meet and join of two propositions is a proposition, and independent observers can always read off the measurements involved by the propositions and combine the results logically.

In quantum mechanics, however, not all propositions can be simultaneously measured to any accuracy (e.g., the momentum and the position of an electron), which means that not all measurements are always compatible. Thus, distributivity does not always hold (this is "the weakest link"). Birkhoff and von Neumann suggested that distributivity be replaced by a weaker property, namely modularity. Thus, they proposed that the model to study the structure of propositions in quantum mechanics be an orthocomplemented modular lattice (which they identified in the set of closed subspaces of the Hilbert space).

Subsequent research, however, has shown that the modular law is not tenable either, and the law that has since been accepted is orthomodularity. Thus, it is currently commonly accepted that the orthomodular lattice is an appropriate framework for the structure of "yes-no" experiments in quantum mechanics. The projectors of an infinite-dimensional Hilbert space form just such a lattice.

As can be seen, the quest for an adequate formal structure of propositions in quantum mechanics has always involved modeling concerns. At this point, it is instructive to quote the creator of this formalism, John von Neumann, as we think that the modeling concerns encountered in quantum mechanics and vector space retrieval are very similar:

> I do not believe absolutely in Hilbert space anymore. After all … Hilbert space was obtained by generalising Euclidean space … Now we begin to believe that it is not the *vectors* which matter, but the lattice of all linear (closed) subspaces. Because: (1) The vectors ought to represent the physical *states*, … (2) and besides, the states are merely a derived notion, the primitive (phenomenologically given) notion being the qualities which correspond to the *linear closed subspaces*. (Rédei 1996)

Definition 8.2 is in harmony with the above quotation, and may thus throw a new light onto the formal mechanism ("phenomenologically given" quality) of VSR in that it says that retrieval is, in essence, a mapping process between two specially ordered sets (as special formal frameworks), namely between two lattices. While the Boolean algebra character (relative to set union, intersection, and inclusion) of the document lattice $\wp(D)$ may be intuitive, the non-Boolean algebra character of the other lattice, the query lattice $L(Q)$, is not at all intuitive or obvious. This latter characteristic may be due to the fact that $L(Q)$ is not being 'organized' by set inclusion, intersection, or union. In other words, the two lattices $L(Q)$ and $\wp(D)$, between which VSR acts as a very special (nonsubmodular) mapping, have different internal organizations, or orderings; they possess different underlying characters.

This situation may suggest that, symbolically speaking, answers in general are "out there," already given in a nicely organized and well-behaved formal structure, while information needs, as products of imagination and free will, materialize in the form of queries that organize themselves into an ordered formal structure that is not well behaved or symmetric in its underlying operations (nondistributive).

Further, comparing the Boolean algebra structure of logical propositions and the non-Boolean algebra of quantum mechanical propositions, von Neumann wrote:

> The main difference seems to be that whereas logicians have usually assumed that ... negation was the one least able to withstand a critical analysis, the study of mechanics points to the distributive identity as the weakest link in the algebra of logic. ... Our conclusion agrees perhaps more with those critiques of logic which find most objectionable the assumption that $(a^{\perp} \vee b = \mathbf{1}) \Rightarrow (a \Rightarrow b)$, or dually, $(a \wedge b^{\perp} = \mathbf{0}) \Rightarrow (b \Leftarrow a)$; the assumption that to deduce an absurdity from the conjunction of a and $\neg b$ justifies one in inferring that a implies b." (Birkhoff and von Neumann 1936)

We note that $(a^{\perp} \vee b = \mathbf{1}) \Rightarrow (a \Rightarrow b)$ is equivalent to the distributive law. Based on these results as well as on Theorem 8.5, we may say that:

- The underlying algebraic structure of mathematical logic and Newtonian mechanics is a Boolean algebra.
- That of quantum mechanics is a non-Boolean (nonmodular) lattice.
- That of VSR is a nonsubmodular mapping to a Boolean lattice from a non-Boolean lattice.

Perhaps the main difference between VSR and quantum mechanics lies not so much in their being characterized by modular and nonmodular lattices, respectively (although that is an important difference), as in the fact that retrieval has an added ingredient: a nonsubmodular mapping (between two different kind of lattices). The algebraic framework of retrieval is not one lattice, as in mechanics, but two lattices (having different types!) together with a special mapping between them. Thus, retrieval seems to have a much more sophisticated algebra than quantum mechanics.

8.7 Exercises

1. Implement a small VSR system using a collection of real documents of your choice. Using the similarity measures from Theorem 8.1 (but keeping the weighting scheme constant), compare the hit lists obtained for real queries Q_j ($j = 1,\ldots,p$).

2. Using a set D of real documents of your choice, experiment with answering a real query **q** having the form $\mathbf{q} = \mathbf{a} \neg \mathbf{b}$ using Eqs. (8.1)–(8.3). Compare and comment the results.

3. Implement a small VSR system using a collection of real documents of your choice. Let Q_j ($j = 1,\ldots,p$) be equivalent queries. Fixing a weighting scheme and a similarity measure, compare and discuss the hit lists for length-normalized as well as not normalized versions of the queries.

4. Study the queries of the test collections ADI, CACM, CISI, MEDLINE, TREC, etc., from the point of view of their being equivalent or not equivalent to one another.

5. Let D denote a set of real documents (of your choice), and Q_j ($j = 1,\ldots,p$) denote pairwise different (real) queries of your choice. Construct the corresponding question lattice.

9 Fuzzy Algebra-Based Retrieval

Ask them what they think not about the truth of theorems but about their importance.
(Évariste Galois)

This chapter explains how fuzzy algebras can be used to provide new or novel retrieval methods.

After presenting the necessary elements of tensor algebra, we show that when the formal framework of information retrieval is a linear space of terms, the scalar product of the space is not necessarily a similarity measure—contrary to the widely held belief.

Then, we present the required notions and results from fuzzy set theory and show that the set of all fuzzy sets in [0; 1] is a fuzzy algebra. Documents and queries are elements of this algebra. By introducing the principle of invariance, latent semantic indexing, vector space retrieval, and generalized vector space retrieval acquire a correct formal framework with which they are consistent (as opposed to the linear space as a framework).

Based on the notion of fuzzy algebra, the fuzzy entropy method and the fuzzy probability method are discussed, together with experimental results as to their relevance effectiveness.

The chapter ends with exercises and problems that are designed to enhance understanding of the mechanism and application possibilities of the concepts and methods presented.

9.1 Elements of Tensor Algebra

Any vector **v** of n-dimensional Euclidean space E_n can be represented as a linear combination of basis vectors $\mathbf{b}_i$, $i = 1,\ldots,n$:

$$\mathbf{v} = p_1\mathbf{b}_1 + \ldots + p_n\mathbf{b}_n = \sum_{i=1}^{n} p_i\mathbf{b}_i \,,\; p_1,\ldots,p_n \in \mathbb{R}. \tag{9.1}$$

As seen in Chapter 7, basis vectors $\mathbf{b}_i$ need not be orthogonal or normal, i.e., they may form a general Cartesian basis of the space. Let $\mathbf{v} = [v_1 \ldots v_n]^\mathrm{T}$ denote a vector in the orthonormal basis $\mathbf{e}_1,\ldots,\mathbf{e}_n$. Further, let $\mathbf{g}_i$ denote the matrix obtained from general basis vectors $\mathbf{b}_i$:

$$\mathbf{g}_i = [\mathbf{b}_1 \ldots \mathbf{b}_i \ldots \mathbf{b}_n] =$$

$$\begin{bmatrix} b_{11} & \ldots & b_{i1} & \ldots & b_{n1} \\ b_{12} & \ldots & b_{i2} & \ldots & b_{n2} \\ . & \ldots & . & \ldots & . \\ b_{1n} & \ldots & b_{in} & \ldots & b_{nn} \end{bmatrix}. \tag{9.2}$$

Matrix $\mathbf{g}_i$ is called a *basis tensor* (Simmonds 1982). As vectors $\mathbf{b}_1,\ldots,\mathbf{b}_n$ are basis vectors, rank($\mathbf{g}_i$) = n. Hence, $\mathbf{g}_i$ has an inverse, denoted by $\mathbf{g}_i^{-1}$, that is called its *reciprocal basis tensor* and is denoted by $\mathbf{g}^i$, i.e., $\mathbf{g}^i = \mathbf{g}_i^{-1}$. Vector **v** (in an orthonormal basis) can also be written in the general basis $\mathbf{g}_i$. Let the coordinates of vector **v** in basis $\mathbf{g}_i$ be $p^1,\ldots,p^n$. Thus, recalling that the vector is invariant with respect to the change of basis, we have $\mathbf{g}_i \times [p^1 \ldots p^n]^\mathrm{T} = \mathbf{v}$, from which, by multiplying by $\mathbf{g}_i^{-1}$ on the left, we obtain

$$\mathbf{g}_i^{-1} \times \mathbf{g}_i \times [p^1 \ldots p^n]^\mathrm{T} = \mathbf{g}_i^{-1} \times \mathbf{v},\; [p^1 \ldots p^n]^\mathrm{T} = \mathbf{g}_i^{-1} \times \mathbf{v}, \tag{9.3}$$

since $\mathbf{g}_i^{-1} \times \mathbf{g}_i = I$. This means that the reciprocal basis tensor can be used to compute the coordinates of any vector **v** in general basis $\mathbf{g}_i$:

$$\mathbf{p}^i = \mathbf{g}^i\mathbf{v} = \mathbf{g}_i^{-1}\mathbf{v}, \tag{9.4}$$

where $\mathbf{p}^i = [p^1 \ldots p^n]^\mathrm{T}$.

Given now two vectors $\mathbf{u} = [u^1\; u^2 \ldots u^n]^\mathrm{T}$ and $\mathbf{v} = [v^1\; v^2 \ldots v^n]^\mathrm{T}$ in a general basis $\mathbf{g}_i$, we compute the scalar product of vectors **u** and **v**:

$$[\mathbf{b}_1 \ldots \mathbf{b}_n] \times \begin{bmatrix} u^1 \\ \ldots \\ u^n \end{bmatrix} \times [\mathbf{b}_1 \ldots \mathbf{b}_n] \times \begin{bmatrix} v^1 \\ \ldots \\ v^n \end{bmatrix} =$$

$$\begin{bmatrix} u^1 & u^2 & \ldots & u^n \end{bmatrix} \times \begin{bmatrix} \langle \mathbf{b}_1, \mathbf{b}_1 \rangle & \langle \mathbf{b}_1, \mathbf{b}_2 \rangle & \ldots & \langle \mathbf{b}_1, \mathbf{b}_n \rangle \\ & \ldots & & \\ \langle \mathbf{b}_n, \mathbf{b}_1 \rangle & \langle \mathbf{b}_n, \mathbf{b}_2 \rangle & \ldots & \langle \mathbf{b}_n, \mathbf{b}_n \rangle \end{bmatrix} \times \begin{bmatrix} v^1 \\ \ldots \\ v^n \end{bmatrix}. \tag{9.5}$$

The matrix of the scalar products of the basis vectors in Eq. (9.5) is called the *metric tensor* and is denoted by $\mathbf{g}_{ij}$, and $\mathbf{g}_{ij} = \mathbf{g}_i^{\mathsf{T}}\mathbf{g}_j$. Thus, a compact expression for the scalar product is

$$<\mathbf{u}, \mathbf{v}> = (\mathbf{u}^i)^{\mathsf{T}}\mathbf{g}_{ij}\mathbf{v}^j. \tag{9.6}$$

As vector magnitude and direction are invariant with respect to the choice of basis, the scalar product of two vectors is also invariant, i.e., the scalar product is the same regardless of the basis of the space.

Example 9.1

Consider the three-dimensional Euclidean space E_3 with the orthonormal basis

$$\mathbf{e}_1 = [1\ 0\ 0]^{\mathsf{T}},\ \mathbf{e}_2 = [0\ 1\ 0]^{\mathsf{T}},\ \mathbf{e}_3 = [0\ 0\ 1]^{\mathsf{T}}.$$

Let $\mathbf{u} = [12\ {-6}\ \ 9]^{\mathsf{T}}$ and $\mathbf{v} = [3\ 3\ 6]^{\mathsf{T}}$ be two vectors in this basis, and let

$$\mathbf{g}_i = \begin{bmatrix} 1 & 0 & -1 \\ -1 & 1 & -2 \\ 2 & 1 & 1 \end{bmatrix}$$

be a new (general) basis. The coordinates $\mathbf{u}^i$ and $\mathbf{v}^j$ of vectors $\mathbf{u}$ and $\mathbf{v}$, respectively, in the new basis $\mathbf{g}_i$ are

$$\mathbf{u}^i = \mathbf{g}^i\mathbf{u} = [8.5\ {-4.5}\ {-3.5}]^{\mathsf{T}},$$

$$\mathbf{v}^j = \mathbf{g}^j\mathbf{v} = [2\ 3\ {-1}]^{\mathsf{T}}.$$

Their scalar product (in an orthonormal basis) is

$$<\mathbf{u}, \mathbf{v}> = [12\ {-6}\ 9] \times [3\ 3\ 6]^{\mathsf{T}} = 72.$$

The metric tensor of space E_3 is

$$\mathbf{g}_{ij} = \mathbf{g}_i^{\mathsf{T}}\mathbf{g}_j = \begin{bmatrix} 6 & 1 & 3 \\ 1 & 2 & -1 \\ 3 & -1 & 6 \end{bmatrix}.$$

The scalar product of vectors **u** and **v** in the new basis $\mathbf{g}_i$ is

$$\langle \mathbf{u}, \mathbf{v} \rangle = (\mathbf{u}^i)^{\mathsf{T}}\mathbf{g}_{ij}\mathbf{v}^j = 72,$$

i.e., it is the same (as expected). □

9.2 Similarity Measure and Scalar Product

Let us now consider, in detail, the following example in the orthonormal Euclidean space of dimension two, E_2. Its unit length and perpendicular basis vectors are $\mathbf{e}_1 = (1, 0)$ and $\mathbf{e}_2 = (0, 1)$. Let us assume that we have the following two terms: t_1 = "computer" and t_2 = "hardware," which correspond to the two basis vectors (or, equivalently, to coordinate axes) $\mathbf{e}_1$ and $\mathbf{e}_2$, respectively (**Fig. 9.1**). Consider a document D being indexed by the term "computer," and having the weights vector $\mathbf{D} = (3, 0)$. Let a query Q be indexed by the term "hardware" and have the weights vector $\mathbf{Q} = (0, 2)$. The dot product $\langle \mathbf{D}, \mathbf{Q} \rangle$ is $\langle \mathbf{D}, \mathbf{Q} \rangle = 3 \times 0 + 0 \times 2 = 0$, which means that document D is not retrieved in response to query Q.

In a thought-provoking theory paper, Wong and Raghavan (1984) argue that:

> The notion of vector in the vector space retrieval model merely refers to data structure… the scalar product is simply an operation defined on the data structure…The main point here is that the concept of a vector was not intended to be a logical or formal tool.

They then show why the model conflicts with the mathematical notion of vector space.

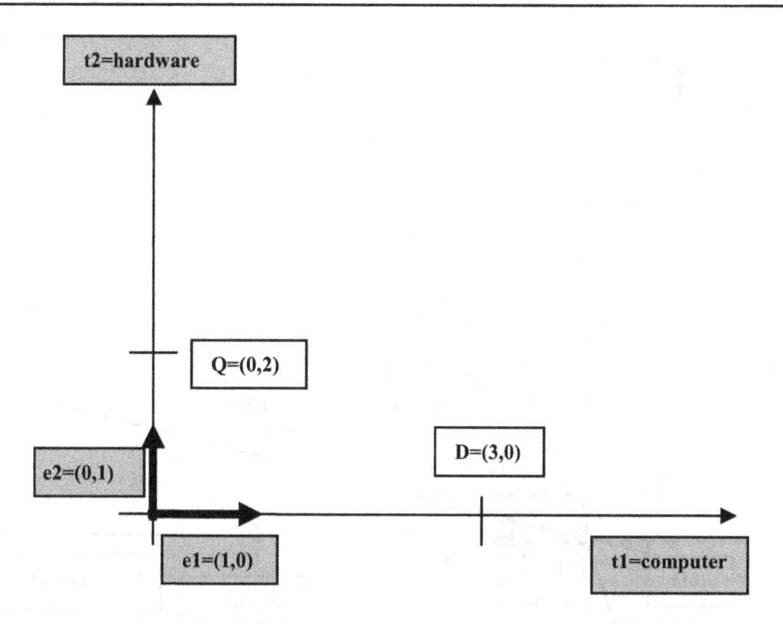

Fig. 9.1. Document and query weight vectors. The document vector **D**(3,0) and query vector **Q**(0,2) are represented in the orthonormal basis (**e**1,**e**2). These basis vectors are perpendicular to each other and have unit lengths. The dot product <**D**,**Q**> is <**D**,**Q**> $= 3 \times 0 + 0 \times 2 = 0$ (which means that document D is not retrieved in response to query Q).

In order to present and illustrate the validity of the concerns with the mathematical modeling as well as of the mathematical subtleties involved, let us enlarge the example of **Fig. 9.1** (Dominich and Kiezer 2007). From the user's point of view, because hardware is part of a computer, he/she might be interested in seeing whether a document D also contains information on hardware. In other words, he/she would not mind if document D would be returned in response to query Q. It is well known that the term independence assumption is not realistic. Terms may depend on each other, and they often do in practice, as in our example. It is also known that the independence assumption can be counterbalanced to a certain degree in practice by, e.g., using thesauri. But can term dependence be captured and expressed in vector space? One possible answer is as follows. Instead of considering an orthonormal basis, let us consider a general basis (**Fig. 9.2**).

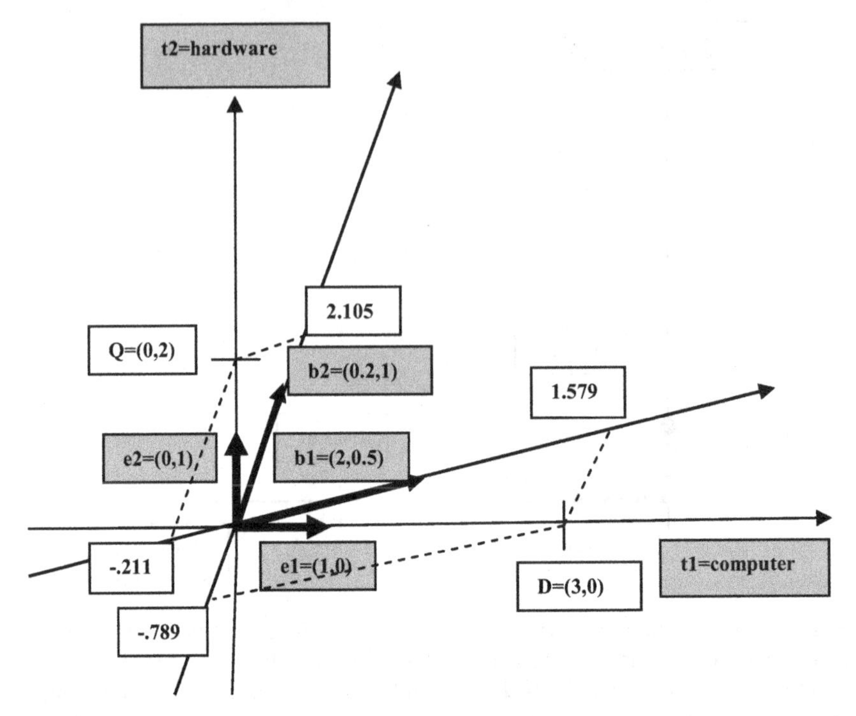

Fig. 9.2. Document and query weight vectors. The document vector **D**(3,0) and query vector **Q**(0;2) are represented in the orthonormal basis (**e**1,**e**2). They are also represented in the general basis (**g**1,**g**2); these basis vectors are not perpendicular to each other, and do not have unit lengths. The coordinates of the document vector in the general basis will be **D**(1.579,–0.789), whereas those of the query vector will be **Q**(–0.211,2.105). The value of the expression <**D**,**Q**> viewed as an inner product between document **D** and query **Q** is always zero, regardless of the basis. But the value of the expression <**D**,**Q**> viewed literally as an algebraic expression is not zero.

The basis vectors of a general basis need not be perpendicular to each other and need not have unit lengths. In our example (**Fig. 9.2**) the term "hardware" is narrower in meaning than the term "computer." If orthogonal basis vectors are used to express the fact that two terms are independent, then a narrower relationship can be expressed by taking an angle smaller than 90° (the exact value of this angle can be the subject of experimentation, but it is not important for the purpose of this example). Thus, let us consider the following two oblique basis vectors: let the basis vector $\mathbf{g}_1$ corresponding to term t_1 be $\mathbf{b}_1 = (2, 0.5)$ and the basis vector $\mathbf{b}_2$ representing term t_2 be $\mathbf{b}_2 = (0.2, 1)$. The coordinates $\mathbf{D}^i$ of the document vector $\mathbf{D}$ in the new (i.e., the general) basis are computed as follows:

$$\mathbf{D}^i = \mathbf{g}_i^{-1} \times \mathbf{D} = [\mathbf{b}_1\ \mathbf{b}_2]^{-1} \times \mathbf{D} = \begin{bmatrix} 2 & 0.2 \\ 0.5 & 1 \end{bmatrix}^{-1} \times [3\ 0]^{\mathrm{T}} =$$

$$= \begin{bmatrix} 0.526 & -0.105 \\ -0.263 & 1.053 \end{bmatrix} \times [3\ 0]^{\mathrm{T}} = [1.579\ -0.789], \tag{9.7}$$

whereas the coordinates $\mathbf{Q}^i$ (in general basis) of query vector $\mathbf{Q}$ are

$$\mathbf{Q}^i = \mathbf{g}_i^{-1} \times \mathbf{Q} = [\mathbf{b}_1\ \mathbf{b}_2]^{-1} \times \mathbf{Q} = \begin{bmatrix} 2 & 0.2 \\ 0.5 & 1 \end{bmatrix}^{-1} \times [0\ 2]^{\mathrm{T}} = \tag{9.8}$$

$$= [-0.211\ 2.105].$$

Now, if dot product is interpreted—as is usual in VSR—as being the expression of similarity between document and query, then the dot product <**D**, **Q**> of document vector **D** and query vector **Q** is to be computed relative to the new, general basis $\mathbf{g}_i$, i.e.,

$$<\mathbf{D}, \mathbf{Q}> = (\mathbf{D}^i)^{\mathrm{T}} \times \mathbf{g}_{ij} \times \mathbf{Q}^j =$$

$$[1.579\ -0.789] \times \begin{bmatrix} 4.25 & 0.9 \\ 0.9 & 1.04 \end{bmatrix} \times [-0.211\ 2.105]^{\mathrm{T}} = 0. \tag{9.9}$$

It can be seen that the dot product of document vector **D** and query vector **Q** is also equal to zero in the new basis (i.e., the document is not retrieved in the general basis either). This should not be a surprise because, as is well known, the scalar product is invariant with respect to the change of basis. Thus, under the inner product interpretation of similarity (i.e., if the similarity measure is interpreted as being the dot product between two vectors), the no-hit case remains valid when using the general basis as well!

The change of basis represents a point of view from which the properties of documents and queries are judged. If the document is conceived as being a vector, i.e., it is the same in any basis (equivalently, its meaning, information content, or properties remain the same in any basis), then the inner product is also invariant, and hence so is the similarity measure.

But then, what is the point of taking a general basis? The orthonormal basis is as good any other basis.

Let us now assume or accept that the meaning or information content of a document and query do depend on the point of view, i.e., on the basis of the space. Then, the properties of documents and queries may be found to be different in different bases. This is equivalent to not interpreting the similarity measure as expressing an inner product, but rather considering it a numerical measure of how much the document and query share. Thus, the similarity measure, which formally looked like the algebraic expression of an inner product, is literally interpreted as a mere algebraic expression (or computational construct) for a measure of how much the document and query share and not as the expression of an inner product.

In this new interpretation, in our example in **Fig. 9.2**, we obtain the following value for the similarity between document and query: $1.579 \times (-0.211) + (-0.789) \times (2.105) = -1.994$, which is different from zero. (Subjectively, a numerical measure of similarity should be a positive number, although this is irrelevant from a formal mathematical, e.g., ranking, point of view). Thus, document D is being returned in response to Q, as intended by:

- Using a general basis to express term dependence.
- Not interpreting similarity as being an inner product.

The Euclidean space as a mathematical/formal framework for VSR is very illustrative and intuitive. But as we have seen, there is no actual and necessary connection between the mathematical concepts used (vector, vector space, scalar product) and the concepts of IR (document, query, similarity). In other words, there is a discrepancy (or inconsistency) between the theoretical (mathematical) model and the effective retrieval algorithm applied in practice. They are not consistent with on another: the algorithm does not follow from the model, and, conversely, the model is not a formal framework for the algorithm.

Sections 9.3 and 9.4 present and discuss the latent semantic indexing (LSI) and general vector space retrieval (GVSR) methods, which exhibit the same inconsistency described above.

9.3 Latent Semantic Indexing Retrieval

9.3.1 Eigenvalue, Eigenvector

Let $A_{n,n}$ be a regular matrix (i.e., $\det(A) \neq 0$). The solutions (roots) of the following n-degree polynomial equation (called a *characteristic equation*; Kurtz 1991),

$$|A - \lambda I| = 0 \tag{9.10}$$

are called *eigenvalues* (*characteristic* or *latent roots*) of A (I denotes the unity matrix).

Example 9.2

Let

$$A = \begin{bmatrix} 1 & 5 \\ 2 & 4 \end{bmatrix}$$

be a regular matrix, det(A) = $|A|$ = –6 ≠ 0. The characteristic equation is

$$\begin{vmatrix} 1-\lambda & 5 \\ 2 & 4-\lambda \end{vmatrix} = 0,$$

i.e., $(1 - \lambda)(4 - \lambda) - 5 \cdot 2 = 0$, which becomes $\lambda^2 - 5\lambda - 6 = 0$. The eigenvalues of A are $\lambda_1 = 6$ and $\lambda_2 = -1$. □

Let λ_i, $i = 1,\ldots,n$, be the eigenvalues of matrix $A_{n,n}$. The vectors (column matrices) $\mathbf{X}_i$ satisfying the simultaneous system of linear equations

$$(A - \lambda_i I)\mathbf{X}_i = 0 \tag{9.11}$$

are called *eigenvectors* (*characteristic* or *latent vectors*) of matrix A. The eigenvectors $\mathbf{X}_1,\ldots,\mathbf{X}_n$ corresponding to distinct eigenvalues $\lambda_1,\ldots,\lambda_n$ are linearly independent of each other, and matrix $S = [\mathbf{X}_1\ldots\mathbf{X}_n]$ has the property $S^{-1}AS = D = \mathrm{diag}(\lambda_1,\ldots,\lambda_n)$, where $\mathrm{diag}(\lambda_1,\ldots,\lambda_n)$ denotes a diagonal matrix (called the *canonical form*) of eigenvalues.

Note: Eigenvalues are useful in many computations, e.g., in computing the powers of matrix A. From the relation $S^{-1}AS = D$ we obtain that $A = SDS^{-1}$ (after multiplying on the left by S and on the right by S^{-1}):

$$SS^{-1}ASS^{-1} = SDS^{-1}, \text{ and } SS^{-1} = SS^{-1} = I.$$

The square of A, i.e., A^2, can now be written as

$$A^2 = AA = (SDS^{-1})(SDS^{-1}) = SD(S^{-1}S)DS^{-1} = SDDS^{-1} = SD^2S^{-1}.$$

In general,

$$A^n = SD^nS^{-1}.$$

Thus, we get a "cost effective" way to compute A^n: S and S^{-1} have to be computed once, and D^n can be calculated in just a few steps using recursion.

9.3.2 Singular Value Decomposition

Given a matrix $A_{m,n}$, $m \geq n$ (albeit that this condition is not necessary; see the second paragraph of Section, 9.3.3, for a justification in IR), and let rank(A) = r. The *singular value decomposition* (SVD) of $A_{m,n}$ is

$$A = USV^{\mathrm{T}}, \tag{9.12}$$

where $U^{\mathrm{T}}U = V^{\mathrm{T}}V = I_{n,n}$ (i.e., matrices U and V are orthogonal), and D is the diagonal matrix $S = \mathrm{diag}(s_1,...,s_n)$, such that $s_i > 0$, $i = 1,...,r$, and $s_j = 0$, $j > r$. The columns of U are called the *left singular vectors*, and those of V the *right singular vectors* of A. The diagonal elements of S are the non-negative square roots of the n eigenvalues of AA^{T}, and are referred to as the *singular values* of A. (In order to obtain the SVD of a matrix, mathematical software or numerical algorithms can be used.)

9.3.3 Latent Semantic Indexing

In principle, LSI derives "artificial concepts" (Deerwester et al. 1990, Berry and Browne 1999) to represent common-meaning components of documents; these are represented by weight vectors indicating a level of association between the documents and these concepts. It is claimed that this representation is computationally economical because the dimension of document vectors can be reduced to a number that is less than the number of terms (the number of terms being equal to the dimension of the term space in which documents are originally represented), and further that LSI better captures common meaning in documents.

Let $D = \{D_1,...,D_j,...,D_m\}$ be a set of elements called *documents* and $T = \{t_1,...,t_i,...,t_n\}$ a set of elements called *terms*. In general, in practical applications there are more documents than terms, i.e., $m \geq n$ (if $n \geq m$, matrices U and V, see below, will be interchanged). Let $W = (w_{ji})_{m\times n}$ be a *weights matrix*, where w_{ji} denotes the weight of term t_i in document D_j. (See Chapter 4 for details on technological aspects on obtaining W.) Let the rank of W be r, i.e., rank(W) = r, and the SVD of W be

$$W = USV^{\mathrm{T}}. \tag{9.13}$$

The SVD (9.13) of W may be viewed as a breakdown of the original relationships, represented by W, between documents and terms. In other words, a set of artificial concepts is obtained that corresponds to a factor value $k = 2, 3,...,r$ (k is the number of selected columns, from left to right, from U, and of selected rows, from top to bottom, from S). Thus,

$$W_k = U_kS_kV^{\mathrm{T}} \tag{9.14}$$

is an approximation of the original matrix W with the weights of artificial concepts (which form a term space of lower dimension). Of course, if $k = r$, then $W_k = W$. Matrix W_k is used for retrieval purposes in that a query q is matched against W_k. Retrieval is performed by computing the value of similarity (e.g., cosine, dot product) between vectors q_k and W_k, e.g., $W_k\, q$.

Example 9.3

Consider the following documents:

D_1 = Bayes's principle: The principle that, in estimating a parameter, one should initially assume that each possible value has equal probability (a uniform prior distribution).

D_2 = Bayesian conditionalization: This is a mathematical procedure with which we revise a probability function after receiving new evidence. Let us say that we have probability function $P(.)$ and that through observation I come to learn that E. If we obey this rule, our new probability function, Q(.) should be such that for all X, $Q(X) = P(X|E)$ we are then said to have "conditionalized on E."

D_3 = Bayesian decision theory: A mathematical theory of decision-making that presumes utility and probability functions, and according to which the act to be chosen is the Bayes act, i.e. the one with highest subjective expected utility. If one had unlimited time and calculating power with which to make every decision, this procedure would be the best way to make any decision.

D_4 = Bayesian epistemology: A philosophical theory that holds that the epistemic status of a proposition (i.e., how well proven or well established it is) is best measured by a probability and that the proper way to revise this probability is given by Bayesian conditionalization or similar procedures. A Bayesian epistemologist would use probability to define concepts such as epistemic status, support, or explanatory power and explore the relationships among them.

Let the terms be

$$t_1 = \textit{Bayes's principle},\ t_2 = \textit{probability},$$
$$t_3 = \textit{Bayesian conditionalization},\ t_4 = \textit{decision-making}.$$

Let the frequency term-document matrix W be

$$W = \begin{pmatrix} 1 & 1 & 0 & 0 \\ 0 & 3 & 1 & 0 \\ 0 & 1 & 0 & 1 \\ 0 & 3 & 1 & 0 \end{pmatrix}.$$

The rank of W is equal to 3, and the singular value decomposition of W is $W = USV^{\mathrm{T}}$, where

$$U = \begin{bmatrix} -0.214 & -0.674 & 0.707 & 0 \\ -0.674 & 0.214 & 0 & -0.707 \\ -0.214 & -0.674 & -0.707 & 0 \\ -0.674 & 0.214 & 0 & 0.707 \end{bmatrix},$$

$$S = \begin{bmatrix} 4.68 & 0 & 0 & 0 \\ 0 & 1.047 & 0 & 0 \\ 0 & 0 & 1 & 0 \\ 0 & 0 & 0 & 0 \end{bmatrix},$$

$$V = \begin{bmatrix} -0.046 & -0.644 & 0.707 & 0.289 \\ -0.955 & -0.062 & 0 & -0.289 \\ -0.288 & 0.409 & 0 & 0.866 \\ -0.046 & -0.644 & -0.707 & 0.289 \end{bmatrix}.$$

Taking the factor value k = 2, we find the approximation of W_k [Eq. (9.14)]) to be

$$W_k = \begin{bmatrix} 0.5 & 1 & 0 & 0.5 \\ 0 & 3 & 1 & 0 \\ 0.5 & 1 & 0 & 0.5 \\ 0 & 3 & 1 & 0.5 \end{bmatrix}.$$

Now consider query q consisting of two terms: q = *probability, decision-making*. The corresponding query matrix is

$$[0\ 1\ 1\ 0],$$

which is to be compared—in terms of a similarity measure—with W_k. For example, $W_k \cdot \mathbf{q} = [1\ 4\ 1\ 4]^{\mathrm{T}}$, where $\mathbf{q} = [0\ 1\ 1\ 0]^{\mathrm{T}}$. □

Retrieval in LSI uses a vector-space-based approach (or framework), just like the traditional VSR method. Originally, documents are represented, using matrix W, as vectors in the term space whose dimension is n. Using the SVD of W, we say that documents are represented in a vector space of artificial concepts, whose dimension is k ($k \leq r \leq n$). The expression $W\mathbf{q}$ for similarity is interpreted as having the meaning of a scalar product. If this is the case, $W\mathbf{q} = W_k\mathbf{q}$ (because documents and queries remain the same vectors even if they are viewed as vectors of a subspace of the original space). Equality occurs when $W = W_k$, i.e., when $k = r = \text{rank}(W)$. Otherwise, the expression $W_k\mathbf{q}$ may not be viewed as having the meaning of scalar product.

9.4 Generalized Vector Space Retrieval

Wong and Raghavan (1984) showed why the vector space model of IR conflicts with the mathematical notion of a vector space. Further on, they rightly observed that the usual similarity functions (dot product, Dice coefficient, and Jaccard coefficient) can also be written in a general basis (not just in the orthonormal basis). They interpret the metric tensor G, which they refer to as the *correlation matrix*, of the space as expressing correlations between terms t_i, $i = 1,\ldots,n$, viewed as basis vectors. G can be used as a model of term dependences: $G = (<\mathbf{t}_i, \mathbf{t}_j>)_{n \times n}$, where $\mathbf{t}_i$ denotes the basis vector corresponding to term t_i.

Subsequently, Wong et al. (1985) proposed an automatic (and very computationally demanding) method to build correlation matrix G. The value of similarity S between a document and a query was computed as the matrix product between:

- Query vector $\mathbf{q}$ expressed in general basis.
- Metric tensor, i.e., G.
- Document vector $\mathbf{d}$ in orthonormal basis.

Thus, $S = \mathbf{q}^{\mathrm{T}} \cdot G \cdot \mathbf{d}$. The method was referred to as the generalized vector space model (GVSM). If $\mathbf{d}$ had been expressed in a general basis, then S would have been the scalar product of $\mathbf{q}$ and $\mathbf{d}$ in that basis (and would have been the same as that in an orthonormal basis). Thus, the expression for S seems to be a mere computational construct rather than the expression of a scalar product in a general basis.

9.5 Principle of Invariance

In this section, a principle is proposed that is designed to deal with the inconsistencies and anomalies discussed in Sections 9.2–9.4, which stem from taking linear space as a formal framework for retrieval and conceiving similarity as a scalar product.

The concepts of position, translation, rotation, velocity, acceleration, force, etc. are primarily physical concepts—not just abstract or mathematical notions (Feynman et al. 1964). They reflect certain aspects of reality and thus possess underlying properties such that the physical laws are the same in any coordinate system regardless of the basis of the space. For example, the position of a physical object in space does not depend on the angle from which we look at it or on the choice of the coordinate axes (i.e., on the choice of the basis of the space). The position of a physical object is invariant with respect to the basis of the space; the same holds true for velocity, force, etc. Such entities are referred to as vectors, for short. In other words, vectors are entities that have an "identity" (namely magnitude and direction), and this identity is preserved in any system or basis; i.e., it is invariant with respect to the change of the basis of the space. An immediate—but very important!—consequence of this is that the quantity called the scalar product of two vectors is also preserved; i.e., it is invariant with respect to the choice of the basis of the space. In other words, apart from vectors, the scalar product is another quantity that is basis-invariant. The mathematical apparatus developed to deal correctly with the physical operations involved (e.g., the addition of velocities) is referred to as vector algebra or tensor calculus; see, e.g., Lánczos (1970) and Simmonds (1982).

As it is well known, one of the basic concepts of IR is that of a document, i.e., that of objects or entities to be searched. (Their physical appearance, such as the language in which they are written or strings of bits on a computer disk, etc., is now irrelevant.) The notion of document is not merely a mathematical or abstract concept. Just as in physics, it is used to reflect certain aspects of reality. But unlike in physics, a document need not have an identity (meaning, content, property) that is basis-invariant, and a user basically operates with the identity. This may depend on the point of view or on the judgment of the user (mathematically, on the basis of the space). As a consequence, and as we have already seen, even if the space is assumed to be, or is related to, a linear space, the similarity measure need not necessarily be viewed as being the expression of an inner product. This is rather an option or hypothesis that we may or may not accept, or accept to a certain extent. Thus, it is reasonable to introduce the following principle:

Principle of Invariance (PI). *In information retrieval, the identities of entities are preserved with probability π.*

The case when $\pi = 1$ means that the identity of entities (documents, queries) remains the same, regardless of any point of view or interpretation. If, however, $\pi < 1$, the identity of entities does depend on a point of view or interpretation.

Based on PI, we may state the following:

- In the classical VSR method, $\pi = 1$, the notion of linear space is used as framework, documents and queries are vectors, and the similarity is the scalar product of the vector space.
- In the LSI IR method, $\pi < 1$, the notion of linear space is used as framework, documents and queries are vectors belonging to different spaces having different dimensions (depending on the *k* factor), and the similarity is the scalar product of the vector space.
- In the GVSR method, $\pi = 1$, the notion of linear space is used as a framework and documents and queries are vectors, but the similarity is not the scalar product of the vector space.

In what follows, new or novel retrieval methods are proposed for the case when $\pi < 1$ that do not use linear spaces as frameworks.

9.6 Elements of Fuzzy Set Theory

9.6.1 Fuzzy Set

Let X be a finite set. A *fuzzy set* $\tilde{A}$ in X is a set of ordered pairs (Zimmerman 1996),

$$\tilde{A} = \{(x, \mu_{\tilde{A}}(x)) \mid x \in X\}, \tag{9.15}$$

where $\mu: X \to [0; a] \subset \mathbb{R}$, $a > 0$, is called a *membership function* (or degree of compatibility or truth function), meaning the degree to which x belongs to $\tilde{A}$. Elements with a zero degree membership are normally not listed.

The fuzzy set $\tilde{A}$ in X for which $\mu_{\tilde{A}}(x) = 0$, $\forall x \in X$ is denoted by o. The fuzzy set $\tilde{A}$ in X for which $\mu_{\tilde{A}}(x) = a$, $\forall x \in X$ is denoted by l.

Example 9.4

Let $X = \{1, 10, 15, 18, 24, 40, 66, 80, 100\}$ be a set denoting possible ages for humans. Then, the fuzzy set $\tilde{A}$ = "ages considered as young" could be the set $\{(10, 1), (15, 1), (18, 1), (24, 1), (40, 0.7)\}$.

The fuzzy set Ã = "real numbers much larger than 10" could be the fuzzy set Ã = {(x, $\mu_{Ã}(x)$) | $x \in X$}, where

$$\mu_{Ã}(x) = \begin{cases} 0, & x \le 10 \\ \dfrac{1}{1+\dfrac{1}{(x-10)^2}} & otherwise. \end{cases}$$

A graphical representation of the fuzzy set Ã = "real numbers much larger than 10" is illustrated in **Fig. 9.3**.

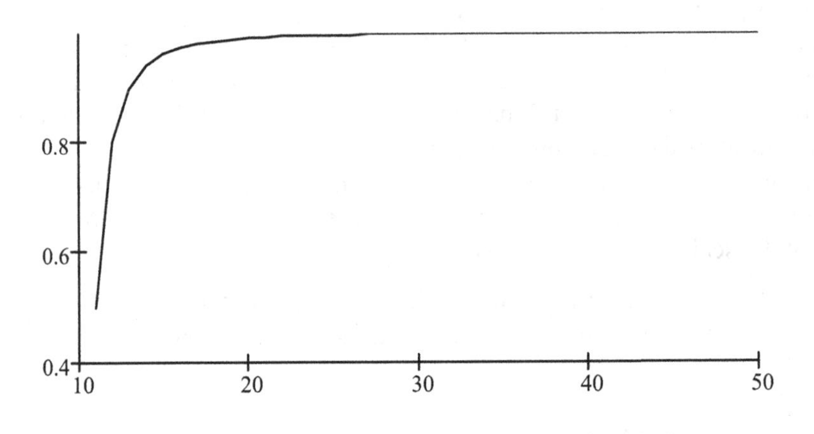

Fig. 9.3. Graphical representation of the fuzzy set Ã = "real numbers much larger than 10."

If the membership function can only take on two values, 0 and a, the fuzzy set becomes a (*classical* or *crisp*) set: an element either belongs to the set (the membership function is equal to a) or not (the membership function is equal to 0). If the membership function takes on values in the interval [0; 1], i.e., $a = 1$, the fuzzy set is called a *normalized* fuzzy set. In the rest of this chapter, we consider term weights as values of some membership function. Any membership function can be normalized (e.g., by division by a). While it is true that not all weighting schemes result in weights between 0 and 1 (e.g., the inverse document frequency weighting scheme may yield weights greater than 1), within document weights can always be normalized (e.g., by division by the largest weight) so as to be between 0 and 1 (while keeping the relative importance of terms within the document). Thus, in what follows we are considering normalized fuzzy sets.

9.6.2 Fuzzy Intersection

Given two fuzzy sets $\tilde{A}_1$ and $\tilde{A}_2$ in X with membership functions μ_1 and μ_2, respectively, the membership function μ of *fuzzy intersection* $\tilde{A} = \tilde{A}_1 \sqcap \tilde{A}_2$ can be defined in several ways. The usual definitions are

$$\begin{aligned} \text{Standard: } \mu(x) &= \min(\mu_1(x), \mu_2(x)), \forall x \in X, \\ \text{Algebraic product: } \mu(x) &= \mu_1(x)\mu_2(x), \forall x \in X. \end{aligned} \tag{9.16}$$

9.6.3 Fuzzy Union

Given two fuzzy sets $\tilde{A}_1$ and $\tilde{A}_2$ in X with membership functions μ_1 and μ_2, respectively, the membership function μ of the *fuzzy union* $\tilde{A} = \tilde{A}_1 \sqcup \tilde{A}_2$ can be defined in several ways. The usual definitions are

$$\begin{aligned} \text{Standard: } \mu(x) &= \max(\mu_1(x), \mu_2(x)), \forall x \in X, \\ \text{Algebraic product: } \mu(x) &= \mu_1(x) + \mu_2(x) - \mu_1(x)\mu_2(x), \forall x \in X. \end{aligned} \tag{9.17}$$

Example 9.5

Let $\tilde{A}_1 = \{(10, 0.5), (15, 1), (18, 1), (24, 1), (40, 0.4)\}$ and $\tilde{A}_2 = \{(24, 0.1), (40, 0.3), (70, 0.9)\}$ be two fuzzy sets. Their standard fuzzy union is $\tilde{A}_1 \sqcup \tilde{A}_2 = \{(10, 0.5), (15, 1), (18, 1), (24, 1), (40, 0.4), (70, 0.9)\}$, and their standard fuzzy intersection is $\tilde{A}_1 \sqcap \tilde{A}_2 = \{(24, 0.1), (40, 0.3)\}$. □

9.6.4 Fuzzy Complement

The membership function $\mu_{¢\tilde{A}}(x)$ of *fuzzy complement* $¢\tilde{A}$ of fuzzy set $\tilde{A}$ in X is defined as

$$\mu_{¢\tilde{A}}(x) = 1 - \mu_{\tilde{A}}(x), \forall x \in X. \tag{9.18}$$

9.6.5 Fuzzy Subset

Given two fuzzy sets $\tilde{A}_1$ and $\tilde{A}_2$ in X with membership functions μ_1 and μ_2, respectively, fuzzy sets $\tilde{A}_1$ and $\tilde{A}_2$ are equal to each other, i.e., $\tilde{A}_1 = \tilde{A}_2$, if $\mu_1(x) = \mu_2(x)$, $\forall x \in$ X. If $\mu_1(x) \le \mu_2(x)$ $\forall x \in X$, we say that fuzzy set $\tilde{A}_1$ is a fuzzy subset of fuzzy set $\tilde{A}_2$, i.e., $\tilde{A}_1 \sqsubseteq \tilde{A}_2$.

9.7 Retrieval Using Linear Space

Consider a set

$$T = \{t_1, t_2, \ldots, t_i, \ldots, t_n\} \quad (9.19)$$

of *terms*, and let

$$D = \{D_1, D_2, \ldots, D_j, \ldots, D_m\} \quad (9.20)$$

denote a set of *documents* indexed by T. As is usual in IR, let w_{ij} denote the weight of term t_i in document D_j. Let Q denote a query and q_i the weight of term t_i in Q, $i = 1,\ldots,n$. We do not make any assumption as to whether documents and queries are elements of a linear space or not. They simply form a collection. Each document is represented by a sequence of numeric weights: D_j is represented by the sequence of weights $w_{1j},\ldots,w_{ij},\ldots,w_{nj}$. Likewise Q is represented by the sequence $q_1,\ldots,q_i,\ldots,q_n$. In the traditional VSR model, the expression $\sum_{i=1..n} q_i w_{ij}$ is conceived as the scalar product of the space of terms whose vectors represent the documents. As we have already seen, this view leads to inconsistencies, so let us drop it and accept that $\pi < 1$. Thus, documents and queries may have multiple identities (they are not vectors), and the expression $\sum_{i=1..n} q_i w_{ij}$ is interpreted simply as a numerical measure of similarity between document D_j and query Q. We may assume, without restricting generality, that $0 \leq w_{ij} \leq 1$. Under these conditions, any document D_j (and any query Q) may be identified with (described by) a fuzzy set $\tilde{A}_j$:

$$\tilde{A}_j = \{(t_i, \mu_j(t_i)) \mid t_i \in T, i \in \{1,\ldots,n\}, \mu_j(t_i) = w_{ij}\}. \quad (9.21)$$

Let $\mathcal{T} = [0; 1]^T$ denote the set of all possible fuzzy sets in T. Then, in general, any conceivable document or query is an element of $\mathcal{T}$. Similarity is defined in the following way:

Definition 9.1. Function $\sigma: \mathcal{T} \times \mathcal{T} \rightarrow \mathbb{R}$ defined as

$$\sigma(\tilde{A}_j, \tilde{A}_k) = \sum_{i=1}^{n} \mu_j(t_i)\mu_k(t_i)$$

is called a *similarity measure.* □

We now use a linear space, but not as a framework, but rather as a tool or operator to design and propose a retrieval method that is based on $\mathcal{T}$ as a formal framework.

It is possible to relate any document $\tilde{A}_j = \{(t_i, \mu_j(t_i))\}$ to an n-dimensional real linear space L (having basis $\mathbf{b}_1,\ldots,\mathbf{b}_n$):

- The values $\mu_j(t_1),\ldots,\mu_j(t_n)$ of the membership function can be used to form a vector $\mathbf{v}_j$ of space L as the linear combination $\mathbf{v}_j = \mu_j(t_1)\cdot\mathbf{b}_1 +\ldots+ \mu_j(t_n)\cdot\mathbf{b}_n$.
- Thus, every document $\tilde{A}_j$ may be viewed as corresponding (not being identical) to vector $\mathbf{v}_j \in$ L with coordinates $\mu_j(t_i)$, i.e., $\mathbf{v}_j = (\mu_j(t_1), \ldots,\mu_j(t_n))$.

(Documents are related to vectors, but this does not mean that they become vectors.) The following retrieval method can be formulated in a natural way:

1. Let $T = \{t_1,\ldots,t_i,\ldots,t_n\}$ denote a set of terms.
2. Let $\tilde{A}_j = \{(t_i, \mu_j(t_i) \mid t_i \in T, i = 1,\ldots,n\}$ denote documents as elements of $\mathcal{T} = [0; 1]^T, j = 1,\ldots,m$.
3. Let L be a real linear space.
4. Let any document $\tilde{A}_j$ correspond to vector $\mathbf{v}_j$ of space L such that $\mu_j(t_i)$, $i = 1,\ldots,n$, corresponds to the ith coordinate of vector $\mathbf{v}_j$, i.e., $\mathbf{v}_j = [\mu_j(t_1), \ldots,\mu_j(t_n)]$.
5. Let $Q = \{(t_i, \mu_Q(t_i) \mid t_i \in T, i = 1,\ldots,n\}$ denote a query, $Q \in \mathcal{T}$, and let $\mathbf{q} = (\mu_Q(t_1), \ldots,\mu_Q(t_n)) \in$ L denote the corresponding query vector.
6. The documents retrieved in response to query Q are obtained using Definition 9.1.

Steps 2–5 can be expanded in the following way.

(a) Let $TD_{n\times m} = (f_{ij})_{n\times m}$ denote the frequency term-document matrix, i.e., f_{ij} denotes the number of times term t_i occurs in document D_j.

(b) Let $Q = (f_1,\ldots, f_i,\ldots, f_n)$ denote a query, where f_i is the number of times term t_i occurs in query Q.

(c) Compute, using some weighting scheme, a term-document weight matrix $W_{n\times m} = (\mathbf{d}_j)_m = (w_{ij})_{n\times m}$ for documents $w_{ij} = \mu_j(t_i)$, and one for query $\mathbf{q} = (q_1, \ldots,q_i, \ldots,q_n)$.

(d) Let us consider a general basis $\mathbf{g}_i$ of Euclidean space E_n:

$$\mathbf{g}_i = (\mathbf{b}_1 \ldots \mathbf{b}_n) = \begin{pmatrix} b_{11} & \ldots & b_{n1} \\ . & \ldots & . \\ b_{1n} & \ldots & b_{nn} \end{pmatrix}.$$

A general basis $\mathbf{g}_i$ can be obtained as follows (Silva et al. 2004): Correlations are determined among terms, and the corresponding axes are rotated in space. Thus, "proximity" between basis vectors is related to the degree of correlation (dependence) between the respective terms. The closer the vectors are, the greater the dependence. A *confidence index* c_{ij} is computed between any two terms t_i and t_j: :

$$c_{ij} = \frac{|\{D \mid t_i, t_j \in D\}|}{m}.$$

(e) The coordinates b_{ik}, $k = 1,\ldots,n$, of basis vector $\mathbf{b}_i$ are given by

$$b_{ik} = \begin{cases} \sin\theta_{ij} & if \quad k = i \\ \cos\theta_{ij} & if \quad k = j \\ 0 & otherwise \end{cases}$$

where $\theta_{ij} = 90 \cdot (1 - c_{ij})$.

f) Compute the coordinates $\mathbf{D}'^j = (w'_{1j},\ldots,w'_{ij},\ldots, w'_{nj})$ of every document D_j in a general basis as follows:

$$\mathbf{D}'^j = \mathbf{g}_i^{-1} \cdot \mathbf{d}_j.$$

g) Similarly, the coordinates $\mathbf{q}' = (q'_1,\ldots,q'_i,\ldots,q'_n)$ of the query in a general basis are

$$\mathbf{q}' = \mathbf{g}_i^{-1} \cdot \mathbf{q}.$$

h) The similarity σ_j between document D_j and query Q is computed using Definition 9.1:

$$\sigma_j = \sum_{i=1}^{n} q'_i \cdot w'_{ij}.$$

The retrieval method was tested by Silva et al. (2004) on the test collections CACM, CISI, and TREC-3. Relevance effectiveness was higher than that of the traditional VSR method by 7, 14, and 16%, respectively.

We note that the correlation c_{ij} between terms may be computed in other ways as well, e.g., with the EMIM (expected mutual information measure) method (van Rijsbergen 1979, Savoy and Desbois 1991):

Term Correlation Using the EMIM Method

1. Let $t_1,\ldots,t_i,\ldots,t_j,\ldots,t_n$ be terms. For every pair of terms t_i and t_j perform the following steps:
2. Determine S_{ij}: the number of documents indexed by both t_i and t_j.
3. Determine S_j: the number of documents indexed by t_j but not by t_i.
4. Determine S_i: the number of documents indexed by t_i but not by t_j.
5. Determine S: the number of documents not indexed by either t_i or t_j.
6. Compute the EMIM I_{ij} (= c_{ij}) for t_i and t_j as follows:

$$I_{ij} = S_{ij} \times \ln \frac{S_{ij}}{(S_{ij}+S_i)\times(S_{ij}+S_j)} + S_j \times \ln \frac{S_j}{(S_j+S)\times(S_{ij}+S_j)} +$$

$$S_i \times \ln \frac{S_i}{(S_{ij}+S_i)\times(S_i+S)} + S \times \ln \frac{S}{(S_j+S)\times(S_i+S)}$$

The EMIM values can be used to construct a tree of term dependencies known as the maximum spanning tree of the graph whose nodes are the terms and weighted edges that correspond to the respective EMIM values (i.e., I_{ij} is the weight of the edge between the nodes corresponding to terms t_i and t_j). The tree can be constructed using the following method:

1. Sort descendingly the EMIM values to obtain a sequence $I_1,\ldots,I_N$.
2. Initialize the tree with the nodes as well as the edge connecting them corresponding to I_1.
3. Take the next value I_j from the sequence, and 'grow' the tree with the nodes and edge corresponding to I_j if this does not yield a cycle. If this does yield a cycle, repeat step 3.

9.8 Fuzzy Algebra-Based Retrieval Methods

In this section, we propose new retrieval methods based on the notions of algebra and measure, which stem from the following observations. The similarity measure $\sigma(\tilde{A}_i, \tilde{A}_j) = \sum_{t\in T} \mu_i(t)\mu_j(t)$ of Definition 9.1 may be viewed as the fuzzy cardinality of the fuzzy intersection of $\tilde{A}_i$ and $\tilde{A}_j$ based on an algebraic product, i.e., $\tilde{A}_i \sqcap \tilde{A}_j = \{(t, \mu(t)) \mid \mu(t) = \mu_i(t)\mu_j(t)\}$. (The cardinality of a fuzzy set is equal to the sum of the values of its membership function.) Further, if the fuzzy sets $\tilde{A}_i$ and $\tilde{A}_j$ are disjoint, the cardinality of their fuzzy union (based on the algebraic product) is equal to the sum of their cardinalities. This property—called additivity—is characteristic of the mathematical notion of measure. These observations suggest looking for or analyzing other mathematical measures (other than fuzzy cardinality) that may then be used as similarity functions.

In the rest of this chapter, the above ideas are presented and discussed in detail. Two retrieval methods are proposed using new measures as similarities. Experiments on the relevance effectiveness offered by these measures are also reported.

9.8.1 Fuzzy Jordan Measure

The notion of mathematical measure is an abstraction of length, area, volume, etc. The mathematical theory of measure, known as measure theory, offers a theoretical and formal basis for integration theory and probability theory. In words, a measure is a nonnegative function of subsets of a set such that the measure of the union of a sequence of mutually disjoint sets is equal to the sum of the measures of the sets. Formally, a concept of measure can be defined in the following way.

First, we introduce the notion of algebra as an abstract space (Kiyosi 2000):

Definition 9.2. Let X denote a set. A collection C of sets from $\wp(X)$, $C \subseteq \wp(X)$, is called an *algebra* (equivalently *clan*, *field*, or *finitely additive class*) if the following conditions hold:

a) $X \in C$.

b) $\forall A, B \in C \Rightarrow A \cup B \in C$.

c) $\forall A \in C \Rightarrow \complement_X A \in C$.

From Definition 9.2 it follows that:

- $\emptyset \in C$ because $\complement_X X = \emptyset$.
- The union of a finite number of sets A_i (i=1,...,m) from C belongs to C, i.e., $A_1 \cup ... \cup A_m \in C$.

The notion of measure is defined over an algebra as a set function that 'measures' a set (Kiyosi 2000):

Definition 9.3. A function $m : C \rightarrow \mathbb{R}$ is called a *Jordan measure* over an algebra C if the following conditions hold:

1. $\mathcal{m}(A) \geq 0, \forall A \in \mathcal{C}$ (nonnegativity).
2. $\mathcal{m}(\varnothing) = 0$.
3. $\forall A, B \in \mathcal{C}, A \cap B \in \varnothing \Rightarrow \mathcal{m}(A \cup B) = \mathcal{m}(A) + \mathcal{m}(B)$ (additivity).

As an analogue of the notion of algebra (Definition 9.2), we introduce its fuzzy counterpart, i.e., a generalization to arbitrary values of the interval [0; 1]:

Definition 9.4. Let X denote a set. A collection $\mathcal{C}$ of fuzzy sets in $[0; 1]^X$ is called a *fuzzy algebra* (equivalently *fuzzy clan*, *fuzzy field*, or *fuzzy finitely additive class*) if the following conditions hold:

- $\mathbf{I} \in \mathcal{C}$.
- $\forall\, \tilde{A}_1, \tilde{A}_2 \in \mathcal{C} \Rightarrow \tilde{A}_1 \sqcup \tilde{A}_2 \in \mathcal{C}$.
- $\forall\, \tilde{A} \in \mathcal{C} \Rightarrow ¢\tilde{A} \in \mathcal{C}$. □

From Definition 9.4 it follows that $\mathbf{o} \in \mathcal{C}$ because $\mathbf{o} = ¢\mathbf{I}$. We now define our fuzzy algebra as a formal theoretical framework for retrieval: Let T denote a set of terms: $T = \{t_1,\ldots,t_i,\ldots,t_n\}$ and $\mathcal{T} = [0; 1]^T$ denote the set of all fuzzy sets in T. Then, $\mathcal{T}$ may be taken as a general framework:

Theorem 9.1. *$\mathcal{T}$ is a fuzzy algebra with respect to an algebraic product.*

Proof. We have to show that the conditions of Definition 9.4 hold. Obviously, we have $\mathbf{I} \in \mathcal{T}$. Let us now consider two arbitrary fuzzy sets $\tilde{A}_i, \tilde{A}_j \in \mathcal{T}$. Their algebraic product union is $\tilde{A}_i \sqcup \tilde{A}_j = \{(t, \mu(t)) \mid \mu(t) = \mu_i(t) + \mu_j(t) - \mu_i(t)\mu_j(t)\}$. Because $\mu_i(t) + \mu_j(t) - \mu_i(t)\mu_j(t) \in [0; 1]$, the fuzzy union $\tilde{A}_i \sqcup \tilde{A}_j$ belongs to $\mathcal{T}$. Further, for any fuzzy set $\tilde{A} \in \mathcal{T}$ the fuzzy complement $¢\tilde{A} = \{(t, 1 - \mu(t))\}$ belongs to $\mathcal{T}$ because $1 - \mu(x) \in [0; 1]$. □

As an analogue of the Jordan measure (Definition 9.3), we introduce the notion of a fuzzy Jordan measure on a fuzzy algebra:

Definition 9.5. A *fuzzy Jordan measure* on a fuzzy algebra $\mathcal{C}$ is a function $\mathrm{m} : \mathcal{C} \rightarrow \mathbb{R}$ such that:

- $\mathrm{m}(\tilde{A}) \geq 0, \forall \tilde{A} \in \mathcal{C}$ (nonnegativity).
- $\mathrm{m}(\mathbf{o}) = 0$.
- $\tilde{A}_i, \tilde{A}_j \in \mathcal{C}, \tilde{A}_i \sqcap \tilde{A}_j = \mathbf{o} \Rightarrow \mathrm{m}(\tilde{A}_i \sqcup \tilde{A}_j) = \mathrm{m}(\tilde{A}_i) + \mathrm{m}(\tilde{A}_j)$ (additivity).

Let us now define a concept for 'how many elements' there are in a fuzzy set (Zimmerman 1996).

Definition 9.6. The *fuzzy cardinality* κ of a fuzzy set $\tilde{A}_i$ in T is the sum of the values of its membership function:

$$\kappa(\tilde{A}_i) = \sum_{j=1}^{n} \mu_i(t_j) . \square$$

We now prove that fuzzy cardinality is a fuzzy Jordan measure on the fuzzy algebra $\mathcal{T}$.

Theorem 9.2. *Fuzzy cardinality is a fuzzy Jordan measure on the fuzzy algebra $\mathcal{T}$.*

Proof. We obviously have that $\kappa(\tilde{A}_i) \geq 0$ for every $\tilde{A}_i$ from $\mathcal{T}$, and that $\kappa(\text{o}) = 0$. Further, we have to show that the cardinality of two disjoint fuzzy sets is equal to the sum of their cardinalities. Let

$$\tilde{A}_i \sqcap \tilde{A}_j = \{(t, \mu(t)) \mid \mu(t) = \mu_i(t)\mu_j(t)\} = \varnothing \Leftrightarrow$$
$$\Leftrightarrow \mu_i(t)\mu_j(t) = 0.$$

Hence,

$$\kappa(\tilde{A}_i \sqcup \tilde{A}_j) =$$
$$= \kappa[\{(t, \mu(t)) \mid \mu(t) = \mu_i(t) + \mu_j(t) - \mu_i(t)\mu_j(t)\}] =$$
$$= \kappa[\{(t, \mu(t)) \mid \mu(t) = \mu_i(t) + \mu_j(t)\}] =$$
$$= \sum_{t \in T} (\mu_i(t) + \mu_j(t)) = \sum_{t \in T} \mu_i(t) + \sum_{t \in T} \mu_j(t) =$$
$$= \kappa(\tilde{A}_i) + \kappa(\tilde{A}_j). \square$$

Thus, the following method for designing new or novel similarity measures for retrieval may be given:

- Look for/design new fuzzy Jordan measures.
- Perform experiments to measure the relevance effectiveness of a retrieval system whose similarity is based on the thus defined fuzzy Jordan measure.
- Depending on the results obtained, reject, accept, or fine-tune the similarity function.

In the next three sections, the following retrieval methods will be proposed and tested (Dominich and Kiezer 2007):

- Fuzzy entropy retrieval method.
- Fuzzy probability retrieval method.

9.8.2 Fuzzy Entropy Retrieval Method

Let us first introduce the concept of fuzzy entropy (Zimmerman 1996):

Definition 9.7. The *fuzzy entropy* of a fuzzy set $\tilde{A} = \{(x, \mu(x)) \mid \forall x \in X\}$ in a finite set X is given by: $H(\tilde{A}) = -\sum_{x \in X} \mu(x) \cdot \log \mu(x)$. □

We now show that fuzzy entropy is a measure in our framework of documents and queries.

Theorem 9.3. *Fuzzy entropy H is a fuzzy Jordan measure on the fuzzy algebra $\mathcal{T}$.*

Proof. We know from mathematical analysis that $\lim_{y \to 0}(y \log y) = 0$. From this, we have that $H(\text{o}) = 0$. Obviously, $H(\text{o}) \geq 0$. Further, we have to show that the entropy of two disjoint fuzzy sets is equal to the sum of their fuzzy entropies. Let $\tilde{A}_i$ and $\tilde{A}_j$ denote two disjoint fuzzy sets, i.e., $\tilde{A}_i \sqcap \tilde{A}_j = \{(t, \mu(t)) \mid \mu(t) = \mu_i(t)\mu_j(t), \forall t \in T\} = \varnothing \Leftrightarrow \mu_i(t)\mu_j(t) = 0$ (in other words it cannot happen that both $\mu_i(t)$ and $\mu_j(t)$ are different from zero; i.e., either both are zero or one of them is zero and the other is not zero). We can write that

$$H(\tilde{A}_i \sqcup \tilde{A}_j) =$$

$$= H[\{(t, \mu(t)) \mid \mu(t) = \mu_i(t) + \mu_i(t) - \mu_i(t)\mu_j(t)\}] =$$

$$-\sum_{t \in T}(\mu_i(t) + \mu_j(t) - \mu_i(t)\mu_j(t)) \times \log(\mu_i(t) + \mu_j(t) - \mu_i(t)\mu_j(t))$$

$$= -\sum_{t \in T}(\mu_i(t) + \mu_j(t)) \times \log(\mu_i(t) + \mu_j(t)) =$$

$$= -\sum_{t \in T}\mu_i(t)\log(\mu_i(t) + \mu_j(t)) - \sum_{t \in T}\mu_j(t)\log(\mu_i(t) + \mu_j(t)) =$$

$$= H(\tilde{A}_i) + H(\tilde{A}_j). \square$$

The following retrieval method, based on fuzzy entropy as similarity, can now be formulated:

Fuzzy Entropy Retrieval Method

1. Given terms $T = \{t_1,\dots,t_n\}$, and documents D_j, $j = 1,\dots,m$.
2. Let $W_{n\times m} = (w_{ij})_{n\times m}$ denote a term-document matrix, where w_{ij} is the weight of term t_i in document D_j (see Chapter 4 for technological aspects).
3. Given a query Q. The query weights are $q_1,\dots,q_i,\dots,q_n$, where q_i denotes the weight of term t_i in Q.
4. The similarity σ_j between document D_j (conceived as a fuzzy set) and query Q (conceived as a fuzzy set) is computed as the fuzzy entropy of their intersection, i.e., $H(D_j \sqcap Q)$:

$$\sigma_j = -\sum_{i=1}^{n} q_i \cdot w_{ij} \cdot \log(q_i \cdot w_{ij}).$$

9.8.3 Fuzzy Probability Retrieval Method

Let $p(t_i)$ denote a frequency-based probability of term $t_i \in T$, $i = 1, \dots, n$. The fuzzy probability $P(\tilde{A}_j)$ of a fuzzy set $\tilde{A}_j = \{(t_i, \mu_j(t_i)) \mid t_i \in T, i = 1,\dots,n\}$ in T is defined as (Zimmerman 1996)

$$P(\tilde{A}_j) = \sum_{i=1}^{n} \mu_j(t_i) \cdot p(t_i). \tag{9.22}$$

We now show that fuzzy probability is a measure in our framework of documents and queries.

Theorem 9.4. *Fuzzy probability P is a fuzzy Jordan measure on the fuzzy algebra* $\mathcal{T}$.

Proof. Obviously, $P(\tilde{A}_j)$ is nonnegative for any fuzzy set. The fuzzy probability of the empty fuzzy set is equal to zero. This is immediately

$$P(\{(t_i, 0) \mid \forall t_i \in T\}) = \sum_{i=1}^{n} 0 \cdot p(t_i) = 0.$$

Further, the fuzzy probability of two disjoint fuzzy sets is equal to the sum of their fuzzy probabilities. We have

$$\tilde{A}_i \sqcap \tilde{A}_j = \{(t, \mu(t)) \mid t \in T, \mu(t) = \mu_i(t)\mu_j(t)\} = \mathbf{o} \Leftrightarrow \mu_i(t)\mu_j(t) = 0.$$

Hence,

$$P(\tilde{A}_i \sqcup \tilde{A}_j) =$$

$$= P[\{(t, \mu(t)) \mid t \in T, \mu(t) = \mu_i(t) + \mu_j(t) - \mu_i(t)\mu_j(t)\}] =$$

$$= P[\{(t, \mu(t)) \mid t \in T, \mu(t) = \mu_i(t) + \mu_j(t)\}] =$$

$$= \sum_{i=1}^{n} (\mu_i(t_i) + \mu_j(t_i)) p(t_i) = \sum_{i=1}^{n} \mu_i(t_i) p(t_i) + \sum_{i=1}^{n} \mu_j(t_i) p(t_i) =$$

$$P(\tilde{A}_i) + P(\tilde{A}_j).$$

In the language model (Ponte and Croft 1998), the conditional probability $P(Q|D)$ of a document D generating a query Q is considered and used as a similarity measure σ:

$$\sigma = P(Q \mid D) = \frac{P(Q \cap D)}{P(D)}. \quad (9.23)$$

There are several ways to fuzzify (or relax) Eq. (9.23)—as a starting point—depending on how set intersection is defined and on what measures are used in the numerator and denominator. We consider a form in which the numerator is the fuzzy cardinality of the fuzzy intersection (based on an algebraic product) between query Q and document D (both viewed as fuzzy sets), whereas the denominator is the fuzzy probability of a document as a fuzzy set:

$$\sigma = \frac{\sum_{i=1}^{n} q_i \cdot w_{ij}}{\sum_{i=1}^{n} w_{ij} \cdot p(t_i)}. \quad (9.24)$$

The following retrieval method may now be formulated:

Fuzzy-Probability-Based Retrieval Method

1. Given terms $T = \{t_1,\ldots,t_n\}$ and documents D_j, $j = 1,\ldots,m$.
2. Let $TD_{n \times m} = (f_{ij})_{n \times m}$ denote the term-document frequency matrix, where f_{ij} is the number of occurrences of term t_i in document D_j.
3. The frequency-based probability $p(t_i)$ of any term t_i may be calculated as follows:

$$p(t_i) = \frac{\sum_{i=1}^{n} f_{ij}}{\sum_{i=1}^{n} \sum_{j=1}^{m} f_{ij}}.$$

4. Let $W_{n \times m} = (w_{ij})_{n \times m}$ denote a term-document weight matrix, where w_{ij} is the weight of term t_i in document D_j.
5. Given a query Q. The query weights are $q_1, \ldots, q_i, \ldots, q_n$, where q_i denotes the weight of term t_i in Q.
6. The similarity σ_j between a document D_j and query Q is as follows:

$$\sigma = \frac{\sum_{i=1}^{n} q_i \cdot w_{ij}}{\sum_{i=1}^{n} w_{ij} \cdot p(t_i)}.$$

9.8.4 Experimental Results

Experiments were performed to estimate the relevance effectiveness of the following retrieval methods:

- Fuzzy entropy method.
- Fuzzy probability method.

The standard test collections ADI, MED, TIME, and CRAN were used. These collections were subjected to the usual Porter stemming and stoplisting (using computer programs written in the C++ language). **Table 9.1** lists the statistics for these test collections.

Table 9.1. Statistics of the Test Collections Used in Experiments

Test collection.	Number of. documents (d)	Number of. queries (q)	Number of. terms (t)	Avg. number (t/d)	Std. dev (t/d)	Avg. number. (t/q)	Std. dev (t/q)
ADI	82	35	791	21	7	6	2
MED	1033	30	7744	45	20	9	5
TIME	423	83	13479	193	140	8	3
CRAN	1402	225	4009	49	21	8	3

For each test collection, the normalized term frequency weighting scheme was used. The classical VSR method (i.e., in an orthonormal basis) was also implemented and used as a baseline. All three retrieval methods as well as the evaluation of retrieval effectiveness were performed using computer programs written in MathCAD. The standard 11-point precision-recall values were computed for all the test collections and for all documents and queries. **Table 9.2** shows the mean average precision values.

Table 9.2. Mean Average Precision Obtained on Standard Test Collections (E: entropy method; H: probability method; VSM: traditional vector space method, used as baseline)

Test Collection	VSM	E	E over VSM	H	H over VSM
ADI	0.33	0.33	0 %	0.35	+6 %
MED	0.44	0.48	+9 %	0.50	+14 %
TIME	0.52	0.56	+8 %	0.58	+12 %
CRAN	0.18	0.20	+11 %	0.20	+11 %
			Avg. = +7%		Avg. = +11%

Table 9.3 compares the results obtained in experiments with those obtained by Deerwester et al. (1990) using LSI with normalized term frequency.

Table 9.3. Comparison of Retrieval Effectiveness Obtained with the Methods E, H, and LSI (Baseline)

Test collection	LSI	E over LSI	H over LSI
ADI	0.30	+10 %	+17 %
MED	0.48	0 %	+4 %
TIME	0.32	+75 %	+81 %
CRAN	0.25	−25 %	−25 %
		Avg =+15 %	Avg = +19 %

9.9 Discussion

9.9.1 More on Measures

Some authors (Doob 1994) define the notion of fuzzy measure as an analogue of the notion of measure over a σ-algebra (Borel field), which is completely additive, i.e., closed under countable unions [condition (b) of Definition 9.2 holds for countable—even infinitely countable—many sets). In this sense, $\mathcal{T}$ is completely additive with respect to standard union, but it is not, in general, completely additive with respect to algebraic product union (because the series $\sum_{j=1}^{\infty} \mu_j(t)$ does not necessarily converge).

A measure $\mathcal{m}$ is monotonic, i.e., if $A \subseteq B$, and A, B and $B \setminus A$ belong to the algebra, then $A \cap (B \setminus A) = \varnothing$, and so $\mathcal{m}(A \cup (B \setminus A) = \mathcal{m}(B) = \mathcal{m}(A) +$

m $(B \setminus A)$, from which it follows that m $(A) \leq m$ (B). Hence, the monotonicity property of a measure follows from the additivity property. Typically, a fuzzy measure is defined as a monotonic function with respect to fuzzy inclusion [i.e., from $\tilde{A}_i \sqsubseteq \tilde{A}_j$ it follows that m $(\tilde{A}_i) \leq$ m $(\tilde{A}_j)$], without requiring that it also be additive (monotonicity does not necessarily imply additivity). The reason for requiring monotonicity rather than additivity can be illustrated by a simple example: the degree to which a house is/looks white is not the mere sum of the color of its entrance, windows, walls, and roof. We note that there are several types of measures used in the theory of fuzzy sets, e.g., the Sugeno measure, the Klement measure, the belief measure, the possibility measure, and the necessity measure (Zimmerman 1996, Wang and Klir 1991).

In this book, the notion of the Jordan measure (which is finitely additive) and its fuzzy counterpart are being used, rather than the notions of measure that are infinitely additive and fuzzy measure that is monotonic. The monotonicity of a Jordan fuzzy measure is debatable. If, e.g., the difference $\tilde{A}_2 \setminus \tilde{A}_1$, where $\tilde{A}_1 \sqsubseteq \tilde{A}_2$, is defined as $\mu_2(x) - \mu_1(x)$, then the intersection $\tilde{A}_1 \sqcap (\tilde{A}_2 \setminus \tilde{A}_1) \neq \mathbf{o}$ (as one would normally expect). Thus, just as a Jordan measure is not a measure (in the modern and widely accepted mathematical sense today), the Jordan fuzzy measure is not a fuzzy measure (in the usual and widely accepted sense). Despite these concerns and mathematically arguable aspects, the validity of the use of such a concept of measure in this book is supported by the very good experimental results obtained with retrieval methods that are based on it.

9.9.2 More on Algebra, Entropy, and Probability

$\mathcal{T}$ is a fuzzy algebra also with respect to standard fuzzy union. In this book, we are using algebraic union instead of standard union, owing to the fact that similarity functions are based on the sum of products rather than maxima or minima.

Relative to Theorem 9.3, one may object that condition $H(\mathrm{o}) = 0$ only holds in the limit rather than exactly (the logarithm is not defined at point 0). This is obviously true. However, two reasons for accepting the theorem, at least in principle, can be formulated. One is that the falsity of the condition $H(\mathrm{o}) = 0$ only means that H is not a continuous function at point 0, but this does not invalidate the behavior or tendency toward zero of H, which may be accepted from a practical point of view in retrieval (the closer the value of the membership function to zero, the closer the entropy to zero). The other reason is that good experimental results (relevance effectiveness) were obtained using fuzzy entropy as similarity.

Fuzzy probability can, of course, be used as a similarity function on its own, but the experimental results as to its relevance effectiveness are weak. This is the reason that it is not used directly as similarity; rather it is used to build a better similarity function on it. The case in which the numerator in Eq. (9.24) is also a fuzzy probability was also tested, but relevance effectiveness was low. Further, the basic equation for the similarity function used in the probability retrieval model, i.e., $P(D|Q) = P(D \cap Q) / P(Q)$, was also fuzzified and tested. The results were also weak. Equation (9.24) gave the best results, which is why we propose only this version here.

9.9.3 Information Retrieval and Integration Theory

This section is based on Dominich and Kiezer (2007) and is basically intended as a preparation Section 9.9.4. The notion of a definite integral can be defined in a general way using the concepts of σ-algebra, measure μ on it, and a simple function s. A simple function s on a set X is defined as

$$s: X \rightarrow \{a_1,\ldots,a_n\} \subset [0; \infty). \tag{9.25}$$

Function s can also be written as

$$s(x) = \sum_{i=1}^{n} a_i \chi_{A_i}, \tag{9.26}$$

where $A_i = \{x \in X | \ s(x) = a_i\}$ and $\chi_{A_i} = 1$ if $x \in A_i$, while $\chi_{A_i} = 0$ if $x \notin A_i$. Now let E be an element of a σ-algebra. Then, the integral of function s over E is defined as

$$\int_E s d\mu = \sum_{i=1}^{n} a_i \mu(A_i \cap E). \tag{9.27}$$

In a retrieval system, the retrieved documents D_i in response to a query Q are presented to the user rank ordered, i.e., sorted descendingly on their relevance score based on the values of a similarity function (as some measure): $\mu(o_i \cap q)$. In other words, there is a ranking function (method, procedure) r that rank orders documents $D_1,\ldots,D_i,\ldots,D_n$. Each document receives its own rank, which is different from any other rank. The ranking function r means computing relevance scores and then sorting documents in descending order according to their relevance scores. In procedural terms, the ranking function r can be expressed in pseudocode as follows:

```
r()
    FOR i = 1 TO n compute similarity μ(oi∩q)
    SORT descendingly (D1,...,Dn)
    FOR i = 1 TO n PRINT(i, Di).
```

Formally, ranking function r maps each document to the set of positive integers $\{1, 2,\ldots,n\}$, i.e., $r: O \rightarrow \{1, 2,\ldots,n\}$. Thus, function r may be conceived as being a simple function. It is then possible to construct the sum

$$R = \sum_{i=1}^{n} r(D_i)\mu(D_i \cap Q). \tag{9.28}$$

Sum R may be referred to as the integral of the ranking function r over query Q, i.e.,

$$R = \sum_{i=1}^{n} r(D_i)\mu(D_i \cap Q) = \int_Q r d\mu . \tag{9.29}$$

In terms of integration theory, a retrieval system is computing the integral of its own ranking function r over query Q.

Dominich and Kiezer (2007) give an example from the ADI test collection for the first query: Q_1 (using normalized frequency weights, dot product similarity measure). The first five elements of the ranked hit list are the documents with numbers 69, 47, 17, 46, 71 (the corresponding similarity values are 0.186, 0.16, 0.158, 0.155, 0.126). Thus, $R = 1.726$.

As any automatic retrieval system is completely defined by giving the documents, the query, and the retrieval algorithm (including ranking), the retrieval system is completely defined by giving its integral:

$$\int_Q r d\mu . \tag{9.30}$$

Different retrieval methods are obtained (e.g., vector space, fuzzy entropy, fuzzy probability) depending on how measure μ is defined (κ, H, P, respectively).

9.9.4 Principle of Invariance and String Theory

This section is designed, above all, to be thought provoking.

String theory[1] is a formal construct in theoretical physics designed to unifying quantum theory and general relativity. It is based on replacing the

[1] B. Schellekens: Introduction to String Theory. http://www.nikhef.nl/~t58/lectures.html (4 Nov 2007).

basic principle of pointlike particles, which underlies our intuition for quantum theory and general relativity, with the assumption that the elementary building blocks of the universe are not particles but strings, which are conceived as tiny line segments ("strings," "pieces of rope") of energy. There is no elementary constituent smaller than strings. Strings can be at most 10^{-15} meters in length; otherwise they could be seen in particle accelerators. But just because they are so small, they look like points.

When strings move in (Minkowski) space-time they sweep out surfaces (ribbons or cylinders). Such a surface S can be described by choosing a time coordinate (parameter) on it. The evolution in time (motion, excitation) of a string from its initial to its final state is given mathematically by a surface integral with the general form

$$\iint_S L dS , \tag{9.31}$$

where L denotes the Lagrangian (expressing kinetic and potential energy, which is usually taken to be $L = -T$; T denotes string tension). Strings may have many excitation modes that look like particles and are perceived as particles. On the other hand, the principle of invariance (Section 9.5) proposes that entities can change their identities: they may actually be conceived as being different from different points of view. This suggests the following parallel between information retrieval and string theory:

String theory	**Information retrieval**
String	Document
Excitation modes (perceived as particles)	Identities (perceived as meanings, information, representations)
Mathematical description of evolution (in time): $\iint_S L dS$	Mathematical description of evolution (in relevance): $\int_\Omega r d\mu$

If the parallel given above is indeed possible or acceptable, then we may ask whether there is any basic principle or concept according to which identities are what we normally call the meaning or information content or representation of documents. It may be that what we normally call a document is/should be a counterpart of a string, as an abstract notion. A further option for a counterpart of the notion of a string would perhaps be the abstract concept of "infon" as a basic building block of information as a further physical notion next to energy or mass.

9.10 Exercises and Problems

1. Write down the details of the proof that the scalar product <**u**, **v**> of the vectors **u** and **v** in the general basis $\mathbf{g}_i$ is equal to $(\mathbf{u}^i)^T \mathbf{g}_{ij} \mathbf{v}^j$.

2. Given the following term-document matrix (whose columns represent the coordinates of documents in the orthonormal basis of space E_4):

$$W = \begin{bmatrix} 2 & 3 & 0 & 5 \\ 1 & 1 & 7 & 0 \\ 2 & 0 & 1 & 0 \\ 0 & 1 & 1 & 1 \end{bmatrix},$$

write the term-document matrix in the following orthogonal basis:

$$\begin{bmatrix} 2 \\ 0 \\ 0 \\ 0 \end{bmatrix}, \begin{bmatrix} 0 \\ 1 \\ 0 \\ 0 \end{bmatrix}, \begin{bmatrix} 0 \\ 0 \\ 3 \\ 0 \end{bmatrix}, \begin{bmatrix} 0 \\ 0 \\ 0 \\ -2 \end{bmatrix}.$$

3. Compute the scalar product between the documents of Exercise 2 in both the orthonormal basis and the orthogonal basis.

4. Given the fuzzy sets $\tilde{A}_1$ = "real numbers close to 10" = $\{(x, \mu_1(x)) \mid x \in X\}$, where

$$\mu_1(x) = \frac{1}{1+(x-10)^2},$$

and $\tilde{A}_2$ = "real numbers much larger than 10" = $\{(x, \mu_2(x)) \mid x \in X\}$, where

$$\mu_2(x) = \frac{1}{1+\dfrac{1}{(x-10)^2}},$$

calculate the standard and algebraic product fuzzy union and fuzzy intersection of $\tilde{A}_1$ and $\tilde{A}_2$.

5. Implement the VSR algorithm on a standard test collection and measure its relevance effectiveness.
6. Implement the fuzzy entropy retrieval method on a collection of your choice and measure its relevance effectiveness.
7. Implement the fuzzy probability retrieval method on a collection of your choice, and measure its relevance effectiveness.
8. Given the matrix

$$A := \begin{pmatrix} 1 & 0 & 3 & 5 \\ 0 & 4 & 2 & 2 \\ 1 & 1 & 0 & 1 \\ 1 & 1 & 0 & 0 \end{pmatrix},$$

use the canonical form to compute A^3.

9. Give other ways to compute the probability $p(t_i)$ in the fuzzy probability retrieval method. Measure the relevance effectiveness of your method.

10 Probabilistic Retrieval

The logic of the world is comprised in computing probabilities.
(James Clerk Maxwell)

After reviewing the necessary notions and results from probability theory (probability measure, event space, relative frequency, independent events, conditional probability, Bayes's theorem), probabilistic retrieval methods are presented (probability ranking principle, Bayes's decision rule, nonbinary method, language model) together with examples.

After discussing various formal frameworks for probabilistic retrieval, we propose a framework using lattices (distributive lattice of logical implications).

The notion of a Bayesian network is defined as a special kind of distributive lattice, and the inference retrieval method is described as an application of Bayesian networks.

Exercises and problems are provided to enhance understanding of the concepts as well as the application possibilities of the different methods.

10.1 Elements of Probability Theory

The notion of probability is usually related to the intuition of chance or degree of uncertainty, which develops at a relatively early age. However, a precise and formal definition of probability, universally accepted, seems to be a delusion.

A widely accepted mathematical definition of the notion of probability is the following. Let $\Im = \wp(\Omega)$ denote a Boolean algebra (i.e., a complemented and distributive lattice), and let its elements be called *events*. A *probability measure* (in short, *probability*) P is defined as (Kolmogorov 1956):

- $P: \Im \rightarrow [0, 1] \subset \mathbb{R}$.
- $P(\Omega) = 1$.
- $A \cap B = \varnothing \Rightarrow P(A \cup B) = P(A) + P(B)\ A, B \in \Im$.

The triple $(\Omega, \Im, P)$ is referred to as an *event space* (Kurtz 1991). In practice, the probability of an event is computed based on frequencies (number of occurrences). Let us assume that a trial has n possible equally likely outcomes. If any one of r outcomes produces an event E, the *relative frequency* $f_{rel}(E)$ of E is calculated as $f_{rel}(E) = r/n$.

In probability theory, it is demonstrated that the following relationship between the probability $P(E)$ of an event E and its relative frequency $f_{rel}(E)$ holds:

$$\lim_{n \to +\infty} f_{rel}(E) = P(E). \tag{10.1}$$

In other words, relationship (10.1) makes it possible to interpret probability as being the relative frequency in the long run. A consequence of Eq. (10.1) is that, in practice, the probability of an event can be established empirically by performing a large number of trials and equating relative frequency with probability.

Example 10.1

What is the probability that when tossing a die, the outcome is odd and greater than 2? If the die is fair (i.e., all sides are equiprobable), this event is satisfied by two ($r = 2$) of the six possible outcomes ($n = 6$), namely 3 and 5. Therefore, the probability that this event will occur is equal to $2/6 = 1/3$. □

We note that there are other ways to compute the probability of an event (equivalently, the degree of uncertainty we associate with it). Thus, in the so-called frequentist view, the probability of an event is the proportion of times we would expect the event to occur if the experiment were repeated many times. In the subjectivist view, the probability of an event is one's degree of belief in the occurrence of that event.

Two events A and B are said to be *independent* if

$$P(A \cap B) = P(A) \times P(B). \tag{10.2}$$

In words, two trials are said to be independent if the outcome of one trial does not influence the outcome of the other.

Example 10.2

Tossing a die twice means two independent trials. The outcomes of two flips of a fair coin are independent events. □

Two events that have no outcomes in common are called *mutually exclusive* (i.e., it is impossible for both to occur in the same experiment). For example, in die tossing, the events "I toss a 2" and "I toss a 1" are mutually exclusive.

The *union* of two events means that, in a trial, at least one of them occurs. It is an interesting result that for any two events, say, A and B, we have $P(A \cup B) = P(A) + P(B) - P(A \cap B)$. Then, if A and B are mutually exclusive, $P(A \cup B) = P(A) + P(B)$.

Let $P(A) > 0$. Then, the quantity denoted by $P(B|A)$ and defined as

$$P(B|A) = \frac{P(AB)}{P(A)} \tag{10.3}$$

is called the *conditional probability* of event B relative to event A. Alternatively, assume that there are two trials, the second dependent on the first. The probability $P(AB) = P(A \cap B)$ that the first trial will yield an event A and the second trial will yield an event B (which thus is dependent on A) is the product of their respective probabilities, where the probability $P(B|A)$ of B is calculated on the premise that A has occurred [the probability of A is $P(A)$]:

$$P(AB) = P(A) \times P(B|A) \tag{10.4}$$

We note briefly that, fairly counterintuitively, the notion of dependence (independence) is not transitive. Transitivity of events would mean that if event A depends on event B, and event B depends on event C, then event A depends on event C.

Let $A_1, A_2, \ldots, A_n$ be a mutually disjoint and complete system of events, i.e.,

$$A_1 \cup A_2 \cup \ldots \cup A_n = \Im,\ A_i \cap A_j = \varnothing,\ i, j = 1, 2, \ldots, n,\ i \neq j, \tag{10.5}$$

and B an arbitrary event. If event B occurs under any A_i with probability $P(B|A_i)$, then the probability $P(B)$ of event B is given by the *total probability equation*:

$$P(B) = P(B|A_1)P(A_1) + \ldots + P(B|A_n)P(A_n). \tag{10.6}$$

The conditional (*posterior*) probability $P(A_i|B)$, given that event B has happened is calculated using Bayes's theorem (or Bayes's equation):

$$P(A_i|B) = \frac{P(B \mid A_i)P(A_i)}{\sum_{i=1}^{n} P(B \mid A_i)P(A_i)}, \tag{10.7}$$

where $P(A_i)$ is called the *prior* probability.

10.2 Principles of Probabilistic Retrieval

Probabilistic retrieval is based on whether a probability of relevance (relative to a query) of a document is higher than that of irrelevance (and exceeds a threshold value).

Maron and Kuhns (1960) argued that since a retrieval system cannot predict with certainty which document is relevant, we should deal with probabilities. The relevance of a term to a document was defined as the probability of user satisfaction if that term would be used as a query. Relevance was taken as a dichotomous variable, i.e., the user either is satisfied or is not satisfied with a retrieved document. Then, it seems obvious that documents that are most likely to satisfy the information need should be presented first. This has become known as the *probability ranking principle* (Cooper 1971, Robertson 1977):

> *If the retrieved documents (in response to a query) are ranked decreasingly on their probability of relevance, then the effectiveness of the retrieval system will be the best that is obtainable.*

Note: However, counterexamples can be given (Robertson 1977).

Indeed, the probability ranking principle yields an optimal solution under certain conditions. Let A and B denote two events. By Bayes's theorem (10.7), we have $P(A|B)P(B) = P(B|A)P(A)$, and $P(\neg A|B)P(B) = P(B|\neg A)P(\neg A)$. Thus, where $\neg A$ is the negated event A, we obtain

$$\frac{P(A \mid B)}{P(\neg A \mid B)} = \frac{P(B \mid A)P(A)}{P(B \mid \neg A)P(\neg A)}. \tag{10.8}$$

Using the *logit* logistic (log-odds) transformation, defined as (Robertson 1977):

$$\text{logit } P(X) = \log \frac{P(X)}{1-P(X)} = \frac{P(X)}{P(\neg X)}, \tag{10.9}$$

yields

$$\text{logit } P(A|B) = \log \frac{P(B \mid A)}{P(B \mid \neg A)} + \text{logit } P(A). \tag{10.10}$$

Let us define the following probabilities:

P(document retrieved | document relevant),
P(document retrieved | document irrelevant),
P(document relevant | document retrieved).

Let A = "document relevant," and B = "document retrieved." As seen in Chapter 4, recall is a measure of relevance effectiveness. It is defined as the proportion of retrieved documents out of those relevant. Thus, we may view recall as an estimate of $P(B|A)$. Using Eq. (10.10), we obtain

$$P(B|A) = P(B|\neg A) \times \exp(\text{ logit } P(A|B) - \text{logit } P(A) \text{ }), \tag{10.11}$$

from which it follows that recall is monotonic with respect to $P(A|B)$, which is the probability of relevance of a retrieved document.

Now let:

- D be a set of documents.
- Q be a query.
- $\alpha \in \mathbb{R}$ a cut-off value.
- $P(R|(Q, d))$ and $P(I|(Q, d))$ the probabilities that document $d \in D$ is relevant (R) and irrelevant (I), respectively, to query Q.

The retrieved documents in response to query Q belong to the set $\Re(Q)$ defined as (van Rijsbergen 1979):

$$\Re(Q) = \{d \mid P(R|(Q, d)) \geq P(I|(Q, d)), P(R|(Q, d)) > \alpha\}. \tag{10.12}$$

The elements of set $\Re(Q)$ are shown to the user ranked in descending order on their $P(R|(Q, d))$ values (on the basis of the probability ranking principle). The inequality

$$P(R|(Q, d)) \geq P(I|(Q, d)) \tag{10.13}$$

is referred to as *Bayes's decision rule*.

The rest of this chapter includes a description of probabilistic retrieval methods and a proposal for a lattice theoretical framework for probabilistic retrieval.

10.3 Probabilistic Retrieval Method

In what follows, we describe a probabilistic retrieval method (Yu et al. 1989).

Given a set D of elements called *documents*:

$$D = \{D_1,\ldots,D_i,\ldots,D_n\}, \tag{10.14}$$

a set T of elements called *terms*:

$$T = \{t_1,\ldots,t_k,\ldots,t_m\}, \tag{10.15}$$

and a set F_i of nonnegative integers:

$$F_i = \{f_{i1},\ldots,f_{ik},\ldots,f_{im}\}, \tag{10.16}$$

where f_{ik} represents the number of occurrences of term t_k in document D_i, a *weights vector*

$$\mathbf{w}_i = (w_{i1},\ldots,w_{ik},\ldots,w_{im}), \tag{10.17}$$

where w_{ik} is the *weight* (significance) of term t_k in document D_i, is calculated as follows:

$$w_{ik} = \log\frac{P(f_{ik} \mid R)}{P(f_{ik} \mid I)}, \tag{10.18}$$

where $P(f_{ik} \mid R)$ and $P(f_{ik} \mid I)$ denote the probability that a relevant or irrelevant, respectively, document D_i has f_{ik} occurrences of term t_k.

It is assumed that (optimal retrieval hypothesis) an optimal way to retrieve documents is in descending order of relevance, i.e., for any two documents D_i and D_j we have $P(R|D_i) \geq P(R|D_j)$, where $P(\,|\,)$ denotes conditional probability (i.e., that a document is relevant). By Bayes's theorem [Eq. (10.7)], we have

$$P(R|D_i) = \frac{P(D_i \mid R)P(R)}{P(D_i)} \text{ and } P(I|D_i) = \frac{P(D_i \mid I)P(I)}{P(D_i)}. \tag{10.19}$$

Q is to be understood next to D, or simultaneous with D. However, it may be omitted as it is a constant (during its own retrieval).

Thus, Bayes's decision rule [Eq. (10.13)] becomes ($P(D_i) \neq 0$):

$$P(R|D_i) \geq P(I|D_i) \qquad \Leftrightarrow$$

$$P(D_i|R)P(R) \geq P(D_i|I)P(I) \qquad \Leftrightarrow$$

$$\frac{P(D_i \mid R)}{P(D_i \mid I)} \geq \frac{P(I)}{P(R)}.$$

$P(R)$ and $P(I)$ denote the probability that a randomly chosen document is relevant and irrelevant, respectively. They may be viewed as being constant for a given collection D of documents. By making use of the *term independence assumption* (i.e., any term in a document occurs independently of every other term), event $D_i|R$ means the simultaneous occurrence of the independent events $f_{ik}|R$ for every term in D_i. Thus, we may write that

$$P(D_i \mid R) = \prod_{k=1}^{m} P(f_{ik} \mid R). \tag{10.20}$$

In a similar way, we have

$$P(D_i \mid I) = \prod_{k=1}^{m} P(f_{ik} \mid I). \tag{10.21}$$

Thus, Bayes's decision rule becomes:

$$\frac{P(D_i \mid R)}{P(D_i \mid I)} \geq \frac{P(I)}{P(R)} \qquad \Leftrightarrow$$

$$\frac{\prod_{k=1}^{m} P(f_{ik} \mid R)}{\prod_{k=1}^{m} P(f_{ik} \mid I)} \geq \frac{P(I)}{P(R)} = c \qquad \Leftrightarrow$$

$$\prod_{k=1}^{m} \frac{P(f_{ik} \mid R)}{P(f_{ik} \mid I)} \geq c.$$

If we take the logarithm of both sides and use Eq. (10.18), we get

$$\sum_{k=1}^{m} w_{ik} \geq \log c = C. \tag{10.22}$$

Hence, the optimal retrieval hypothesis can be rewritten as

$$P(R|D_i) \geq P(R|D_j) \Leftrightarrow \sum_{k=1}^{m} w_{ik} \geq \sum_{k=1}^{m} w_{jk} . \tag{10.23}$$

Let $Q = (q_1,\ldots,q_k,\ldots,q_m)$ denote a query (q_k are binary weights), i.e., q_k = 1 if t_k occurs in Q, and $q_k = 0$ otherwise. Using the dot product similarity measure, we find that the similarity σ_i between query Q and document D_i is equal to

$$\sigma_i = \sum_{k=1}^{m} w_{ik} q_k = \sum_{k=1}^{m} w_{ik} , \tag{10.24}$$

which means that the optimal retrieval hypothesis is rewritten as

$$P(R|D_i) \geq P(R|D_j) \Leftrightarrow \sum_{k=1}^{m} w_{ik} \geq \sum_{k=1}^{m} w_{jk} \Leftrightarrow \sigma_i \geq \sigma_j. \tag{10.25}$$

The following method can be used to apply this model in practice:

Probabilistic Retrieval Method

1. Let q denote a query.
2. Let query q be the one-element set T [given in (10.15)], $|T| = m$.
3. In order for a document D_i to be retrieved in response to q, the following condition can be used:

$$\sum_{k=1}^{m} f_{ik} \geq K,$$

where K is a threshold.
4. The retrieved documents are then presented to the user, who judges which are relevant and which are not. (This action is called *relevance feedback*).
5. From the retrieved and relevant documents, the following table is constructed first for each term t_k in T:

T_k =	0	1	...	j	...
	b_0	b_1	...	b_j	...

where T_k is a variable associated with term t_k and takes on the values 0, 1,..., j,... (which can be interpreted as the number of occurrences), and b_j is the number of relevant and retrieved documents having j occurrences of term t_k. The probabilities $P(f_{ik}|R)$ are calculated as:

$$P(f_{ik}|R) = \frac{b_j}{b_0 + b_1 + ...},$$

for $f_{ik} = j$. The same method is used for the irrelevant documents.

6. Calculate weight vectors $\mathbf{w}_i$ for documents, assign weight 1 to each query term, and use the optimal retrieval hypothesis to retrieve and rank order documents.

Note: This method gives better results if probabilities are (re-)computed using accumulated statistics for many queries.

Example 10.3

Let the set of documents be $D = \{D_1, D_2, D_3\}$, where:

D_1 = Bayes's principle: The principle that in estimating a parameter, one should initially assume that each possible value has equal probability (a uniform prior distribution).

D_2 = Bayesian decision theory: A mathematical theory of decision-making that presumes utility and probability functions, and according to which the act to be chosen is the Bayes's act, i.e., the one with highest subjective expected utility. If one had unlimited time and calculating power with which to make every decision, this procedure would be the best way to make any decision.

D_3 = Bayesian epistemology: A philosophical theory that holds that the epistemic status of a proposition (i.e., how well proven or well established it is) is best measured by a probability and that the proper way to revise this probability is given by Bayesian conditionalization or similar procedures. A Bayesian epistemologist would use probability to define concepts such as epistemic status, support, or explanatory power and explore the relationships among them.

Let query q be $q = \textit{probability}$. $T = \{t_1 = \text{probability}\}$, $k = 1$. In order to retrieve an initial set of documents, f_{i1}, $i = 1, 2, 3$, are calculated first:

$$f_{11} = 1, f_{21} = 1, f_{31} = 3.$$

Taking $K = 1$, we retrieve documents D_1, D_2, and D_3:

$$\Sigma_k f_{1k} = 1,\ \Sigma_k f_{2k} = 1,\ \Sigma_k f_{3k} = 3\ (\geq K).$$

In a relevance feedback, D_3 is judged as relevant, whereas D_1 and D_2 are irrelevant. The probabilities of relevance are

$$P(f_{i1}{=}1 \mid R) = 0,\ \ P(f_{i1}{=}3 \mid R) = 1,$$

and those of irrelevance are

$$P(f_{i1}=1 \mid I) = 1,\ P(f_{i1}=3 \mid I) = 0.$$

The weight vectors for documents are

$$\mathbf{w}_1 = -\infty,\ \mathbf{w}_2 = -\infty,\ \mathbf{w}_3 = \infty.$$

The query vector is $\mathbf{w}_q = (1)$. In the new retrieved rank order, D_3 precedes D_2 and D_1. □

10.4 Language Model Retrieval Method

Instead of computing (estimating) the conditional probability $P(R|(D, Q))$ of relevance R of a given document D with respect to query Q, Croft and Ponte (1998) and Song and Croft (1999) suggested a method for estimating the conditional probability $P(Q|D)$ of a query Q given document D, known as the language model of information retrieval.

Let Q denote a query consisting of (or represented as) a sequence of terms: $t_1,\ldots,t_i,\ldots,t_m$. The occurrence of each term t_i is conceived as being an independent event, i.e., each term is independent of any other term (*term-independence assumption*). Let $P(t_i|D)$ denote the probability of term t_i in document D. Then,

$$P(Q \mid D) = \prod_{i=1}^{m} P(t_i \mid D). \tag{10.26}$$

Probability $P(t_i|D)$ can be estimated as

$$P(t_i \mid D) = \frac{f_{iD}}{N_D}, \tag{10.27}$$

where f_{iD} denotes the number of occurrences of term t_i in document D and N_D the total number of term occurrences in D.

It is known from practice that, in general, many terms may be missing in a document, which means that their probabilities would be zero. If such a term were used in Q, then probability $P(t_i|D)$ would vanish, which is not desirable if another query term is present in D. A solution to such a situation would be to assign some probability to missing terms as well. The method used to perform this assignment is referred to as *smoothing*. The number f_{iD} of occurrences of term t_i in document D is adjusted to a value f'_{iD} according to the following equation (known as the Good-Turing estimate):

$$f'_{iD} = (f_{iD} + 1)\frac{E(N_{f_{iD}+1})}{E(N_{f_{iD}})}, \tag{10.28}$$

where N_x is the number of terms with frequency x in D, and $E(.)$ denotes expected value. Then, probability $P(t_i|D)$ is defined as

$$P(t_i \mid D) = \frac{f'_{iD}}{N_D}. \tag{10.29}$$

Thus, the probability of a missing term will be set to $E(N_1)$ / $[E(N_0)N_D]$. Since the length and content of a document is fixed (in practice), the expected value $E(N_x)$ is almost impossible to obtain. However, one way to get around this problem is to estimate (or approximate) $E(N_x)$ with N_x, albeit that this may create problems. For example, a term t with highest frequency h will have probability zero because N_{h+1} is zero. One solution is to use curve fitting to smooth the observed frequencies to a fitted (smoothing) function S. It is known (Chapter 4) that term occurrences follow a power law. The number of occurrences f is represented on the x-axis, while the number N_f of terms having frequencies f is on the y-axis; e.g.:

f	N_f
0	2134
1	34
. . .	

With a smoothing function S, probability $P(t_i|D)$ becomes

$$P(t_i \mid D) = (f_{iD} + 1)\frac{S(f_{iD} + 1)}{S(f_{iD})N_D}. \tag{10.30}$$

Song and Croft (1999) report that, according to experimental results, an appropriate smoothing function was a particular geometric distribution. The relevance effectiveness of the retrieval method thus obtained was measured on two test databases: The *Wall Street Journal* and TREC4. The MAPs (mean average precisions) obtained were 0.2198 and 0.1905, respectively.

Another smoothing method used to estimate $P(t_i|D)$ is the *Jelinek-Mercer smoothing* (Metzler and Croft 2004). With this method, $P(t_i|D)$ is approximated as

$$P(t_i \mid D) = \lambda \frac{f_{iD}}{|D|} + (1-\lambda)\frac{N_i}{|C|}, \tag{10.31}$$

where:

- $|D|$ is the number of terms in D.
- N_i is the number of times term t_i occurs in the entire collection of documents.
- $|C|$ is the number of terms in the entire collection of documents.
- λ is a smoothing parameter that can be set manually (experimentally or automatically), $0 \leq \lambda \leq 1$; recommended value: $\lambda = 0.6$.

10.5 Lattice Theoretical Framework for Probabilistic Retrieval

We have seen that probabilistic retrieval methods are based on the notion of probability quantified by $P(R|(Q, d))$ and $P(I|(Q, d))$. These quantities are referred to as the conditional probabilities that document d is relevant (R) and irrelevant (I), respectively, to query Q.

Then, we can ask the following question: What is the event space over which the probabilities are defined? Several answers have been given.

Robertson et al. (1982) propose that the event space be generated by the Cartesian product between a document set D and a query set Q, $Q \times D$, i.e., the event space is the Boolean lattice $(Q \times D, \wp(Q \times D), P)$. Relevance is defined as a binary relation $R \subseteq Q \times D$. Then, obviously, $P(R) = 1$ (since R, as the event, is given). But how realistic is this (relevance may also be a posteriori)?

Fuhr (1992) takes a slightly different approach. He proposes the same event space, i.e., $(Q \times D, \wp(Q \times D), P)$. However, in his view, relevance is not an element of this event space, but rather a real number attached a query-document pair (q, d) as follows. The probability $P(R|(q, d))$ is calculated as the proportion of pairs (q, d) that are judged as relevant out of the total number of documents and queries having the same representation d and q, respectively. In Fuhr's view, the set $Q \times D$ may even be infinite. In such a case, it may happen that all representations are also infinitely many (when they are all different from each other). How is probability calculated in this case? On the other hand, if $P(R|(q, d))$ is either 0 or 1, the probability ranking principle may become superfluous: ranking becomes meaningless (the relevance degree of all retrieved documents is equal to 1).

Van Rijsbergen (1992), recognizing these difficulties, examines the notion of probability in probabilistic retrieval. There are two ways to view probability. One is to see it as a measure of chance (of an event) and the other is as a measure of the degree of belief (in a proposition). In probabilistic retrieval, the most thoroughly researched notion was conditional probability $P(R|(q, d))$ within the Bayesian framework. As this probability may leave room for confusion, the question of whether $P(R|(q, d))$ designates a conditional probability or a degree of implication (of the type q implies d) may be raised. In order to get around such problems, van Rijsbergen proposes the application of *Jeffrey's rule of conditioning*. Let X be an event, and let E denote a proposition that signifies the "passage of experience." Then, $P(X)$ is the probability (measure of the degree of belief) of X before observation (i.e., before the "passage of experience"). The passage of experience (represented by E) leads P to a revised P^*. In Bayesian notation: $P^*(X) = P(X|E)$. To give an example: one believes that a piece of cloth is green to the degree of $P(X) = 0.3$; but, after examining it by candlelight, one modifies one's belief as follows: the piece of cloth is blue to the degree $P^*(X) = 0.7$. Van Rijsbergen developed a mathematical formalism that implements Jeffrey's rule of conditioning in IR. With the notation $X =$ relevance, and E is the observation of a query term in a document, the following formula for relevance is proposed:

$$P^*(\text{relevance}) =$$

$$P(\text{relevance} \mid \text{term occurs})P^*(\text{term occurs}) +$$

$$P(\text{relevance} \mid \text{term does not occur})P^*(\text{term does not occur}).$$

Robertson (2002) discusses at length the problems that may arise when the event space is $(Q \times D, \wp(Q \times D))$. Apart from those already mentioned thus far, another major problem that can be raised in the context of IR: In this event space, every event is distinct and every query is paired with every document. How adequate is this in retrieval? Robertson proposes the following solution. For the probabilistic model, the event space is based on a single query, the actual one, paired with documents (i.e., the event space is regenerated with every query).

We now propose a formal framework for probabilistic retrieval based on the distributive lattice of logical propositions ordered by logical implication.

Let $\{d_1,\ldots,d_j,\ldots,d_m\}$ denote a set of documents to be searched in response to a query Q. Let us introduce the function

$$f\colon \{d_1,\ldots,d_j,\ldots,d_m, Q, R, I\} \rightarrow (\{T, F\}, \wedge, \vee, \neg), \qquad (10.32)$$

where $(\{T, F\}, \wedge, \vee, \neg)$ is the Boolean algebra of logical propositions, and

$d_j \mapsto$ "Document d_j is observed," $j = 1,\ldots,m$.
$Q \mapsto$ "Query Q is given."

$R \mapsto$ "Document is relevant."
$I \mapsto$ "Document is irrelevant."

The degree of ordering in a lattice can be measured as follows:

Definition 10.1 (Knuth 2005). In a Boolean lattice, the *degree of ordering* is measured by the function

$$z(x,y) = \begin{cases} 1 & \text{if} \quad x \geq y \\ 0 & \text{if} \quad x \wedge y = \mathbf{0} \\ p & \text{otherwise} \quad 0 < p < 1 \end{cases}$$ □

We use the following result (whose proof is only sketched here, as for us it is the result itself that is important rather than its proof):

Theorem 10.1 (Knuth 2005). *In a distributive lattice L, the measure z of the degree of ordering satisfies the relation*

$$z(y, x \wedge t) = \frac{z(y,t) z(x, y \wedge t)}{z(x,t)}.$$

Proof. Since lattice L is distributive, we have

$$z(x \wedge y, t) = z(x, t) \cdot z(y, x \wedge t).$$

Owing to commutativity, we may rewrite the above as

$$z(y \wedge x, t) = z(y, t) \cdot z(x, y \wedge t).$$

The left-hand sides are equal, so the right-hand sides are equal to each other as well:

$$z(x, t) \cdot z(y, x \wedge t) = z(y, t) \cdot z(x, y \wedge t),$$

which yields

$$z(y, x \wedge t) = \frac{z(y,t) \cdot z(x, y \wedge t)}{z(x,t)}.$$ □

Obviously, Theorem 10.1 also holds for Boolean algebras (because any Boolean algebra is a distributive lattice).

The relationship in Theorem 10.1 reminds us of Bayes's theorem from probability theory. It can be similarly shown (Knuth 2005) that the sum and product rules of probability also hold for function z. (Moreover, these rules are the only ones that are consistent with lattice L.

As we saw earlier, the lattice of propositions can be ordered using logical implication and thus transformed into a distributive lattice (moreover, even into a Boolean algebra). Logical implication in mathematical logic is usually referred to as *material conditional*. Recall that this type of conditional may create paradoxes or strange situations. For example, a contradiction (which is always false) may imply anything (i.e., falsity as well as truth). Or, if the consequent is true, then the value of the material conditional is also true regardless of the truth-value of the antecedent. (In $p \Rightarrow q$, p is the antecedent and q is the consequent; $p \Rightarrow q$ is also denoted by $p \rightarrow q$.)

Another strange situation is that the material conditional allows us to link any propositions that might not even be used together in normal speech (see examples in Chapter 2). Conditionals that allow only propositions that bear on the same subject are called *entailments*.

There are other types of conditionals as well. Thus, the *counterfactual* (or subjective) *conditional* has the form "if p were to happen, then q would." The counterfactual conditional is typically used in science, e.g., "if ice were to be heated, it would melt,' or "if the equipment were to fail, then the lamp would flash." It can be seen that, in general, counterfactual conditionals cannot be represented by material conditionals. (Since the material conditional $p \rightarrow q$ is true whenever p is false, the value of the counterfactual would be indistinguishable.)

Another type of conditional is the *indicative conditional*. An example of an indicative conditional is, "if the accused did not kill the victim, then someone else did," which is true. But the falsity or truth of its counterfactual version, "if the accused had not killed the victim, someone else would have," is questionable.

Reasoning or inference (i.e., the process of deriving conclusions from premises known or assumed to be true) is based, among others things, on the use of conditionals.

In *deductive reasoning*, if the premises are true, then the conclusion must be true. Examples for deductive reasoning are:

- The law of noncontradiction (if p is true, then p cannot be false).
- Modus ponens (if $p \rightarrow q$ and p, then q).

False or inconclusive premises may lead to false or inconclusive conclusions.

In another type of reasoning—*inductive reasoning*—true premises are believed to support the conclusion, albeit that they do not necessarily ensure it. Inductive reasoning is typically used to formulate laws or rules based on limited or uncertain observations or experiments. For example, from "this fire is hot," one infers that "all fires are hot" (which is not necessarily true, since there is also cold fire).

We can thus see that inductive reasoning has a probabilistic (uncertain, belief-based, plausibility) flavor. Inductive reasoning may thus be interpreted as being a material conditional that has a degree of being true attached to it. As Stalnaker (1968) conjectured, and recent experimental research has supported, (Over et al. 2007), the probability of the conditional $p \rightarrow q$ is proportional to the conditional probability $P(q|p)$ of the consequent on the antecedent. Implication induces an order in the lattice of propositions. Thus, the following definition may be introduced:

Definition 10.2. (Knuth 2005) The *conditional probability* of y given x is denoted by $P(y \mid x)$ and defined as

$$z(x, y) = P(x \mid y). \qquad \square$$

We are now formally entitled to write, in probabilistic retrieval, the probability $P(R \mid (Q, d))$, which is now equal to $P(R \mid Q \wedge d)$, and is thus consistent with its formal background. The usual expression (10.19) can now be written as (Theorem 10.1):

$$P(R \mid (d, Q)) = P(R \mid d \wedge Q) = \frac{P(R \mid Q) P(d \mid R \wedge Q)}{P(d \mid Q)}, \tag{10.33}$$

which is the same as Eq. (10.19); Q is constant, so it may be omitted.

Note: At a first look, the derivations in this section may seem strange. We wish to remind the reader that the notion of probability, as defined by Kolmogoroff, is a formal concept: it is a measure on a lattice structure (to be exact, on a σ-algebra). As such, it is consistent with the operations on this lattice, namely the sum, product, and Bayes's rules are satisfied. We have seen that these rules are satisfied by the degree function z defined on a distributive lattice L. Thus, formally, the two concepts, probability and degree function z, are equivalent. Moreover, it is just this equivalence that may elucidate why some measures in science act like probabilities when they are hardly what one would consciously and explicitly call probability (as chance). Yet, this seems to be exactly the case with the quantities P in probabilistic retrieval methods.

10.6 Bayesian Network Retrieval

Let us consider the lattice (P, ⇒) of logical implications. Based on Definitions 10.1 and 10.2, a conditional probability $P(x|y)$ may be assigned to the elements of P.

The notion of a Bayesian network (or inference network) can be defined as a lattice of implications in which the degree function z is given in a very specific way:

Definition 10.3. A *Bayesian network* is a lattice (P, ⇒) in which the conditional probabilities $P(x|y)$ are given as the probability of event x on its supremal events y. □

Usually, a Bayesian network (BN) is defined as follows (Savoy and Desbois 1991, Metzler and Croft 2004): The BN is a directed and acyclic graph (tree if one disregards direction) whose nodes represent events and edges dependence between events. Each nonroot node is assigned conditional probabilities that give the probability of an outcome depending on the outcome of its parent (antecedent) events.

We note that the **0** and **1** of the lattice (P, ⇒) are, in many cases, formal elements that do not have a practical role in applications. However, this does not invalidate Definition 10.3 from a formal point of view.

Given any observed event, referred to as *evidence*, in a BN, it is possible to compute the probability, referred to as *belief*, of an outcome at any node by propagating beliefs through the BN. Since our scope is not a treatment of BNs per se, but rather to show how BNs can be applied in IR, the interested reader is directed to the specialized literature (e.g., Pearl 1988, Cooper 1990). However, a brief example is given below to help clarify the meaning and use of a BN, in general.

Example 10.4

Let the lattice (P, ⇒), i.e., Bayesian network, be defined as follows. The degree functions, i.e., probabilities, are (recall that T = true, F = false):

- $P(\text{Sunny} = T) = 0.5$, $P(\text{Sunny} = F) = 0.5$.
- $P(\text{Warm} = F \mid \text{Sunny} = F) = 0.8$, $P(\text{Warm} = F \mid \text{Sunny} = T) = 0.2$.
- $P(\text{Warm} = T \mid \text{Sunny} = F) = 0.2$, $P(\text{Warm} = T \mid \text{Sunny} = T) = 0.8$.
- $P(\text{Heating} = F \mid \text{Sunny} = F) = 0.5$, $P(\text{Heating} = F \mid \text{Sunny} = T) = 0.8$.
- $P(\text{Heating} = T \mid \text{Sunny} = F) = 0.5$, $P(\text{Heating} = T \mid \text{Sunny} = T) = 0.1$.
- $P(\text{I am warm} = F \mid \text{Heating} = F \wedge \text{Warm} = F) = 1$.
- $P(\text{I am warm} = T \mid \text{Heating} = F \wedge \text{Warm} = F) = 0$.
- $P(\text{I am warm} = F \mid \text{Heating} = T \wedge \text{Warm} = F) = 0.1$.

- P(I am warm = T | Heating = $T \wedge$ Warm = F) = 0.9.
- P(I am warm = F | Heating = $F \wedge$ Warm = T) = 0.1.
- P(I am warm = T | Heating = $F \wedge$ Warm = T) = 0.9.
- P(I am warm = F | Heating = $T \wedge$ Warm = T) = 0.01.
- P(I am warm = T | Heating = $T \wedge$ Warm = T) = 0.99.

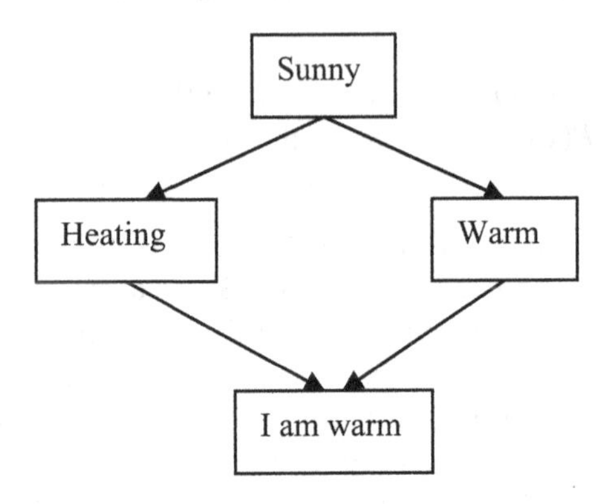

We can use the BN to perform inference:

- For example, if we observe (this is the evidence) that the whether is sunny (i.e., the event "Sunny" is true), then we hardly switch the heating on (i.e., the probability of event "Heating" being true is very low, namely 0.1).
- We can use the BN to infer the probability of causes of an observed event. For example, if the evidence is that "I am warm" (i.e., "I am warm" is true), then the possible causes as well as their probabilities can be obtained. □

In IR, BNs are applied to represent documents and queries as well as to propose similarity (ranking) measures (Turtle and Croft 1991, Savoy and Desbois 1991, Metzler and Croft 2004). **Figure 10.1** shows the basic BN (as a building block) used for retrieval.

Node D_j corresponds to document D_j. Nodes $t_1,\ldots,t_k,\ldots,t_m$ correspond to representations of documents (typically to terms). Node Q corresponds to a query. (We note that the BN of **Fig. 10.1** can be enlarged to encompass all the documents of a whole collection, as well as several queries. In such a case, the lowest node will be one called I, which represents the user's information need.)

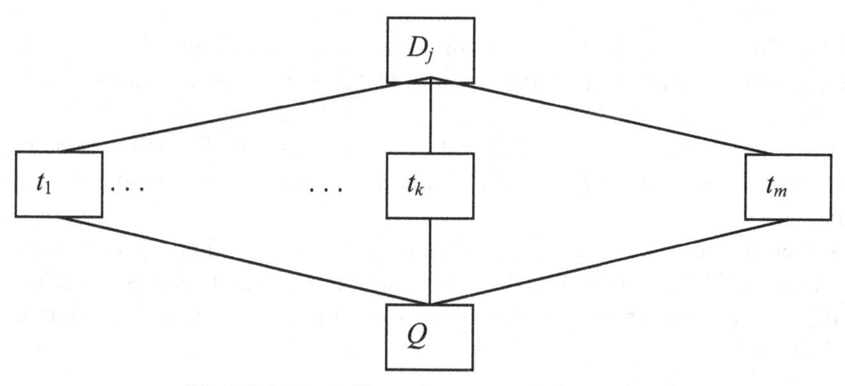

Fig. 10.1. Basic Bayesian network for retrieval.

The document node is binary, i.e., it either happens (i.e., it is being observed) or not.

There are several options to set the belief $bel(t_k)$ of representation nodes t_k. For example, it can be set to

- $bel(t_k) = w_{kj}$ (as defined in Theorem 4.1., Section 4.6),
- $bel(t_k) = P(t_i|D_j)$ [as defined in Eq. (10.27)],
- Okapi inverse document frequency belief score:

$$bel(t_k) = b + (1-b) \times \frac{f_{kj}}{f_{kj} + 0.5 + 1.5\frac{|D_j|}{|D|}} \times \frac{\log\frac{N+0.5}{d_k}}{\log(N+1)}, \tag{10.34}$$

where:

—b is the default belief parameter, $0 \le b \le 1$, typically $b = 0.6$.

—f_{kj} is the number of time term t_k occurs in document D_j.

—$|D_j|$ is thelength of document D_j (typically: the number of its terms).

—N is the total number of documents in the entire collection.

—$|D|$ is the average document length (e.g., $\frac{1}{N}\sum_{j=1}^{N}|D_j|$).

—d_k is the number of documents in which t_k appears.

The choice of $bel(t_k)$ has an influence on the relevance effectiveness of retrieval. Metzler and Croft (2004) provide experimental evidence that the choice $bel(t_k) = P(t_i|D_j)$ as defined in Eq. (10.27) yields good results.

The query node allows us to combine beliefs about representations. Given an evidence, i.e., the observation of a document D_j, belief is being propagated from D_j to Q, which yields a scoring function (used to rank documents).

Given a query $Q = (w_1, q_1; \ldots; w_i, q_i; \ldots; w_n, q_n)$, where q_i is a query term and w_i is the importance that we attach to it, several *scoring functions* $bel(Q)$ have been proposed and tested experimentally (see e.g., Metzler and Croft 2004):

- Weighted sum: $$bel(Q) = \frac{\sum_{i=1}^{n} w_i bel(q_i)}{\sum_{i=1}^{n} w_i}.$$
- Weighted AND: $$bel(Q) = \prod_{i=1}^{n} (bel(q_i))^{w_i}.$$
- Sum: $$bel(Q) = \frac{1}{n} \sum_{i=1}^{n} bel(q_i).$$
- OR: $$bel(Q) = 1 - \prod_{i=1}^{n} (1 - bel(q_i)).$$
- AND: $$bel(Q) = \prod_{i=1}^{n} bel(q_i).$$
- MAX: $$bel(Q) = \max_{1 \le i \le n} bel(q_i).$$

Metzler and Croft (2004) reported experimental results as to the relevance effectiveness of BN retrieval. The test databases used were TREC 4, 6, 7, and 8.[1] Each query was Porter stemmed and stoplisted. The scoring function used was the weighted sum. The average precision obtained with the BN method was 9% higher than that obtained with the language model on TREC 4.

---

[1] www.nist.gov (TREC).

10.7 Exercises

1. Given a blog on the World Wide Web (or a presentation on radio on or television) that contains a section called "Frequently Asked Questions" (FAQ), note that FAQ contains the following questions (in chronological order): $q_1,\ldots,q_i,\ldots,q_n$. The number of times question q_i was asked is f_i, $i = 1,\ldots,n$. What is the probability of selecting q_i?

2. Let A and B be two independent events. Prove that

$$\text{logit}\ (A \cup B) = \log(1 / (1 - P(A) - P(B)) - 1).$$

3. Implement the probabilistic retrieval method on a document collection of your choice. In Eq. (10.18), use frequency f_{ik} first, then a weight w_{ik} of your choice (using Theorem 4.1). Observe and discuss the differences in the two rankings.

4. Implement the language model method on a document collection of your choice. Define and experiment with several smoothing functions. Observe and discuss the rankings.

5. Implement the inference network retrieval method on a document collection of your choice. Observe and discuss the influence of different belief functions on ranking.

11 Web Retrieval and Ranking

The significant problems we face cannot be solved at the same level of thinking we were at when we created them.
(Albert Einstein)

After introducing the notion of a Web graph and discussing degree distribution, we present the basic methods using link structure analysis (impact factor, connectivity, mutual citation, PageRank, HITS, SALSA, associative-interaction) together with clarifying examples for each. A connection between HITS and LSI is also shown.

Then, an aggregated method for Web retrieval based on lattices is presented that allows one to calculate the importance of pages, taking into account both their link importance (using link analysis) and their intrinsic importance (stemming from page content). Experimental evidence for the relevance effectiveness of this method is also given in terms of comparison with commercial search engines (with Google, Altavista, Yahoo!).

After introducing the notion of Web lattice and chain, we define Web ranking as a lattice-lattice function between a Web lattice and a chain. We show that ranking is not submodular. Then, global ranking is defined as a lattice-lattice function (i.e., a mapping from the direct product of Web lattices to the chain [0; 1]). It is shown that global ranking is not submodular. Based on the concept of global ranking, we present a method that allows us to compute the global importance of a Web page at Web level, taking into account the importance of the site the page belongs to, but without the need to consider the entire Web graph of all pages.

After proving that any tree as well as any document can be transformed into a lattice, we show that the DocBall model and Galois (concept) lattice representations of a document are equivalent to one another.

Based on these results as well as on the fact that the structure of any site is a lattice, we describe a method for computing site importance.

The chapter ends with exercises and problems designed to promote a deeper understanding of the notions introduced and the application possibilities of the results obtained.

11.1 Web Graph

Let $W_1,\ldots,W_i,\ldots,W_N$ denote a set of Web pages. A directed link from page W_i to page W_j is defined by the fact that the URL of page W_j occurs on page W_i, which is expressed as $W_i \rightarrow W_j$ (see Section 4.8).

A graph $G = (V, E)$ is referred to as a *Web graph* if vertex $v_i \in V$ corresponds to page W_i ($i = 1,\ldots,N$), and a directed edge $(v_i, v_j) \in E$ exists if there is a link from page W_i to page W_j. Graph G can be represented, e.g., by an adjacency matrix $M = (m_{ij})_{N \times N}$ defined as (**Fig 11.1**)

$$m_{ij} = \begin{cases} 1 & W_i \rightarrow W_j \\ 0 & otherwise \end{cases} \tag{11.1}$$

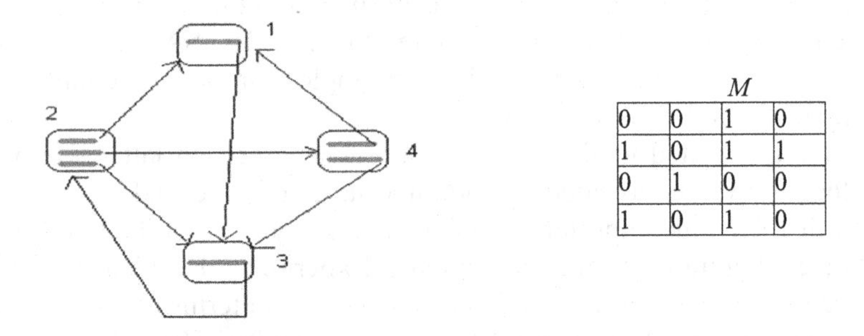

0	0	1	0
1	0	1	1
0	1	0	0
1	0	1	0

Fig. 11.1. A small Web graph G with four pages: 1, 2, 3, and 4. The horizontal bars within each page symbolize URLs indicating links to other pages as shown by the arrows. The corresponding adjacency matrix M is shown on the right.

We note that matrix M is not symmetrical (were the graph undirected, M would be symmetrical). Moreover, it is typically a sparse matrix. This property is important when representing matrix M on computer storage media (e.g., using adjacency lists) and when implementing matrix operations using programming languages.

The number of outgoing links from page W_i is called the *outdegree* of page W_i, and the number of incoming links is called the *indegree* of page W_i. For example, in **Fig 11.1**, the outdegree of page 2 is equal to 3, while its indegree is equal to 1.

Degree Distribution. The experimental discovery by Faloutsos et al. (1999) that the degree distribution for Web pages (and also Internet nodes) follows a power law with a fairly robust degree exponent was a basic milestone toward revealing the properties of the Web graph.

Kahng et al (2002) investigated the question of why the degree exponent, and especially that for indegree, exhibits a fairly robust behavior, just above the value 2. Using a directed network model in which the number of vertices grows geometrically with time and the number of edges evolves according to a multiplicative process, they established the distribution of in- and outdegrees in such networks. They arrived at the result that if the degree of a vertex grows more rapidly than the number of edges, then the indegree distribution is independent of the 'details' of the network.

We recall briefly the notion and technique of a power law (from a slightly different point of view than that used in Chapter 4). Given a discrete random variable $V = V_1, V_2, \ldots, V_n$, if the probability P that the random variable V assumes values equal to or greater than some value v is given by

$$P(V \geq v) = \left(\frac{m}{v}\right)^k, \tag{11.2}$$

where $m > 0$, $k > 0$, and m and k are—problem-dependent—constants, $v \geq m$, then we say that V follows Pareto's law. For example, indivduals' incomes obey Pareto's law (Guilmi et al. 2003); m represents a minimal income. It follows from Eq. (11.2) that

$$P(V < v) = 1 - \left(\frac{m}{v}\right)^k, \tag{11.3}$$

which is the distribution function $F(v)$ of V. A function P as defined in Eq. (11.3) for real values of v is differentiable with respect to v, and the derivative is continuous. Thus, it is absolutely continuous, and hence the random variable V has density function $f(v)$ given by the derivative F', i.e., $f(v) = F'(v) = m^k \cdot v^{-(k+1)}$. The function $f(v)$ is referred to as a power law and is usually written in the following general form:

$$f(v) = C \cdot v^{-\alpha}, \tag{11.4}$$

where C is a—problem-dependent—constant, and α is referred to as the exponent of the power law, or degree exponent. The power law can be used to describe phenomena with frequent small events and rare large ones (Adamic 2003). For visualization purposes, the power law is represented in a log-log plot, i.e., as a straight line obtained by taking the logarithm of Eq. (11.4):

$$\log f(v) = \log C - \alpha \times \log v; \tag{11.5}$$

$\log v$ is represented on the abscissa and $\log f(x)$ on the ordinate; $-\alpha$ is the slope of the straight line and $\log C$ is its intercept.

Given a sequence of values $X = (x_1,\ldots,x_i,\ldots,x_n)$ on the abscissa and another sequence of values $Y = (y_1,\ldots,y_i,\ldots,y_n)$ on the ordinate, if the correlation coefficient $r(X, Y)$ suggests a fairly strong correlation between X and Y at a log scale, then a regression line can be drawn to exhibit a relationship between X and Y. Using the slope and the intercept of the regression line, we can write the corresponding power law.

Thus far, the following values for the degree exponent were obtained experimentally:

1. Faloutsos et al. (1999), using data provided by the National Laboratory for Applied Networks Research between the end of 1997 and end of 1998, arrived at the result that the tail of the frequency distribution of outdegree—i.e., the number of Internet nodes and Web pages with a given outdegree—is proportional to a power law. Their observation was that the values of the exponent seemed to be almost constant: 2.15, 2.16, 2.2, 2.48.

2. Barabási et al. (2000)—using 325,729 HTML pages involving 1,469,680 links from the *nd.edu* domain—confirmed the earlier results obtained for the values of the degree exponent. They obtained the values 2.45 for outdegree and 2.1 for indegree.

3. Broder et al. (2000) and Strogatz (2001) describe two experiments using two Web crawls, one in May and another one in October 1999, provided by Altavista, involving 200 million pages and 1.5 billion links. The results they arrived at were the same in both experiments: the values of the degree exponent were estimated to be 2.1, 2.54, 2.09, 2.67, 2.72 for outlink distribution.

4. The values obtained earlier for the degree exponent were also confirmed by Pennock et al. (2002), who found—using 100,000 Web pages selected at random from 1 billion URLs of Inktomi Corporation Webmap—that the exponent for outdegree was 2.72 and 2.1 for indegree. Similar exponent values were obtained for the indegree distribution for category-specific homepages: 2.05 for companies and newspapers, 2.63 for universities, 2.66 for scientists, and 2.05 for newspapers.

5. Shiode and Batty (2000) assessed the power law for Web country domain names in- and outlink distribution as of 1999. Their results for the power law exponent were 2.91, 1.6, 2.98, 1.46, 2.18, 2.

6. Adamic and Huberman (2000) report on an experiment involving 260,000 sites, each representing a separate domain name. The degree exponent was estimated to be 1.94.

7. Kumar et al. (1998) report that a copy of the 1997 Web from Alexa (a company that archives the state of the Web) was used to estimate the degree exponent of the power law. The data consisted of about 1 terabyte of data representing the content of over 200 million Web pages. It was found that the degree exponent was 2.38.

8. Albert (2000) reports that the value of 2.3 was found for the degree exponent.

9. Experiment 1. Using the Barabási data,[1] the power law for outdegree distribution was assessed (Dominich et al. 2005). The data were provided as a zipped file. After unzipping it, the result was a text file that contained two numbers in each line: the leftmost number was the sequence number of Web pages (0; 1; 2;...; 325,729), and the other number was the sequence number of the Web page pointed to by the page represented by the leftmost number. A noteworthy observation is that the exponent of the Web power law increases slowly from 1 with the number of pages (from a few hundred up to several tens of thousands of pages) and starts to stabilize around the value $\alpha = 2.5$ if the number of Web pages involved is fairly high—above 100,000. Thus, e.g., for 30,000 pages, the correlation—at a log scale—r between outdegree and frequency was only $r = -0.892$, and the fitting of a power law curve $C \cdot x^{-\alpha}$ using MathCAD's built-in curve fitting command *genfit* resulted in $\alpha = 0.867$, with an approximation error of the sum of the absolute values of differences of 3.7×10^6 at 10^{-4} convergence error, whereas using linear regression yielded $\alpha = 1.47$, with an approximation error of 1,589,104 at 10^{-4} convergence error.

   Figure 11.2 shows the results for 256,062 Web pages—involving 1,139,426 links—selected at random from the 325,729 pages that were provided. After processing this file, the X data consisted of the outdegrees of Web pages and the Y data consisted of the corresponding frequencies. For example, there were 2206 pages having outdegree 13, and outdegree 14 had a frequency equal to 1311. The empirical correlation coefficient—taking log scale data—r between outdegree and frequency was $r = -0.94$. The linear regression method yielded the following values: $\alpha = 2.5$ for the exponent and $C = 10^{6.1043}$

[1] Provided at http://www.nd.edu/~networks/database/index.html; downloaded January 2, 2004.

for the constant. The computation was performed using MathCAD's built-in *line* command. The numeric computation used in this command, as well as the fact that we used 69,667 fewer pages may account for the difference of 0.05 in the exponent value compared to the value reported in Barabási et al. (2000). Owing to the strong correlation (see above) as well as to inherently present numeric approximation errors, we believe that the 0.05difference is not important, that the values obtained in our experiment do confirm the earlier results, and further that the power law characterizes the behavior of the Web at very large scale. Hence, our experiment confirmed the earlier results.

10. Experiment 2. The power law for Web country domain names inlink distribution was assessed as of 2004 (Dominich et al 2005). The inlink frequency distribution for country domain names[2] was generated as of January 2004 (**Fig. 11.3**). The domain names *.gov*, *.org, .net, .edu, .us, .com, .mil, .um, .vi* were all considered as representing the United States, whereas the domain names *.ac*, *.uk, .gb* represented the United Kingdom, and *.fr*, *.fx* France. This yielded 238 country domain names (88 domain names more than 5 years earlier). The number of inlinks for every country domain name was identified using Altavista search engine's Webmasters option during January 19–22, 2004. For example, the United Kingdom had a total of 30,701,157 inlinks, and the United States had 271,019,148 inlinks in all. The inlinks were binned into 1000 equally spaced intervals. In this case, the correlation between the number of inlinks and the corresponding number of country domain names was found to be –0.99 (at a log scale). The value for the power law exponent was found to be equal to $\alpha = 1.18$ using MathCAD's *linfit* linear regression command (the approximation error was equal to 14,509).

11. Experiment 3. The inlinks frequency distribution for the 43 state university domain names in Hungary was generated as of January 2004 (Dominich et al 2005). The number of inlinks for every domain name was identified using Altavista's Webmasters option during January 2004. The inlinks were binned into 140 equally spaced intervals. In this case, the correlation between the number of inlinks and the corresponding number of domain names was found to be –0.92 (at a log scale). The value for the power law exponent was found to be equal to $\alpha = 1.15$ using MathCAD's *genfit* curve fitting command

[2] Taken from http://www.webopedia.com/quick_ref/topleveldomains.

to fit the power law curve with an approximation error equal to 14.4 at a convergence error of 10^{-4}.

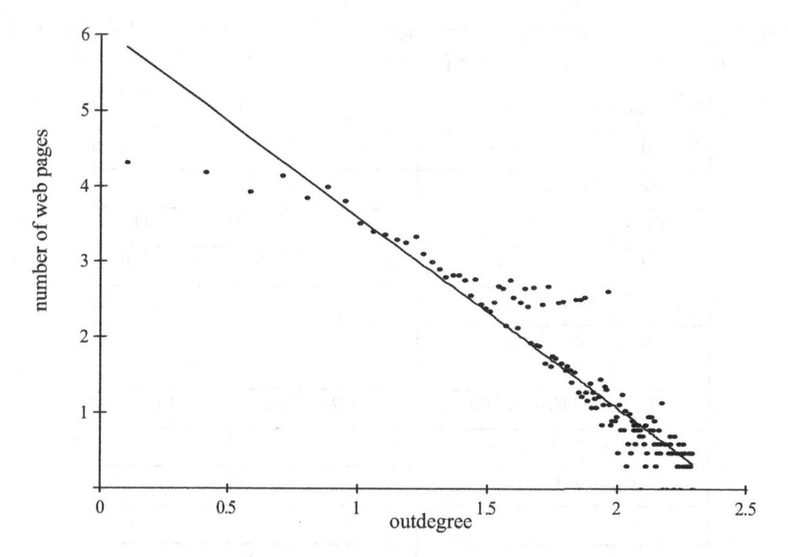

Fig. 11.2. World Wide Web power law. The frequency (i.e., number of Web pages) of the outdegrees of Web pages plotted on a log-log scale. The points represent actual values; the straight line represents the regression line fitted to the real values. The correlation coefficient is equal to $r = -0.94$, the power law exponent is equal to $\alpha = 2.5$.

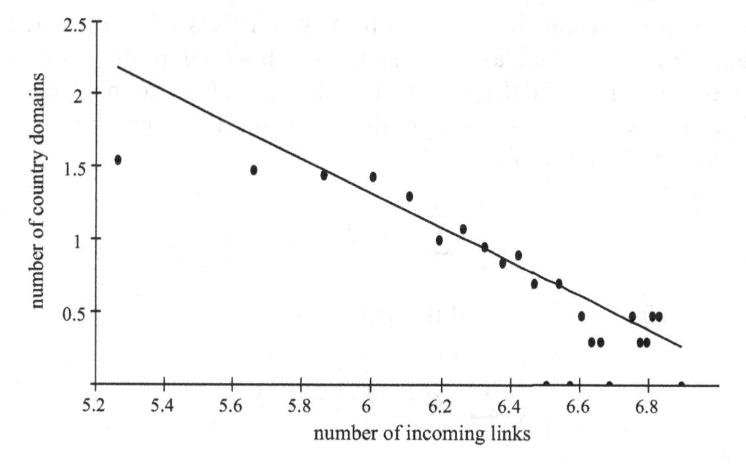

Fig. 11.3. Log-log plot of the power law for the inlinks of country domain names as of January 2004. The correlation between the number of inlinks and the corresponding number of country domain names was found to be –0.99, whereas the value of the power law exponent was 1.18.

The estimated values obtained experimentally for the exponent of the power law for degree distribution in the Web are summarized in **Table 11.1**.

Table 11.1. Estimated Values Obtained Experimentally for the Exponent of the Power Law for Degree Distribution in the World Wide Web

Source (experiment)	**Degree exponent value**
Faloutsos et al. (1999)	2.15; 2.16; 2.2; 2.48
Barabási et al. (2000)	2.1; 2.45
Broder et al. (2001)	2.1; 2.72; 2.09; 2.67; 2.54
Pennock et al. (2002)	2.1; 2.72, 2.05; 2.05; 2.63; 2.66
Kumar et al. (1998)	2.38
Adamic and Huberman (2000)	1.94
Shiode and Batty (2000)	2.91; 1.6; 2.98; 1.46; 2.18; 2
Albert (2000)	2.3
Experiment 1	2.5
Experiment 2	1.18
Experiment 3	1.15

Let us consider the different degree exponent values obtained experimentally as being a sample drawn from a population consisting of degree exponent values. The population may consist, e.g., of the degree exponent values obtained using the data of all Web crawlers (search engines); all domain names, as well as their subsets containing more than 100,000 pages; or a population defined in some other way. Our sample has size $N =$ 29. The following test of the hypothesis for the mean can be performed. The mean M of the sample is

$$M = \frac{1}{N}\sum_{i=1}^{N}\alpha_i = 2.222, \tag{11.6}$$

and the standard deviation s of the sample is

$$s = \sqrt{\frac{1}{N}\sum_{i=1}^{N}(\alpha_i - M)^2} = 0.451. \tag{11.7}$$

Using the χ^2-test (with $\nu = N - 1 = 28$ degrees of freedom), we estimate the population standard deviation σ to lie in the interval

$$\frac{s\sqrt{N}}{\chi_{0.975}} < \sigma < \frac{s\sqrt{N}}{\chi_{0.025}}, \quad \text{i.e., } 0.363 < \sigma < 0.621, \tag{11.8}$$

with 95% confidence. Furthermore, we can check that deviation σ is estimated to lie in the interval $0.35 < \sigma < 0.66$ with 99% confidence. As all of the degree exponent values α_i lie in the open interval (1; 3), i.e., $1 < \alpha_i < 3$, $i = 1,\ldots,N$, the mean, whether sample or population ('true') mean, should also lie in this same interval. We may ask ourselves whether there exist positive integer numbers p such that the null hypothesis H_0, "$\mu = \sqrt{p}$," is supported. Possible values for p are the numbers 4, 5, 6, 7, and 8. Using the z-score (μ)

$$z\text{-score}(\mu) = \left| \frac{M - \mu}{\frac{s}{\sqrt{N}}} \right|, \tag{11.9}$$

we obtain the following z-score(μ) values:

- z-score($\sqrt{4}$) = 2.657.
- z-score($\sqrt{5}$) = 0.163.
- z-score($\sqrt{6}$) = 2.712.
- z-score($\sqrt{7}$) = 5.056.
- z-score($\sqrt{8}$) = 7.238.

As only one of these z-score(μ) values does not exceed 1.96, namely z-score($\sqrt{5}$), we reject the hypothesis H_0 for p = 4, 6, 7, and 8, and accept H_0 for $p = 5$ with 95% confidence. Similarly, only z-score($\sqrt{5}$) is less than 2.58, which means that we may accept H_0 for $p = 5$ with 99% confidence. Thus, we may say that there is statistical support to assume that the sample comes from a population with mean $\mu = \sqrt{5}$. Thus, the power law for the degree distribution in the World Wide Web may be written in the following form:

$$f(x) \approx C \cdot x^{-\sqrt{5}}, \tag{11.10}$$

where $f(x)$ denotes (approximate values of) the frequencies of the nodes with degree x. This demonstrates the "robustness," observed earlier in experiments concerned with the exponent of the power law for the Web.

Note: An interesting property, related to number theory is the connection between Web power law (11.10) and the golden section. As it is known, the golden section (also known as the golden ratio, golden mean, divine proportion) is usually denoted by φ and is defined as the smallest root of the equation $x^2 - x - 1 = 0$, $\varphi = (\sqrt{5} - 1)/2 \approx 0.61803398875$ (the other root is $\Phi = (\sqrt{5} + 1)/2 \approx 1.61803398875$). It is easy to see that the following relationships hold: $\sqrt{5} = 2\varphi + 1$, and $\varphi\Phi = 1$. A direct numerical connection between the degree exponent as defined in Eq. (11.10) and the golden section is $2\varphi + 1 = \sqrt{5}$. Moreover, the connection with Fibonacci numbers can be established. Fibonacci numbers are defined recursively: $F_0 = 0$, $F_1 = 1$, $F_n = F_{n-1} + F_{n-2}$, $n \geq 2$. Thus, their sequence is: 0, 1, 1, 2, 3, 5, 8, 13, 21, 34, 55, 89,... . A noteworthy property of these numbers is that, in general, the ratio of the consecutive numbers has a limit equal to the golden section, namely 5/8 = 0.625, 8/13 = 0.615, 13/21 = 0.619, ...,:

$$\lim_{n \to \infty} \frac{F_n}{F_{n+1}} = \varphi .$$

The golden section and the Fibonacci numbers are related by Binet's equation:

$$F_n = \frac{1}{\sqrt{5}}\left(\Phi^n - (-\varphi)^n\right),$$

from which we can express $\sqrt{5}$.

11.2 Link Structure Analysis

Link structure analysis (*link analysis*, for short) refers to methods used to quantify the importance of networked entities of interest based on the number of links (connections) among them. Entities may be:

- Social objects (e.g., groups of people).
- Written units (e.g., scientific papers).
- Web pages.
- Molecules.
- And so on.

The starting point of link analysis was *citation analysis*, whose principle is as follows: the number of citations a paper gets from other papers is a measure of its importance (Garfield 1955, 1972).

This principle was applied to compute an *impact factor* for journals. For example, the impact factor IF for journal J in 2007 can be calculated as follows:

Impact Factor Method

$$IF = \frac{C}{P},$$

where C is the number of times J's articles published in 2005 and 2006 were cited in other journals during 2007, and P is the number of articles published in J during 2005 and 2006.

The impact factor is based merely on a pure count of links; no other factor (e.g., quality, importance) is taken into account.

The principle of citation analysis was applied for the first time by Carrière and Kazman (1997) for Web retrieval in the following form:

Connectivity Method

1. Using the Boolean retrieval method, we first obtain a list of Web pages (hit list).
2. The Web graph for the hit list is constructed.
3. For each node in the graph, its *connectivity* (i.e., the sum of its indegree and outdegree) is computed.
4. Finally, the hit list is sorted on node connectivity and presented in decreasing order.

Pinski and Narin enhanced the connectivity method by noting that not all citations have equal importance. They argued that a journal is important if it gets citations from other important journals (Geller 1978). The mutual citation method proposed is as follows:

Mutual Citation Method

Let $J_1,\ldots,J_i,\ldots,J_n$ denote entities. A matrix $M = (m_{ij})_{n\times n}$ is constructed:

$$m_{ij} = \frac{c_i^j}{c_i},$$

where c_i denotes the total number of citations in journal J_i, while c_i^j denotes the number of citations journal J_j gets (out of c_i) from journal J_i. The importance vector $\mathbf{w} = (w_1 \ldots w_n)$ of journals is the solution of

$$\mathbf{w} = M^T\mathbf{w}.$$

In other words, the importance vector **w** is the eigenvector corresponding to eigenvalue 1 of matrix M^T.

Example 11.1

Let us consider the following small Web graph:

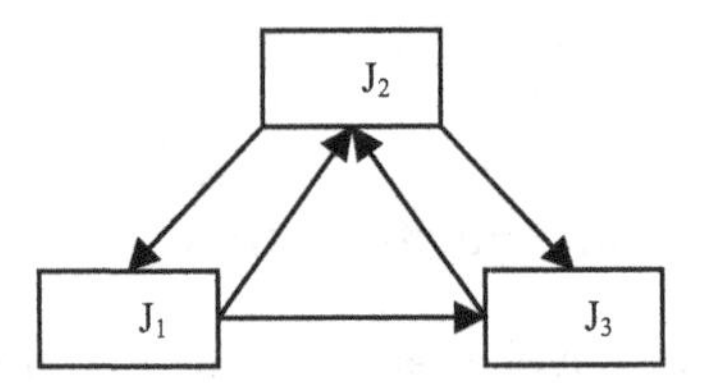

Matrix M is

$$\begin{bmatrix} 0 & \frac{1}{2} & \frac{1}{2} \\ \frac{1}{2} & 0 & \frac{1}{2} \\ 0 & 1 & 0 \end{bmatrix}$$

The importance w_1 of page J_1 is equal to

$$w_1 = 0\cdot w_1 + 0.5\cdot w_2 + 0\cdot w_3,$$

where (0 0.5 0) is the first column of matrix M. Thus, the importance vector $\mathbf{w} = [w_1\ w_2\ w_3]^T$ is given by the equation $\mathbf{w} = M^T\mathbf{w}$ and is equal to $\mathbf{w} = [0.371; 0.743; 0.557]^T$. □

The ideas behind these methods were precursors of other techniques for computing the importance of Web pages. The basic methods are presented in the next sections.

11.3 The PageRank Method

In the PageRank method, a Web page's importance is determined by the importance of Web pages linking to it. Brin and Page (1998) define the PageRank value R_i of a Web page W_i using the equation

$$R_i = \sum_{W_j \in B_i} \frac{R_j}{L_j}, \tag{11.11}$$

where L_j denotes the number of outgoing links (i.e., URLs) from page W_j and B_i denotes the set of pages W_j pointing to page W_i.

Equation (11.11) is a homogeneous and simultaneous system of linear equations in the unknown R_i, $i = 1,\dots,N$, which always has trivial solutions (the null vector, i.e., $R_i = 0$, $i = 1,\dots,N$).

Equation (11.11) also has nontrivial solutions if and only if its determinant is equal to zero. Let $G = (V, A)$ denote (correspond to) a Web graph, where the set $V = \{W_1,\dots,W_j,\dots,W_N\}$ of vertices denotes the set of Web pages. The set A of arcs consists of the directed links (given by URLs) between pages.

Let $M = (m_{ij})_{N\times N}$ denote a square matrix (modified adjacency matrix) attached to graph G such that (**Fig. 11.4**)

$$m_{ij} = \begin{cases} \dfrac{1}{L_j} & W_j \to W_i \\ 0 & otherwise \end{cases}. \tag{11.12}$$

As the elements of matrix M are the coefficients of the right-hand side of Eq. (11.11), this can be rewritten in matrix form:

$$M \times R = R, \tag{11.13}$$

where R denotes the vector (i.e., column matrix) of PageRank values, i.e.,

$$R = \begin{bmatrix} R_1 \\ . \\ R_i \\ . \\ R_N \end{bmatrix} = [R_1, \ldots, R_i, \ldots, R_N]^{\mathrm{T}}. \tag{11.14}$$

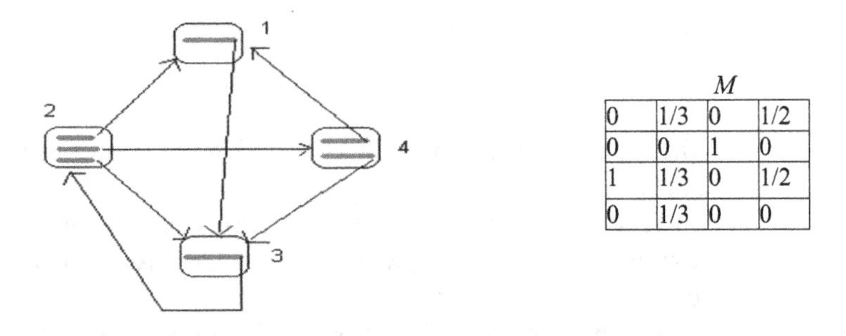

0	1/3	0	1/2
0	0	1	0
1	1/3	0	1/2
0	1/3	0	0

Fig. 11.4. A small Web graph G with four pages: 1, 2, 3, and 4. The elements of matrix M are also shown; they were computed using Eq. (11.12).

If graph G is strongly connected (i.e., every node can be reached from every other node following directed links), the sums of the columns in matrix M are equal to 1. Thus, because matrix M has only zeroes in the main diagonal, in matrix $M - I$ (I denotes the unity matrix), i.e.,

$$M - I = \begin{bmatrix} -1 & \ldots & m_{1N} \\ . & . & . \\ m_{N1} & \ldots & -1 \end{bmatrix}, \tag{11.15}$$

the sums of columns is equal to zero. Let D denote its determinant:

$$D = |M - I|. \tag{11.16}$$

If every element of, e.g., the first line of D is doubled, we obtain a new determinant D', and we have $D' = 2 \times D$. We now add every line to the first line in D'. As the sums of the columns in D are null, it follows that (after these additions) the first row of determinant D' will be equal to the first row of determinant D. Thus, we have

$$D' = 2D = D, \tag{11.17}$$

from which it follows that $D = 0$. Since matrix $M - I$ is exactly the matrix of Eq. (11.11), it follows that it also has nontrivial solutions.

The determinant $|M - I|$ being equal to 0 means that the number 1 is an eigenvalue of matrix M. Moreover, the number 1 is a *dominant eigenvalue* of matrix M, i.e., it is the largest eigenvalue in terms of absolute value (Farahat et al 2006). **Figure 11.5** shows an example for the Web graph of **Fig. 11.4**.

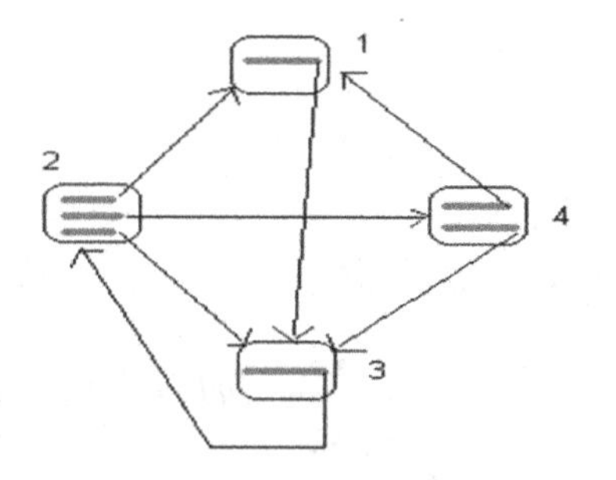

M				R
0	1/3	0	1/2	0.325
0	0	1	0	0.651
1	1/3	0	1/2	0.651
0	1/3	0	0	0.217

Fig. 11.5. A small Web graph G with four pages: 1, 2, 3, and 4. The elements of matrix M are also shown; they were computed by $m_{ij}=1/L_j$. The PageRank values, i.e., the eigenvector corresponding to eigenvalue 1, were computed using the MathCAD command "eigenvec(M,1)."

Owing to the fact that N is large, PageRank values are computed in practice using some numeric approximation procedure by calculating the eigenvector R corresponding to eigenvalue 1. The following approximation method can be used:

$$M \times R_k = R_{k+1},\ k = 0, 1,\ldots,K,$$
$$R_0 = \left[\frac{1}{N} \quad \cdots \quad \frac{1}{N}\right], \tag{11.18}$$

where K is equal to a few tens (typically to 50), or the recursive computation is performed until

$$\max|\, R_{k+1} - R_k \,| < \varepsilon, \tag{11.19}$$

where $\varepsilon \in \mathbb{R}_+$ is some preset error threshold. The approximation obtained is an eigenvector whose elements sum to unity.

Example 11.2

For the Web graph of **Fig. 11.5**, the numerical approximation procedure yields the following PageRank values for $\varepsilon = 0.1$ and after $k = 9$ steps: $[0.177; 0.353; 0.35; 0.12;]^T$. □

Equations (11.18) are derived from the well-known power method used to compute the dominant eigenvector **x** of a matrix M, in general. The steps of the power method are as follows:

POWER METHOD

1. Choose an initial vector $\mathbf{x}_0$.
2. Set $i = 1$.
3. Calculate the next approximation $\mathbf{x}_{i+1}$ as $\mathbf{x}_{i+1} = M\mathbf{x}_i$.
4. Divide $\mathbf{x}_{i+1}$ by its Euclidean norm, i.e., $\mathbf{x}'_{i+1} = \dfrac{\mathbf{x}_{i+1}}{\|\mathbf{x}_{i+1}\|}$. *Note:* One may divide by any nonzero element of $\mathbf{x}_{i+1}$.
5. Repeat steps 3 and 4 until $error(\mathbf{x}'_i, \mathbf{x}'_{i+1}) < \varepsilon$, where $error(\mathbf{x}'_i, \mathbf{x}'_{i+1}) = \|\mathbf{x}'_i - \mathbf{x}'_{i+1}\|$, or $error(\mathbf{x}'_i, \mathbf{x}'_{i+1}) = \max|\mathbf{x}'_i - \mathbf{x}'_{i+1}|$, or some other expression (as more appropriate for the application being considered).
6. The dominant eigenvector can be approximated by the Rayleigh quotient: $\dfrac{\mathbf{x}^T M\mathbf{x}}{\mathbf{x}^T\mathbf{x}}$.

For a real portion of the Web, graph G is not always strongly connected. For example, it may happen that a page W_j does not have any outgoing links (i.e., its outdegree is null). Such a page is referred to as a *dangling page*. In such a case, the jth column—corresponding to page W_j—of matrix M contains only zeroes. The elements of matrix M may be interpreted in the following way: the entry m_{ij} is the probability with which page W_i follows page W_j during a walk on the Web (i.e., the probability with which, during a navigation on the Web, a surfer jumps from page W_j to page W_i). Based on this interpretation, a new matrix, denoted by M', can be constructed:

1. First, the columns corresponding to dangling nodes in matrix M are replaced by columns containing all $1/N$, i.e.,

$$m'_{ij} = \frac{1}{N}, \quad i = 1,\ldots,N, \text{ page } W_j \text{ is a dangling page.} \tag{11.20}$$

2. Second, using matrix M', we have a new matrix M'':

$$M'' = \alpha M' + (1 - \alpha)/N, \quad 0 < \alpha < 1. \tag{11.21}$$

A typical value for α is $\alpha = 0.85$. Thus, the PageRank equation becomes

$$M'' \times R = R. \tag{11.22}$$

Matrix M'' is nonnegative (i.e., its elements are nonnegative numbers); hence it has a nonnegative dominant eigenvalue (Farahat et al 2006). The corresponding eigenvector is the PageRank vector: it is unique, its entries are nonnegative, and it can be calculated using the approximation (or power) method given by Eq. (11.18).

Example 11.3

Let us assume that in **Fig. 11.5**, page W_3 is a dangling page. Then,

$$M' = \begin{pmatrix} 0 & 0.333 & 0.25 & 0.5 \\ 0 & 0 & 0.25 & 0 \\ 1 & 0.333 & 0.25 & 0.5 \\ 0 & 0.333 & 0.25 & 0 \end{pmatrix}.$$

Application of the PageRank Method in Web Retrieval. The PageRank method is being used by the Web search engine Google. **Figure 11.6** shows the query interface and a portion of the hit list for the query "lattice information retrieval" (as of the May 2, 2007).

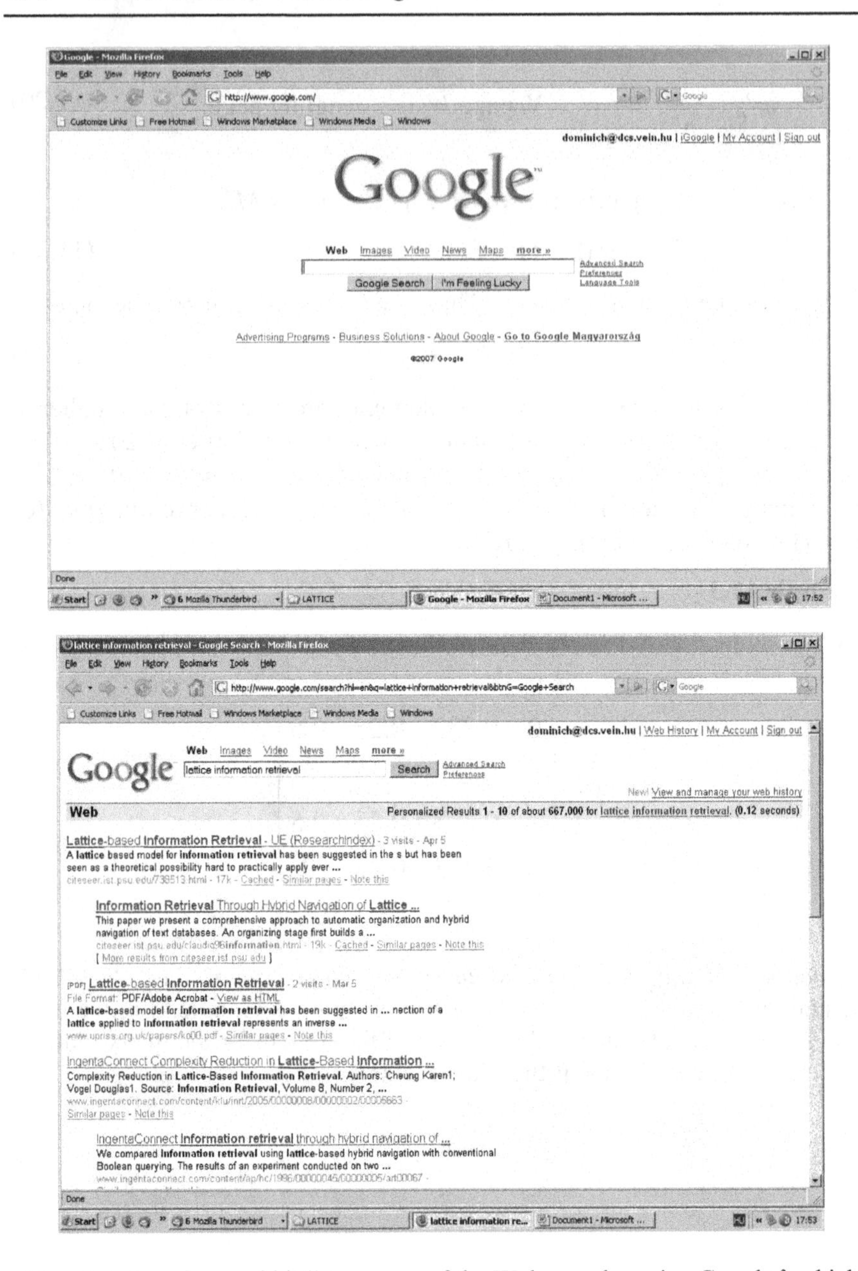

Fig. 11.6. Interface and hit list screens of the Web search engine Google,[3] which is using the PageRank method.

[3] http://www.google.com.

11.4 The HITS Method

A method called HITS for computing hubs and authorities was proposed by Kleinberg (1999). Two types of Web pages are defined: hubs and authorities. They obey a mutually reinforcing relationship, i.e., a Web page is referred to as

- An *authority* if it is pointed to by many *hub* pages.
- A *hub* if it points to many *authoritative* pages (**Fig. 11.7a**).

Given a page p, an authority weight $x^{<p>}$ and a hub weight $y^{<p>}$ is assigned to it. If p points to pages with large x-values, then it receives large y-values, and if p is pointed to by pages with large y-values, then it should receive a large x-value.

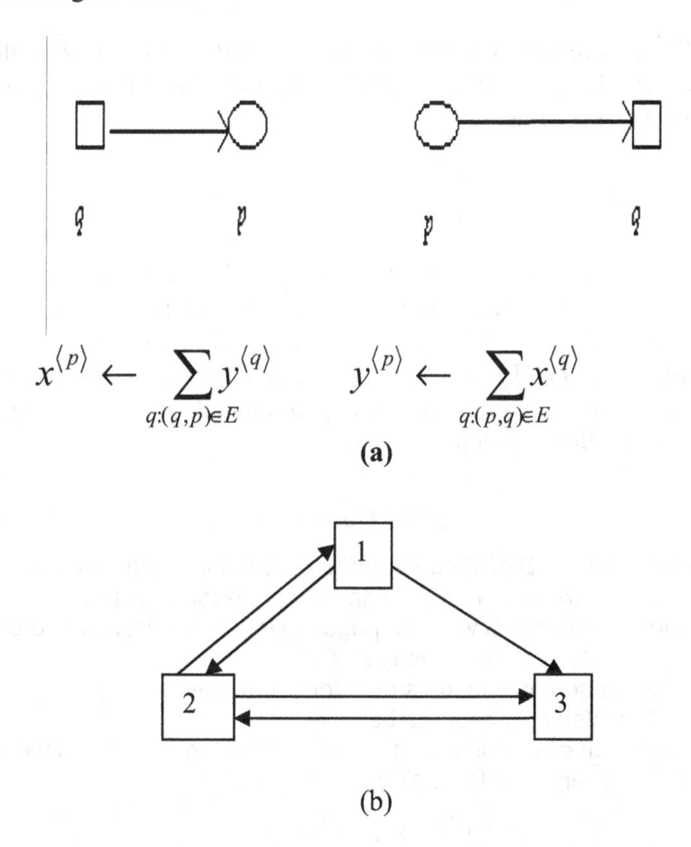

Fig. 11.7(a) Illustration of operations for computing hubs and authorities. **(b)** Mini-Web (example).

The following iterative operations are defined:

$$x^{\langle p\rangle} \leftarrow \sum_{q:(q,p)\in E} y^{\langle q\rangle}$$
$$y^{\langle p\rangle} \leftarrow \sum_{q:(p,q)\in E} x^{\langle q\rangle}, \qquad (11.23)$$

where E denotes the set of arcs of the Web graph. Let M denote the adjacency matrix of the Web graph of interest. Equations (11.23) can then be written in matrix form:

$$x^{\langle k\rangle} = M^{\mathrm{T}} M x^{\langle k-1\rangle},$$
$$y^{\langle k\rangle} = M M^{\mathrm{T}} y^{\langle k-1\rangle}. \qquad (11.24)$$

Matrix $M^{\mathrm{T}}M$ is referred to as the *hub matrix*, while matrix MM^{T} is the *authority matrix*. Thus, the HITS method is equivalent to solving the following eigenvector problems:

$$M^{\mathrm{T}}Mx = \lambda x,$$
$$MM^{\mathrm{T}}y = \lambda y, \qquad (11.25)$$

where λ denotes the dominant eigenvalue of $M^{\mathrm{T}}M$ (MM^{T}). An entry m_{ij} in matrix MM^{T} is equal to the number of pages to which both pages i and j point. An entry m_{ij} in matrix $M^{\mathrm{T}}M$ is equal to the number of pages that point to both pages i and j. A diagonal entry (i, i) in matrix MM^{T} represents the outdegree of page i. In order to compute the authority and hub vectors in practice, the following steps are taken:

HITS Method

1. Define a root set S of Web pages (e.g., submit a query on some topic to a commercial search engine and keep the top L hits).
2. Expand the root set S with the pages given by the inlinks and outlinks of pages in S to obtain a base set T.
3. Eliminate pages having the same domain name.
4. Define the Web graph for the base set T.
5. Repeat a sufficient number of iterations starting with the initial values $x_0 = [1,\ldots,1]^{\mathrm{T}}$ and $y_0 = [1,\ldots,1]^{\mathrm{T}}$ for both x and y:

$$x_{i+1} = M^{\mathrm{T}} y_i,\ \ y_{i+1} = M x_{i+1};$$

(After each iteration, vectors x and y are normalized such that the squares of their entries sum to 1; this operation is called length normalization.)

It can be shown that x is the dominant eigenvector of M^TM, and y is the dominant eigenvector of MM^T (Farahat et al. 2006).

Example 11.4

Let

$$M = \begin{pmatrix} 0 & 1 & 1 \\ 1 & 0 & 1 \\ 0 & 1 & 0 \end{pmatrix},$$

denote the adjacency matrix of the mini-Web graph of **Fig. 11.7b**. We then have:

$$MM^T = \begin{pmatrix} 2 & 1 & 1 \\ 1 & 2 & 1 \\ 1 & 0 & 1 \end{pmatrix}$$

and

$$M^TM = \begin{pmatrix} 1 & 0 & 1 \\ 0 & 2 & 1 \\ 1 & 1 & 2 \end{pmatrix},$$

with the following eigenvalues: 1.5, 0.2, 3.2. Perform the operations $x_{i+1} = M^Ty_i$ and $y_{i+1} = Mx_{i+1}$ until vectors x and y do not change significantly (convergence). In this example, after three steps, the following values are obtained: $x = [0.309; 0.619; 0.722]^T$ and $y = [0.744; 0.573; 0.344]^T$.

11.4.1 Application of the HITS Method in Web Retrieval

The HITS method is being used in the Web search engine Teoma (Ask).[4] **Figure 11.8** shows the interface and hit list screens for the query "lattice information retrieval" (as of May 2, 2007).

[4] http://www.ask.com.

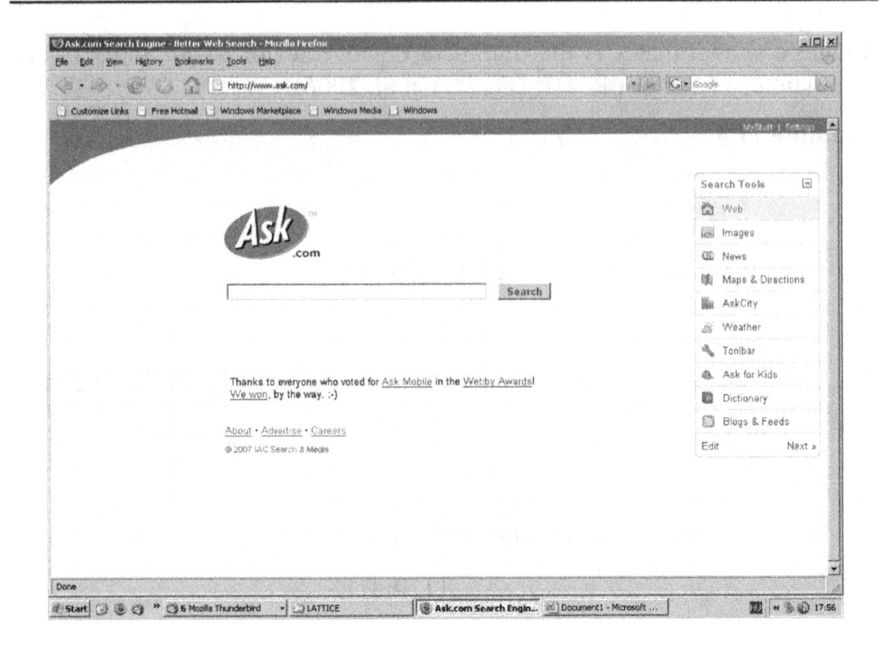

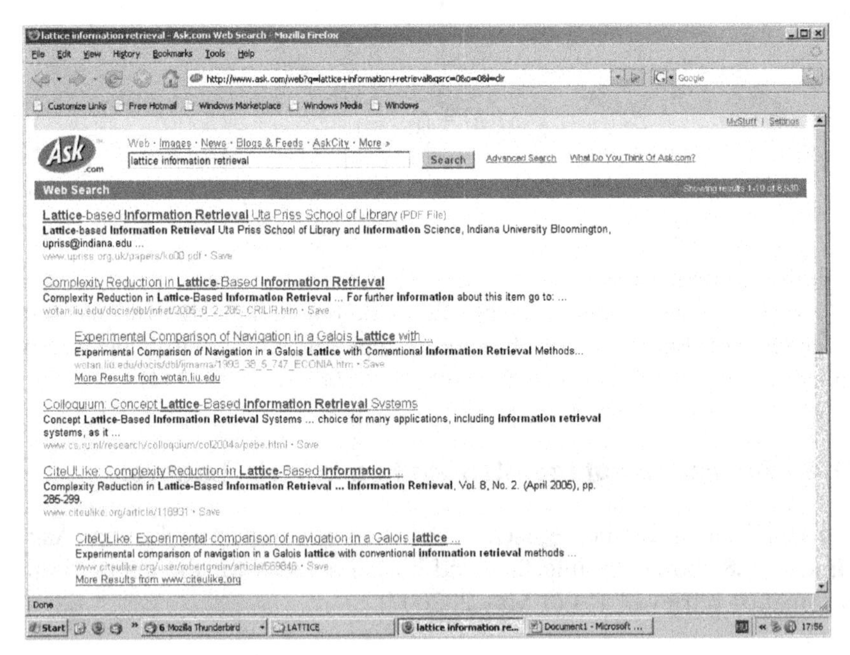

Fig. 11.8. Interface screen and a hit list of the Web search engine Ask (Teoma), which uses the HITS method.

11.4.2 Latent Semantic Indexing and HITS

We recall that the *singular value decomposition* (SVD) of a matrix $A_{m,n}$ (rank(A) = r) is defined as

$$A = UDV^{\mathrm{T}}, \tag{11.26}$$

where $U^{\mathrm{T}}U = V^{\mathrm{T}}V = I_{n,n}$, D is a diagonal matrix $D = \mathrm{diag}(d_1,...,d_n)$, $d_i > 0$, $i = 1,...,r$, and $d_j = 0, j > r$. Matrices U and V are orthogonal, and their first r columns define the orthonormal eigenvectors associated with the r nonzero eigenvalues of AA^{T} and $A^{\mathrm{T}}A$, respectively. The columns of U are called the *left singular vectors*, and those of V the *right singular vectors*. The diagonal elements of D are the nonnegative square roots of the n eigenvalues of AA^{T}, and are referred to as the *singular values* of A.

SVD and HITS are connected in the following way (Ng et al. 2001). Let $D = \{d_1,\ldots,d_j,\ldots,d_m\}$ be a set of documents and $T = \{t_1,\ldots,t_i,\ldots,t_n\}$ be a set of terms. A graph $G = (V, E)$ is constructed:

- The set $V = \{v_1,\ldots,v_i,\ldots,v_n, v_{n+1},\ldots,v_{n+j},\ldots,v_{n+m}\}$ of vertices is such that to every term t_i and every document d_j there is a corresponding vertex v_i and v_{n+j}, respectively, in graph G.
- The set $E = \{e_1,\ldots,e_{ij},\ldots,e_p\}$ of directed edges is constructed so that there is a directed edge from a term vertex v_i to a document vertex v_{n+j} if document d_j contains term t_i.

Let M be the adjacency matrix of the graph G thus constructed. The hub weight vector **y** given by HITS has nonzero elements only for term vertices, whereas the authority weight vector **x** has nonzero elements only for document vertices (since no document vertex links to any vertex; only term vertices link to document vertices). The hub weight vector **y** is equal to the first left singular vector (i.e., the first column from left to right) of matrix U.

Example 11.5

Let $D = \{d_1, d_2\}$ be two documents and $T = \{t_1, t_2, t_3\}$ be three terms such that $t_1, t_2 \in d_1$ and $t_1, t_2, t_3 \in d_2$. The adjacency matrix is

$$M = \begin{bmatrix} 0 & 0 & 0 & 1 & 1 \\ 0 & 0 & 0 & 1 & 1 \\ 0 & 0 & 0 & 0 & 1 \\ 0 & 0 & 0 & 0 & 0 \\ 0 & 0 & 0 & 0 & 0 \end{bmatrix}.$$

The hub weight vector is

$$\mathbf{y} = [-0.657\ -0.657\ -0.369\ 0\ 0]^{\mathrm{T}},$$

while the authority weight vector is

$$\mathbf{x} = [0\ 0\ 0\ -0.615\ -0.788]^{\mathrm{T}}.$$

The SVD of matrix M is $M = UDV^{\mathrm{T}}$, where

$$U = \begin{pmatrix} -0.657 & -0.261 & -0.707 & 0 & 0 \\ -0.657 & -0.261 & 0.707 & 0 & 0 \\ -0.369 & 0.929 & 0 & 0 & 0 \\ 0 & 0 & 0 & 1 & 0 \\ 0 & 0 & 0 & 0 & 1 \end{pmatrix}$$

$$D = \begin{pmatrix} 2.136 & 0 & 0 & 0 & 0 \\ 0 & 0.662 & 0 & 0 & 0 \\ 0 & 0 & 0 & 0 & 0 \\ 0 & 0 & 0 & 0 & 0 \\ 0 & 0 & 0 & 0 & 0 \end{pmatrix}$$

$$V = \begin{pmatrix} 0 & 0 & 1 & 0 & 0 \\ 0 & 0 & 0 & -0.707 & -0.707 \\ 0 & 0 & 0 & -0.707 & 0.707 \\ -0.615 & -0.788 & 0 & 0 & 0 \\ -0.788 & 0.615 & 0 & 0 & 0 \end{pmatrix}$$ □

11.5 The SALSA Method

The SALSA method (Lempel and Moran 2001) offers another computation of authorities and hubs. Let $M = (w_{ij})_{N \times N}$ denote the adjacency matrix of the Web graph of interest. Let $M_r = (r_{ij})$ and $M_c = (c_{ij})$ be the following matrices (Langville and Meyer 2005):

$$r_{ij} = \frac{w_{ij}}{\sum_{j=1}^{n} w_{ij}}; \quad \sum_{j=1}^{n} w_{ij} \neq 0; \quad i, j = 1, \ldots, N$$
$$c_{ij} = \frac{w_{ij}}{\sum_{i=1}^{n} w_{ij}}; \quad \sum_{i=1}^{n} w_{ij} \neq 0; \quad i, j = 1, \ldots, N \tag{11.27}$$

Two matrices, H (*hub matrix*) and A (*authority matrix*), are then introduced:

$$H = M_r \times M_c^{\mathsf{T}}, \quad A = M_c^{\mathsf{T}} \times M_r. \tag{11.28}$$

The hub weights and authority weights are the elements of the dominant eigenvectors of H and A, respectively.

Example 11.6

Using **Fig. 11.7**, we have:

$$M = \begin{pmatrix} 0 & 1 & 1 \\ 1 & 0 & 1 \\ 0 & 1 & 0 \end{pmatrix}, \; M_r = \begin{pmatrix} 0 & 1/2 & 1/2 \\ 1/2 & 0 & 1/2 \\ 0 & 1 & 0 \end{pmatrix}, \; M_c = \begin{pmatrix} 0 & 1/2 & 1/2 \\ 1 & 0 & 1/2 \\ 0 & 1/2 & 0 \end{pmatrix}.$$

Then,

$$H = \begin{pmatrix} 0.5 & 0.25 & 0.25 \\ 0.25 & 0.75 & 0 \\ 0.5 & 0 & 0.5 \end{pmatrix}, \text{ and } A = \begin{pmatrix} 0.5 & 0 & 0.5 \\ 0 & 0.75 & 0.25 \\ 0.25 & 0.25 & 0.5 \end{pmatrix}.$$

The dominant eigenvalue of H is 1 and the hub vector is the corresponding eigenvector: $[0.577\ 0.577\ 0.577]^{\mathsf{T}}$. The dominant eigenvalue of A is 1 and the authority vector is the corresponding eigenvector: $[0.577\ 0.577\ 0.577]^{\mathsf{T}}$.

Originally, the computation method of H and A was as follows (Lempel and Moran 2001). The Web graph $G = (V, E)$ was used to construct a bipartite graph $G' = (V_h, V_a, E')$ where (**Fig. 11.9**):

- $V_h = \{s \mid s \in V, \text{outdegree}(s) > 0\}$, *hub side*.
- $V_a = \{s \mid s \in V, \text{indegree}(s) > 0\}$, *authority side*.
- $E' = \{(s, r) \mid (s, r) \in E\}$.

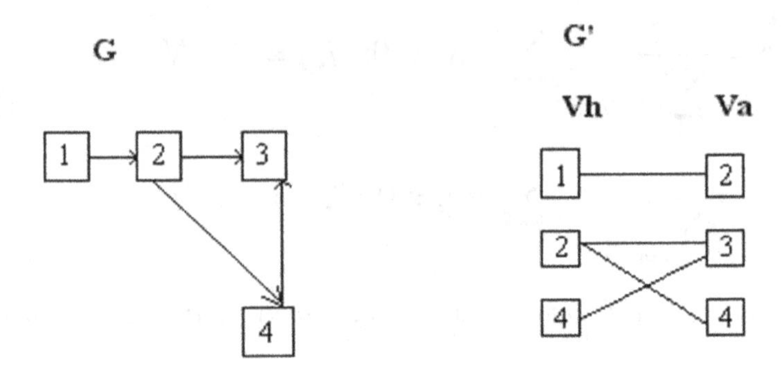

Fig. 11.9. Graph G, bipartite graph G' in the SALSA method.

It was assumed that graph G was connected. (If it is not connected, then the graph G' is constructed for every connected subgraph of G). Matrix $A = (a_{ij})$ was defined as

$$a_{ij} = \sum_{\{k|(k,i),(k,j)\in E'\}} \frac{1}{\deg(i \in V_a)} \times \frac{1}{\deg(k \in V_h)}, \tag{11.29}$$

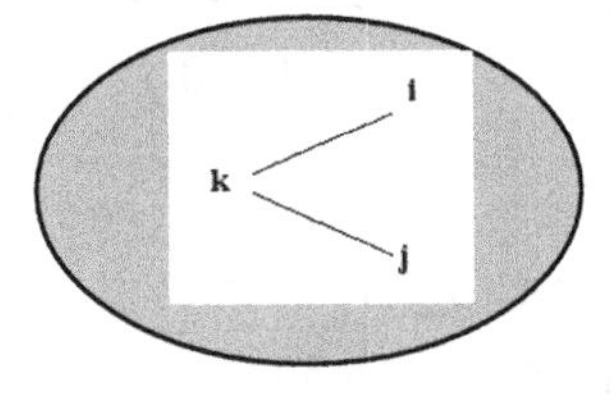

and matrix $H = (h_{ij})$ as

$$h_{ij} = \sum_{\{k|(i,k),(j,k)\in E'\}} \frac{1}{\deg(i \in V_h)} \times \frac{1}{\deg(k \in V_a)}. \tag{11.30}$$

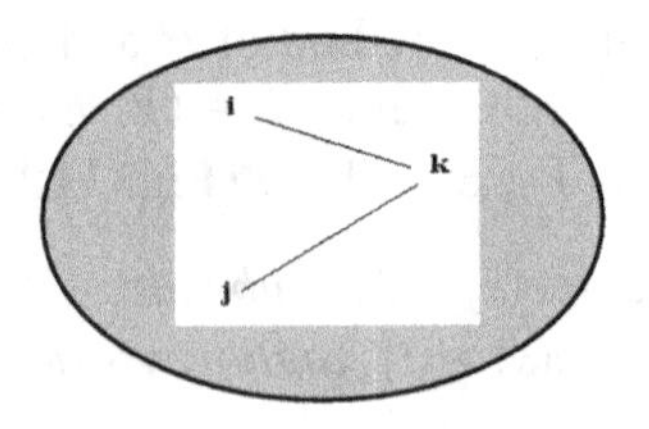

11.6 The Associative Interaction Method

Before we describe the method, we discuss the relevant concepts and results from the theory of artificial neural networks (ANNs).

11.6.1 Artificial Neural Networks

The fundamental principle of artificial neural networks (ANNs), states that the amount of activity of any neuron depends on (James, 1988):

- Its weighted input.
- The activity levels of artificial neurons connecting to it.

An *artificial neuron* is a formal processing unit abstracted from real, biological neurons (**Fig. 11.10a**) (Feldman and Ballard 1982, Grossberg 1976, Hopfield, 1984).

An artificial neuron ν has inputs $I_1,\ldots,I_n$, which can be weighted by the weights $w_1,\ldots,w_n$. The total input I depends on inputs and their weights. The typical form of I is a linear combination of its inputs:

$$I = \sum_{i=1}^{n} I_i w_i \,. \tag{11.31}$$

As a result of total input I, the neuron can take on a *state* (also called an *activation level*) z. State z is a function g of I, $z = g(I)$. For example:

- Threshold function: $z = 1$ if $I > k$, and $z = 0$ if $I \le k$, where k is a threshold value.
- Identity function: $z = I$.

The artificial neuron produces an output O via its *transfer function* f depending on its state z, i.e., $O = f(z)$, e.g.:

- Identity function: $O = f(z) = z$.
- Sigmoid function: $O = f(z) = \dfrac{1}{1+e^{-z}}$.

Artificial neurons can be connected to form an ANN (**Fig. 11.10b**). Given two interconnected neurons ν_i and ν_j in an ANN, the output $f_j(z_j)$ of ν_j can be transferred to ν_i via the connection between them, which can alter $f_j(z_j)$ by a weight w_{ij}. The quantity $w_{ij} \times f_j(z_j)$ reaches artificial neuron ν_i, for which it is an input.

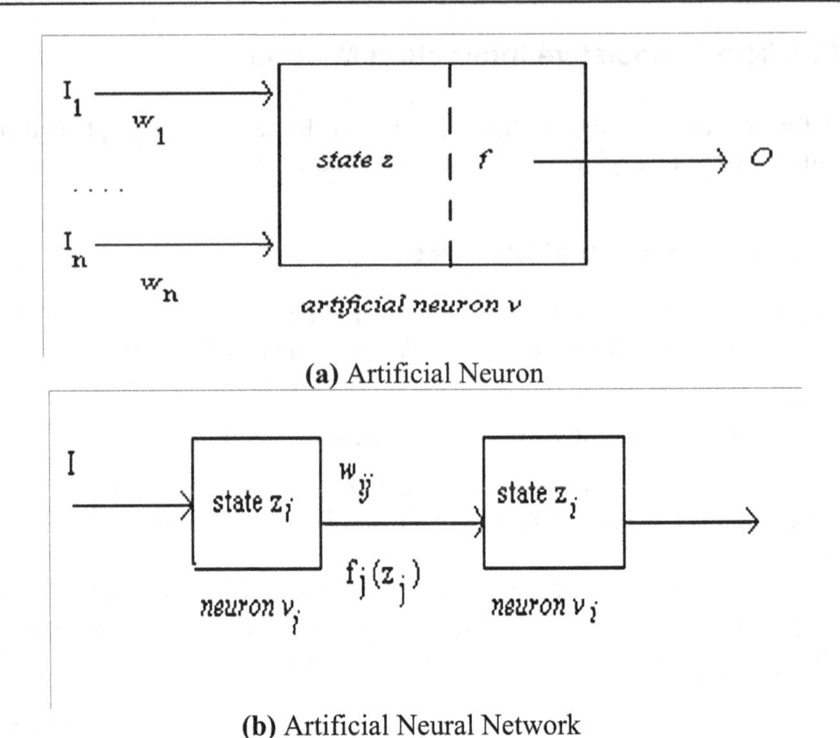

Fig. 11.10(a) An artificial neuron:.a linear combination of the weighted (w_i) inputs (I_i) activates neuron v, which takes on a state z and produces an output O via its transfer function f. **(b)** ANN: interconnected artificial neurons (ν_j, ν_i); I is an input to neuron ν_j and $f_j(z_j)$ is its output, which is an input to neuron ν_i weighted by the quantity w_{ij}, i.e., $w_{ij} \cdot f_j(z_j)$.

The state z_i of neuron ν_i can be described by the following generic differential equation (DeWilde, 1996):

$$\frac{dz_i(t)}{dt} = -z_i(t) + \sum_{j=1}^{n} f_j(w_{ij}, z_j(t), z_i(t)) + I_i(t), \tag{11.32}$$

where:

- t denotes time.
- $z_i(t)$ denotes the activity level of neuron ν_i.
- w_{ij} denotes the weight of a link from neuron ν_j to neuron ν_i.
- $I_i(t)$ denotes external input to neuron ν_i.
- $f_j(z_j(t), w_{ij}, z_i(t))$ denotes the influence of neuron ν_j on neuron ν_i.

Equation (11.32) is a generic equation and can have different forms depending on the choice of I_i, f_j, w_{ij} corresponding to the particular case or application where the ANN is being used. For example, when applied to real (i.e., biological) neurons, then:

- z_i denotes membrane voltage.
- I_i means an external input.
- w_{ij} is interpreted as a weight associated to the synapse.
- f_j takes the form of a product between weight and z_j.

For analogue electric circuits:

- z_i denotes the potential of a capacitor.
- The left-hand side of the equation is interpreted as a current charging a capacitor to potential z_i.
- The summed terms mean potentials weighted by conductance.

As Eq. (11.32) can be written for every $i = 1, 2,...,n$, we have a *system of differential equations*. The study of an ANN is carried out by assuming that *initial states* z_0 are known at some initial point t_0. It can be shown that in a small enough vicinity $|z-z_0|$ of z_0 and $|t-t_0|$ of t_0, system (11.32) has a unique solution. From a practical point of view, the question as to the existence of solutions of Eq. (11.32) can be answered positively owing to the Cauchy-Lipschitz theorem (Martin and Reissner 1961). It is stated here without proof (as it is well-known in the theory of differential equations, and because it is the result of the theorem rather than the proof that is important for us now in IR):

Theorem 11.1. *Given the following system of differential equations:*

$$F(t,z) = \frac{1}{\mu_i}(I_i(t) - z_i(t) + \sum_j f_j(z_j(t), w_{ij}, z_i(t))),$$

where μ_i is a coefficient, consider the initial condition $z(t_0) = t_0$. If function $F(t, z)$ is continuous in a region $\Omega \subset \mathbb{R}^2$ ($\mathbb{R}^2$ denotes the real plane), and function $F(t, z)$ is a local Lipschitz contraction, i.e.,

$$\forall P \in \Omega \exists K \subset \Omega \text{ and } \exists L_K > 0 \text{ constant such that}$$

$$|F(t,z_1) - F(t,z_2)| \leq L_K |z_1 - z_2|, \ \forall (t,z_1), (t,z_2) \in K,$$

then there exists a vicinity $V_0 \subset \Omega$ of point (t_0, z_0) in which the equation has a unique solution satisfying the initial condition $z(t_0) = t_0$, which can be obtained by successive numeric approximations.

Equation (11.32) gives the state of every neuron at time t. By allowing time t to evolve, a sequence $z_i(t)$, $i = 1,...,n$, of states is obtained. This is referred to as the *operation* of ANN. Normally, an ANN evolves in time toward a state that does not change any further. This is called an *equilibrium* and is given by

$$\frac{dz_i}{dt} = 0, \ i = 1, 2,...,n. \tag{11.33}$$

An important mode of operation of an ANN is known as the *winner-take-all* (WTA) strategy, which reads as follows: only the neuron with the highest state will have output above zero; all the others are "suppressed." In other words, WTA means selecting the neuron that has maximum state and deactivating all the others. Formally, the WTA can be expressed as

$$(z_i = 1 \text{ if } z_i = \max_j z_j) \wedge (z_k = 0 \text{ if } z_k \neq \max_j z_j).$$

11.6.2 Associative Interaction Method

Let (**Fig. 11.11**):

- $\Delta = \{O_1, O_2,\ldots,O_i,\ldots,O_N\}$ denotes a set of Web pages of interest. Each page O_i is assigned an artificial neuron $\aleph_i$, $i = 1,\ldots,N$. Thus, we may write $\Delta = \{\aleph_1, \aleph_2,\ldots, \aleph_i,\ldots, \aleph_N\}$.
- $\Phi_i = \{\aleph_k \,|\, k = 1,\ldots,n_i\}$ denotes the set of artificial neurons that are being influenced (i.e., synapsed, pointed to by) by $\aleph_i$, $\Phi_i \subseteq \Delta$.
- $\mathrm{B}_i = \{\aleph_j \,|\, j = 1,\ldots,m_i\}$ denotes the set of artificial neurons that influence (i.e., synapse to, point to) $\aleph_i$, $\mathrm{B}_i \subseteq \Delta$.

$$\begin{array}{ccc} \mathbf{B}_i\mathbf{:} & & \mathbf{\Phi}_i\mathbf{:} \\ \aleph_j \longrightarrow & \aleph_i & \longrightarrow \aleph_k \end{array}$$

Fig. 11.11. $\aleph_1, \aleph_2,\ldots, \aleph_i,\ldots, \aleph_N$ form an artificial neural network. $\Phi_i=\{\aleph_k \,|\, k=1,\ldots,n_i\}$ denotes the set of artificial neurons that are being influenced by $\aleph_i$. $\mathrm{B}_i=\{\aleph_j \,|\, j=1,\ldots,m_i\}$ denotes the set of artificial neurons that influence $\aleph_i$

The associative interaction method is derived from the generic equation (11.32) (Dominich et al 2006). As the objects to be searched are Web pages, no external input (i.e., from outside the Web) can be assumed, so we take $I_i(t) = 0$. One way to define f_j is to consider the influence of a page j on another page i as being determined by the strengths of the connections that convey this influence, i.e., weights w_{ij} of the links between them. Equation (11.31) thus reduces to

$$\frac{dz_i(t)}{dt} = -z_i(t) + \sum_{\aleph_j \in B_i} w_{ij} \tag{11.34}$$

In order to simplify the expression, we introduce the notation: $\sum_{\aleph_j \in B_i} w_{ij} = \Sigma^{(i)}$.

It is known from the theory of differential equations (Theorem 11.1) that the solution of Eq. (11.34) has the general form

$$z_i(t) = Ce^{-t} + \Sigma^{(i)}, \tag{11.35}$$

where C is a constant that depends on the initial condition.

When the network operates for retrieval, activation spreading is taking place according to a WTA strategy. At any time step t_u, $u = 0, 1,\ldots,$ exactly one neuron $k \in \{1,\ldots,N\}$, i.e., the winner, is active; all the other neurons $s \in \{1,\ldots,k-1, k+1,\ldots,N\}$, $s \neq k$, are deactivated, i.e., $z_s(t_u) = 0$. Taking this initial condition into account, we express the activity level of any nonwinner neuron s as

$$z_s(t) = (1 - e^{t_u - t})\Sigma^{(s)}. \tag{11.36}$$

If time t is allowed to increase, activity level $z_s(t)$ tends to stabilize on the total input value $\Sigma^{(s)}$ of that neuron s:

$$\lim_{t \to \infty} z_s(t) = \Sigma^{(s)}. \tag{11.37}$$

At the next time step t_{u+1} of these neurons s, the winner will be the neuron p whose activity level z_p exceeds the activity level z_s of any other neuron s, i.e., $z_p \geq z_s$, expressed as

$$(1 - e^{t_u - t})\Sigma^{(p)} \geq (1 - e^{t_u - t})\Sigma^{(s)}. \tag{11.38}$$

As $t > t_u$, we have $e^{t_u - t} < 1$, and so $(1 - e^{t_u - t})$ is strictly positive. Hence, the winner condition $z_p \geq z_s$ becomes equivalent to $\Sigma^{(p)} \geq \Sigma^{(s)}$. In other words, the neuron with the highest total input will be the winner.

Thus, from a practical point of view, the associative interaction method can be applied in the following way.

Each Web page W_i is viewed as an artificial neuron and is associated with an n_i-tuple of weights corresponding to its terms (obtained after stemming and stoplisting) t_{ik}, $k = 1,\ldots,n_i$. Given another page W_j, if term t_{jp},

$p = 1,\ldots,n_j$, occurs f_{ijp} times in W_i, then there is a link from W_i to W_j, and this may have the weight (normalized frequency weighting scheme)

$$w_{ijp} = \frac{f_{ijp}}{\sum_k f_{ik}}. \tag{11.39}$$

If identifier t_{ik} occurs f_{ikj} times in W_j and df_{ik} denotes the number of pages in which t_{ik} occurs, then there is another link from W_i to W_j, and this may have the weight (inverse document frequency weighting scheme)

$$w_{ikj} = f_{ikj} \cdot \log \frac{2N}{df_{ik}}. \tag{11.40}$$

The total input to W_j is then

$$\sum_{k=1}^{n_i} w_{ikj} + \sum_{p=1}^{n_j} w_{ijp}. \tag{11.41}$$

The other two connections—in the opposite direction—have the same meaning as above:

- w_{jik} corresponds to w_{ijp}.
- w_{jpi} corresponds to w_{ikj}.

A query Q is considered a page; i.e., it is interlinked with pages (referred to as *interaction* between query and pages). The process of retrieval is as follows (**Fig. 11.12**):

- A spreading of activation takes place according to a WTA strategy.
- The activation is initiated at the query $Q = o_j$, and spreads along the strongest total connection, thus passing onto another page, and so on.
- After a finite number of steps, the spreading of activation reaches a page that was a winner earlier, giving rise to a loop (known as a *reverberative circle*) This is analogous to a "local memory" recalled by the query. (This process may be conceived as a process of *association*: some pages are associated to the given query). The pages that are retrieved are those that belong to the same reverberative circle.

Figure 11.13 shows sample—and typical—plots of activity levels $z_s(t)$ for four neurons. It can be seen how activity levels asymptotically reach their limit, which is equal to the corresponding total input value: 1, 5, 3, 6.

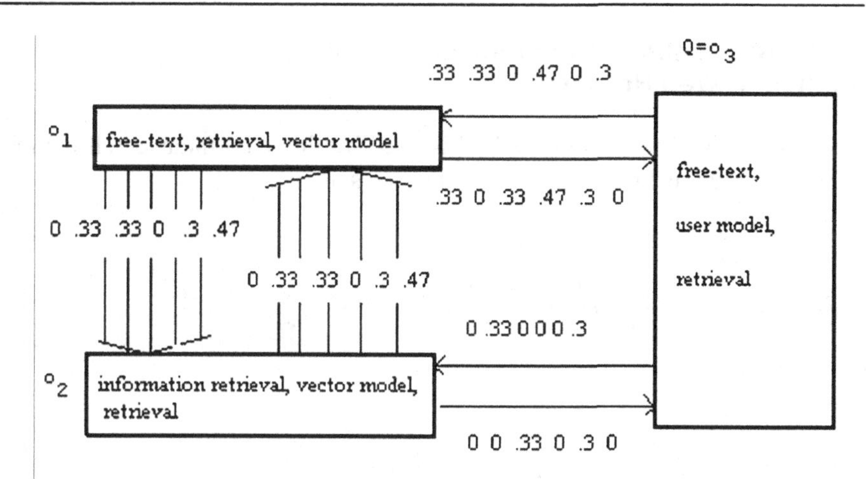

Fig. 11.12. Associative interaction retrieval method (example). All links having the same direction between Q and o_1 and Q and o_3 are shown as a single arrow to simplify the drawing. The activation starts at Q and spreads over to o_1 (total weight = 0 .33 + 0.33 + 0.47 + 0.3 = 1.43), then to o_2, and then back to o_1. o_1 and o_2 form a reverberative circle; hence o_1 and o_2 will be retrieved in response to Q.

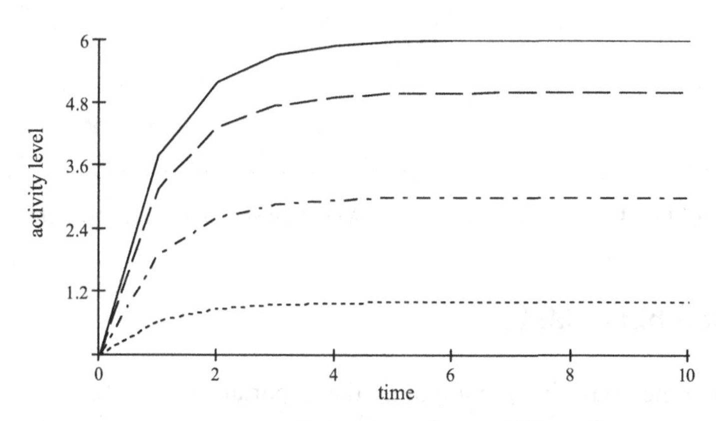

Fig. 11.13. Sample plots of activity levels for four neurons in the associative interaction method during the operation for retrieval. It can be seen how activity levels asymptotically reach their limit, which is equal to the corresponding total input. The highest will be the winner.

11.6.3 Application of the Associative Interaction Method in Web Retrieval

The associative interaction method is used by the Web metasearch engine I²Rmeta.[5] **Figure 11.14** shows the interface screen.

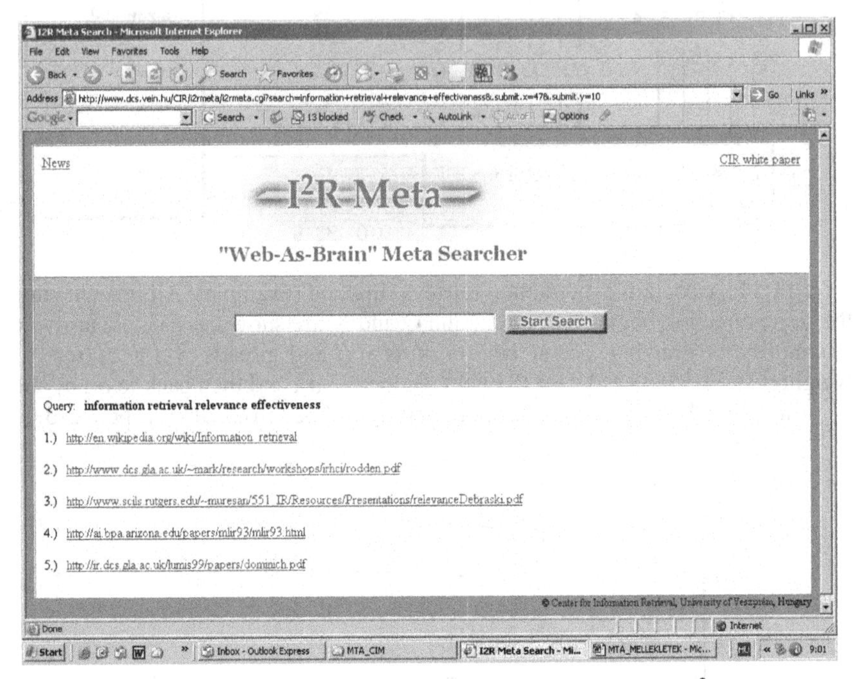

Fig. 11.14. Interface screen of the Web metasearch engine I²Rmeta.

11.7 Combined Methods

Combined methods aim at computing the importance (used for ranking) of Web pages as a combination of the following:

- Their importance stemming from their belonging to a network of pages (link importance or link-based evidence),
- Their importance given by their content (referred to as intrinsic importance or content-based evidence).

In Tsikrika and Lalmas (2004), an overview of combined methods is given. It is also reported that, according to experimental results, considering

[5] www.dcs.vein.hu/CIR (Meta Search).

link importance is beneficial for finding a homepage but not for searching for a topic. The impact factor, connectivity, mutual citation, PageRank, HITS, SALSA, and associative-interaction (or other) methods can be used to compute a link importance for a Web page.

In what follows, we describe major combined importance methods that are based on lattices.

11.7.1 Similarity Merge

It is worth recalling first an early method that combines link-based and content-based evidence as the sum of normalized deviations from minimum (Fox and Shaw 1994).

The combined importance Ψ_j for a Web page W_j of interest is computed using the equation

$$\Psi_j = \sum_{i=1}^{m} \frac{s_i - s_{min}}{s_{max} - s_{min}},$$

where

- s_i is the importance of page W_j given by method i.
- $m > 1$ is the number of methods used to calculate importance.
- $s_{min} = \min_i s_i$, $s_{max} = \max_i s_i$

The combined importance Ψ_j can be weighted (i.e., multiplied) by a factor M denoting the number of methods for which $s_i \neq 0$.

In experimental results using the similarity merge method with vector space retrieval and HITS on the TREC WT10 test collection, the similarity merge method did not outperformed the vector space method. This might have been due to the characteristics of the WT10 collection, or to other causes (which were not analyzed).

11.7.2 Belief Network

The probabilistic retrieval method based on Bayesian networks can be used to elaborate a combined method using content importance based on terms and authority/hub importance (Calado et al. 2003).

The basic Bayesian network is constructed in three versions: one for content-based evidence and two for two link-based evidence (one for hub and another for authority evidence) of a page W_j of interest (**Fig. 11.15**).

The combined importance Ψ_j for a Web page W_j of interest is computed as follows:

$$\Psi_j = \eta \cdot (1 - (1 - C_j) \cdot (1 - H_j) \cdot (1 - A_j)),$$

where:

- η is a normalizing parameter.
- C_j is a content-based importance (given by the cosine similarity measure) of page W_j given a query Q.
- H_j is a hub-based importance of page W_j.
- A_j is an authority-based importance of page W_j.

Experimental results were reported as to the relevance effectiveness of the above method. A total of 5,939,061 pages from the Brazilian Web (domain.br) were automatically collected and indexed to obtain inverted file structures (number of terms: 2,669,965; average number of terms per page: 413). Fifty queries were selected from the most frequently asked 100,000 queries[6] (average number of terms per query: 1.8). Of these 28 queries were general (e.g., movies), 14 queries were specialized (e.g., transgenic food), and 8 queries concerned music band names. The relevance lists were compiled manually by human experts. A precise description of which page was relevant to which query was given to the assessors (e.g., for the query "employment" only pages with employment ads were considered to be relevant). When applied separately, the cosine measure outperformed both link-based methods. When content-based evidence was combined with authority-based evidence, the content-based only and the authority-based only methods were both outperformed. When content-based evidence was combined with hub-based evidence, the content-based only and the hub-based only methods were both outperformed. The combination of all three methods outperformed each method used separately.

---

[6] www.todobr.com.br.

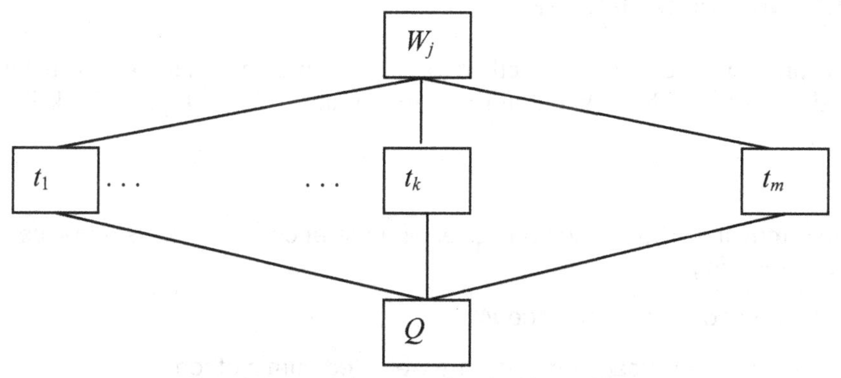

Bayesian network for content-based evidence of page W_j

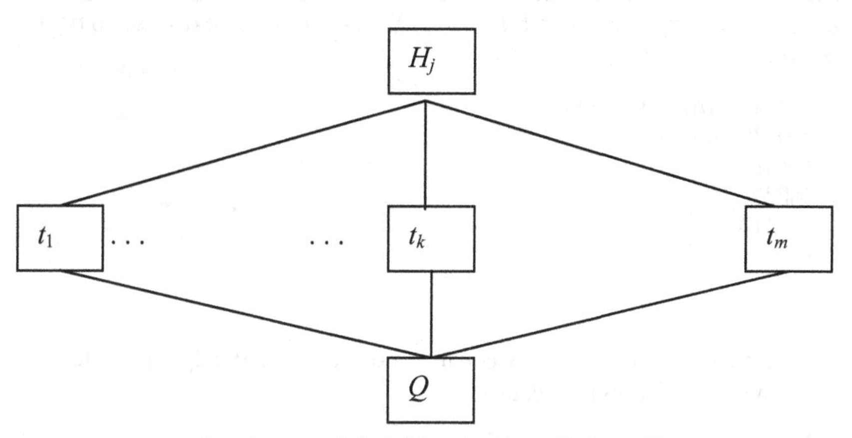

Bayesian network for hub-based evidence H_j of page W_j

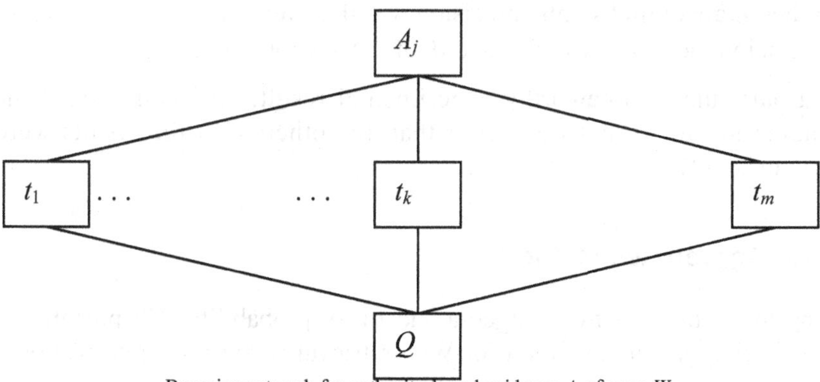

Bayesian network for authority-based evidence A_j of page W_j

Fig. 11.15. Bayesian network types used in a belief network.

11.7.3 Inference Network

The probabilistic retrieval method based on inference networks can be used to combine the link and intrinsic importance of a Web page (Tsikrika and Lalmas 2004).

The retrieval method based on Bayesian (inference) networks is enhanced in the following way:

- D_j corresponds to Web page W_j.
- Probability is propagated using the weighted sum method.
- An anchor window (i.e., the text inside a window of W bytes around the anchor text, typically N = 50, N = 100) is used as text describing the page to which it refers.
- t_k (k = 1,...,m) may denote
  - word
  - term
  - inlink
  - outlink

Extensive experimentation was carried out (on the WT2g TREC test database) with the following results:

- When t_k are terms only, precision was highest.
- When t_k are inlinks only, precision was second highest.
- When t_k are outlinks only, precision was third highest.
- Precision increases with the size W of the anchor window.

As a particularly noteworthy experimental result, they emphasized the combination for t_k that was better than the others (but the results were not consistent).

11.7.4 Aggregated Method

Using the notions of fuzzy algebra and fuzzy probability (Chapter 9), we present an aggregated method for Web retrieval (mainly for retrieval on a given site).

11.7.4.1 Content-Based Importance

A Web page W_j may be interpreted as being a fuzzy set in a set $T = \{t_1,\ldots,t_n\}$ of terms, i.e., $W_j = \{(t_i, \varphi_j(t_i)) \mid t_i \in T, i = 1,\ldots,n\}$. Then, its fuzzy probability P_j is

$$P_j = P(W_j) = \sum_{i=1}^{n} \varphi_j(t_i) p(t_i), \qquad (11.42)$$

where $p(t_i)$ denotes a frequency-based probability of term t_i. One way to compute it is as follows ("favorable cases over all cases"):

$$p(t_i) = \frac{\sum_{j=1}^{N} f_{ij}}{\sum_{i=1}^{n} \sum_{j=1}^{N} f_{ij}}, \; i = 1,\ldots,n, \qquad (11.43)$$

where:

- n = number of terms.
- N = number of Web pages.
- $\varphi_j(t_i)$ = membership function (e.g., weight of term t_i for page W_j).
- f_{ij} = number of occurrences of term t_i in W_j.

P_j may be interpreted as being proportional to (or an indication of) the chance that the page is being selected or occurs (e.g., in a hit list) based on its content. The fuzzy probability of a page is equal to zero if the page does not have any content (is without meaning, the weights of its terms all being zero). The fuzzy probabilities $\Pi = [P_1,\ldots,P_j,\ldots,P_N]^T$ of all pages are given by the following matrix multiplication: $\Pi = \Phi \times P$, where $\Phi = (\varphi_{ji})_{N \times n}$, $P = [p(t_1),\ldots,p(t_i),\ldots,p(t_n)]^T$.

11.7.4.2 Combined Importance Function

A combined importance function Ψ of a Web page W is defined as being a function F of its link importance L (stemming from the link structure of the Web graph) and its intrinsic importance given by the fuzzy probability P:

$$\Psi = \Psi(P, L). \qquad (11.44)$$

From a practical point of view, an analytic form for the combined importance function Ψ should be given. In this regard, the following assumptions (or axioms) seem reasonable:

Assumption 1. It seems straightforward to require that the combined importance of an isolated page without content be null:

$$\Psi(0, 0) = 0. \tag{11.45}$$

Assumption 2. If a Web page does not carry any meaning (practically it does not have any content), i.e., $P = 0$, then its combined importance should vanish, even if it is highly linked Formally:

$$\Psi(L, 0) = 0, \; L \neq 0. \tag{11.46}$$

Note: This assumption may need further investigation, because, e.g., a hub page may be very useful even if it contains only links.

Assumption 3. Further, from zero link importance ($L = 0$) need not necessarily follow a vanishing combined importance Ψ if the fuzzy probability does not vanish (e.g., this may be the case of a "young" Web page that is an isolated node of the Web graph, but which may carry important meaning). Formally,

$$\Psi(0, P) \neq 0, P \neq 0. \tag{11.47}$$

Assumption 4. It seems natural to require that the combined importance of a page increase with its probability P for the same link importance L; the same should also hold for L. Formally,

$$\begin{aligned} P_1 < P_2 &\Rightarrow F(L, P_1) < F(L, P_2), \\ L_1 < L_2 &\Rightarrow F(L_1, P) < F(L_2, P). \end{aligned} \tag{11.48}$$

One possible and simple analytical form for Ψ that satisfies Assumptions 1–4 is

$$\Psi(L,P) = PL + aP = P(L + a), \tag{11.49}$$

where parameter $a > 0$ is introduced to "maintain" a balance between the probability-based importance P and link-based importance L when P happens to be much larger than L. It can be easily seen that Ψ satisfies all of the Assumptions 1–4. **Figure 11.15** shows the plot (surface) of the combined importance function Ψ defined by Eq. (11.49).

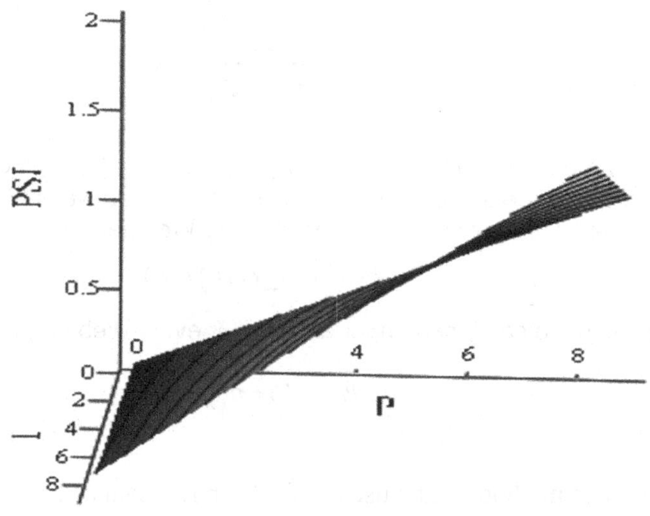

Fig. 11.16. Graphical representation of the combined importance function Ψ = PSI = $PL + aP$ for $a = 1$ (The values on the L and P axes are grid points for scaling purposes; the L and P values are obtained by division by 10).

11.7.4.3 Aggregated Method

The following combined method may be proposed for computing the importance of Web pages:

Combined Method

1. Construct the Web graph G for Web pages of interest, W_j, $j = 1,\ldots,N$.
2. Compute link importance L_j for every such Web page , W_j, $j = 1,\ldots,N$. In principle, any method (connectivity, PageRank, HITS, SALSA, associative, or other) may be used. For example, using the PageRank method, we have

$$L_j = \alpha \cdot (1-d) + \beta \cdot d \cdot \sum_{W_k \in B_j} \frac{L_k}{C_k} + \gamma \cdot E.$$

3. Construct a set of terms $T = \{t_1,\ldots,t_i,\ldots,t_n\}$.
4. Construct the term-page frequency matrix M:

$$M = (f_{ij})_{N \times n}.$$

5. Compute probabilities $p(t_i)$ as follows:

$$p(t_i) = \frac{\sum_{j=1}^{N} f_{ij}}{\sum_{i=1}^{n} \sum_{j=1}^{N} f_{ij}}.$$

6. Define membership functions $\varphi_j(t_i)$, $j = 1,\ldots,N$; $i = 1,\ldots,n$. For example, $\varphi_j(t_i) = w_{ij}$, where the weight w_{ij} is calculated using a weighting scheme.
7. Calculate the fuzzy probability P_j of every Web page W_j, $j = 1,\ldots,N$:

$$P_j = P(W_j) = \sum_{i=1}^{n} \varphi_j(t_i) p(t_i).$$

8. Compute the combined importance Ψ_j for every Web page W_j ($j = 1,\ldots,N$):

$$\Psi_j = L_j P_j + aP_j.$$

The combined method can be used to for Web retrieval as follows:

Aggregated Retrieval Method

1. Given a query Q.
2. Compute similarities between Q and Web pages W_j (j=1,…,N):

$$\rho_j = \frac{\kappa(Q \cap W_j)}{P(Q)} = \frac{\sum_{i=1}^{n} q_i \cdot \varphi_j(t_i)}{\sum_{i=1}^{n} q_i \cdot p(t_i)}.$$

3. Construct the set of pages that matches the query:

$$\{W_j \mid \rho_j \neq 0, j = 1,\ldots, J\}.$$

4. Compute an aggregated importance S_j for Web pages W_j (j=1,…,J):

$$S_j = \alpha\Psi_j + \beta\rho_j, \quad \alpha, \beta \text{ parameters.}$$

5. Rank pages $W_1,\ldots,W_J$ descendingly on their aggregated similarity $S_1,\ldots,S_J$ to obtain a hit list H.
6. Show the entire hit list H or part of it (use cut-off or threshold) to the user.

In order to test the aggregated retrieval method, an experimental Web search engine was developed (in C++, Java, and MathCAD). It consisted of the following modules:

- Crawler
- Indexer
- Matrix generator
- Retrieval and ranking module

All the .html pages on the www.vein.hu domain (which is the site of the University of Pannonia, Veszprém, Hungary) were crawled, and 6551 HTML Web pages were downloaded (as of January 29, 2007). Terms were extracted automatically (anything between two consecutive blank spaces was considered to be a word), and the top 100 words were excluded (stopwords). The majority of the pages were written in Hungarian, while the rest were in English, German, and French. Thus, some 133,405 terms remained (in Hungarian, English, German, and French). The link importance of pages, L, was computed using the PageRank method with $\alpha = 0$, $\beta = 1$, $d = 1$, and $\gamma = 0$. Membership function φ was taken to be the well-known length-normalized term frequency weighting scheme. For the computation of global importance Ψ, parameter a was set to $a = 0.25$, and for the aggregated importance S the parameter values used were $\alpha = 100{,}000$ and $\beta = 0.5$. These values for the parameters were established after several trials so as to obtain nonzero values for combined and aggregated importance (PageRank values were extremely small, and thus the combined importance values would almost vanish, i.e., would be practically zero, otherwise).

Other assumptions as well as combining and aggregating functions may be proposed and experiments done with them. The fine-tuning of the parameters may also be carried out in other ways, e.g., based on user feedback and/or some learning procedure.

It is well known that the in vivo measurement of relevance effectiveness of a Web search engine poses several problems. For example, recall cannot be measured. Neither can precision, in many cases owing to the overly many hits returned, which are practically impossible to assess. Thus, such an evaluation should necessarily follow some reasonable compromise relative to some baseline or benchmark search engine. For example, in Thelwall and Vaughan (2004), users' rankings were compared to those given by Google. In the present measurement, the retrieval and ranking produced by the experimental search engine was compared to those of Google, Altavista, and Yahoo! All the searches were carried out on January 30, 2007. The queries and the results are given below. As our purpose is the comparison of hit lists with one another, it will be sufficient to give only the URLs of the hits returned. The hits (the first five that could be seen on the screen) were assessed for relevance manually. We note that while it was possible to restrict the search to .html format in Altavista and Yahoo!, this was not possible in Google.

Experiment 1. The query was: "nitrálás." The hit lists were are as follows:
Aggregated Method:

www.vein.hu/public_stuff/oik/tematikak/tematikak/2003-04-2/VETKTC2214c.html
www.vein.hu/public_stuff/oik/tematikak/tematikak/2002-03-2/VETKTC2214c.html
www.vein.hu/public_stuff/oik/tematikak/tematikak/2002-03-2/VEMKTC2214c.html
www.vein.hu/public_stuff/oik/tematikak/tematikak/2004-05-2/VEMKTC2214c.html

Google:

www.vein.hu/public_stuff/oik/tematikak/tematikak/2004-05-2/VEMKTC2214c.html

Altavista:

www.vein.hu/public_stuff/oik/tematikak/tematikak/2002-03-2/VETKTC2214c.html
www.vein.hu/public_stuff/oik/tematikak/tematikak/2003-04-2/VEMKTC2214c.html
www.vein.hu/public_stuff/oik/tematikak/tematikak/2002-03-2/VEMKTC2214c.html

Yahoo!:

www.vein.hu/public_stuff/oik/tematikak/tematikak/2002-03-2/VETKTC2214c.html
www.vein.hu/public_stuff/oik/tematikak/tematikak/2003-04-2/VEMKTC2214c.html
www.vein.hu/public_stuff/oik/tematikak/tematikak/2002-03-2/VEMKTC2214c.html

Google returned only one result that was relevant. The same result was also returned by the aggregated method in the fourth position. Altavista and Yahoo! returned the same hits, all of which were relevant. The same three hits were all returned by the aggregated method, but ranked differently. In conclusion, the aggregated method returned the highest number of hits, all were relevant, and some of them were also returned by Google, Altavista, and Yahoo!. According to the results of Experiment 1, the following ranking in terms of relevance effectiveness of the four search engines were:

1. Aggregated method
2. Altavista, Yahoo!
3. Google

Experiment 2. The query was "supercomputer." The hit lists returned were as follows:

Aggregated Method

www.vein.hu/oktatok/egyetemi_szervezetek/szemelyzetio/szem_oszt/szemelyzeti.html

Google (15 hits):

...

www.vein.hu/oktatok/egyetemi_szervezetek/szemelyzetio/szem_oszt/szemelyzeti.html

...

Altavista:

www.vein.hu/oktatok/egyetemi_szervezetek/szemelyzetio/szem_oszt/Publikaciokkal_kapcsolatos_utmutato.doc
www.vein.hu/oktatok/egyetemi_szervezetek/szemelyzetio/szem_oszt/szemelyzeti.html

Yahoo!:
www.vein.hu/oktatok/egyetemi_szervezetek/szemelyzetio/szem_oszt/Publikaciokkal_kapcsolatos_utmutato.doc
www.vein.hu/oktatok/egyetemi_szervezetek/szemelyzetio/szem_oszt/szemelyzeti.html

Google returned 15 hits of which all may be viewed as relevant. Out of these 15, the first six hits had .pdf format. The hit returned by the aggregated method had rank seven in the Google list. Altavista and Yahoo! returned the same hit list; all were relevant, and the first hit was in .doc format. Thus, one may conclude that all four search engines performed practically equally well:

1. Altavista, Yahoo!, Google, Aggregated Method

Experiment 3. The query was: “bizottság” (in English: “committee”). The hit lists returned were as follows:

Aggregated Method
www.vein.hu/belso/2004_2005_tanevi_ertesito/menus/etanacsmenu.html
www.vein.hu/belso/2003_2004_tanevi_ertesito/menus/etanacsmenu.html
http://www.vein.hu/www/intezetek/fdsz/generate.php?file_name=kepvisel.txt
http://www.vein.hu/www/intezetek/fdsz/szak_szerv/menu.php
www.vein.hu/oktatok/egyetemi_szervezetek/fotitkarsag/korlevelek/nre/valasztas.html

Google (1030 hits):
.../index.php
kulugy.vehok.vein.hu
www.vein.hu/library/iksz/serv/dos/szmsz.htm
http://www.vein.hu/www/intezetek/fdsz/szak_szerv/nevsor.html
www.vein.hu/oktatok/szabalyzatok/kozmuv.html
...

Altavista (985 hits):
.../index.php
.../index.php
.../index.php
.../index.php
....

Yahoo! (889 hits):
.../index.php
.../index.php
.../index.php
.../index.php
....

All the hits returned by the aggregated method were relevant. The first three hits returned by Google were not relevant, whereas the fourth and fifth were relevant. The hits returned by Altavista and Yahoo! were not relevant. According to the results of Experiment 3, the following ranking in terms of relevance effectiveness of the four search engines were:

1. Aggregated method
2. Google
3. Altavista, Yahoo!

In conclusion, these experimental results show that a retrieval and ranking based on the aggregated method can outperform commercial search engines such Google, Altavista, and Yahoo! (at least for single-term queries).

11.8 Lattice-Based View of Web Ranking

11.8.1 Web Lattice

Let L_{Web} denote a set $W = \{W_1,\ldots,W_i,\ldots,W_N\}$ of Web pages or a set $S = \{S_1,\ldots,S_i,\ldots,S_N\}$ of Web sites of interest. Then, set L_{Web} can be turned into a lattice as follows:

1. $\mathbf{0} = \varnothing$.
2. $\mathbf{1} = W$ (or S).
3. The meet $\wedge$ and join $\vee$ operations are shown in **Fig. 11.17**.

If the Web lattice is a site lattice L_S, then, in the site graph, there is an edge from site s_i to site s_j if site s_i has a page pointing to a page situated on site s_j.

Matrices M used in link analysis methods are defined for the Web lattice L_{Web}. It can be easily seen that lattice L_{Web} is modular and not distributive.

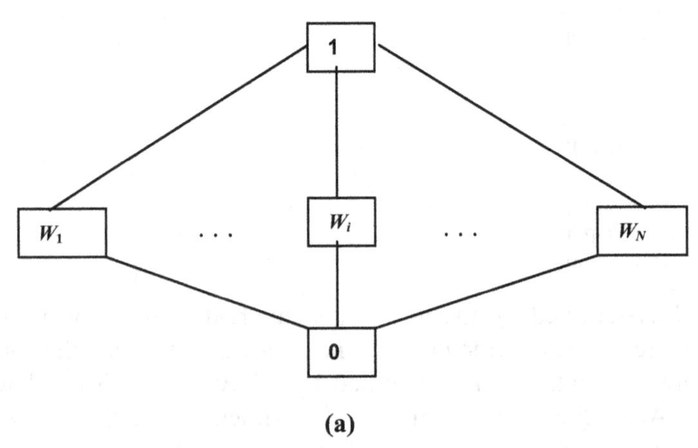

Fig. 11.17. Lattices L_{Web} of **(a)** Web pages and **(b)** of Web sites.

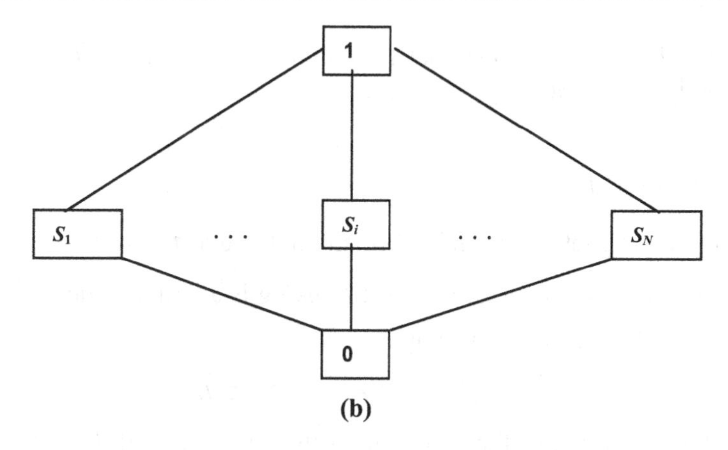

(b)

Fig. 11.17. (Continued)

11.8.2 Chain

Let $(A, \leq)$ be a poset. Then, poset $(A, \leq)$ is said to be *completely ordered* if any two of its elements can be compared with one another, i.e, $\forall a, b \in A$ we have either $a \leq b$ or $a \geq b$ or $a = b$.

Example 11.7

The set $\mathbb{R}$ of real numbers is a completely ordered set with respect to the relation $\leq$.

Let $\mathfrak{C}$ be a subset of a poset A, $\mathfrak{C} \subseteq A$. If $\mathfrak{C}$ is completely ordered, then it is called a *chain*. It can be shown that:

Theorem 11.2. *Any chain is a lattice.*

Proof. We know that the lattice is a poset in which any two elements have an infimum and a supremum. Let $\mathfrak{C}$ be a chain. Then, any two of its elements, a and b, are comparable. Without restricting generality, let $a \leq b$. Then, sup $\{a, b\} = b$, inf $\{a, b\} = a$.

From Theorem 11.2, it follows that any subset R of the set $\mathbb{R}$ of real numbers, $R \subset \mathbb{R}$, is a chain (with respect to $\leq$), and also a lattice.

The methods described in Sections 11.2–11.7 allow for computing page importance, which makes the ranking of a set $W = \{W_1,\ldots,W_i,\ldots,W_N\}$ of Web pages possible. Formally, let I_i denote the importance of page W_i computed by some method ($i = 1,\ldots,N$). Because I_i is a real number, i.e., I_i

$\in \mathbb{R}$, $i = 1,\ldots,N$, it follows that (Theorem 11.2) set $\{I_1,\ldots,I_N\} \subset \mathbb{R}$ is a chain and hence a lattice.

11.8.3 Ranking

Ranking in Web retrieval can be formally defined as follows:

Definition 11.1. Given a lattice L_{Web}. *Ranking* is a lattice-lattice function ρ from a Web lattice L_{Web} to a chain $R \subset \mathbb{R}$:

$$\rho : L_{Web} \to R,\ \rho(A) = r,\ \forall A \in L_{Web},$$

where the lattice-lattice function ρ gives the importance of A. □

We may assume, without restricting generality, that $R = [0; 1]$. It can be shown that ranking has the following property:

Theorem 11.3. *Ranking is not submodular.*

Proof. The submodularity condition is as follows:

$$\rho(x \vee_1 y) \vee_2 \rho(x \wedge_1 y) \leq \rho(x) \vee_2 \rho(y),$$

where $\vee_1$ and $\wedge_1$ are the join and meet defined in lattice L_{Web}, and $\vee_2$ is the join defined in the chain lattice [0; 1]. Let $\rho(W_1) = r_1$ and $\rho(W_2) = r_2$. We can assume that $\rho(\mathbf{1}) = 1$ (i.e., the whole Web is ranked highest) and $\rho(\mathbf{0}) = 0$ (i.e., the empty set is ranked lowest). If we take $x = W_1$ and $y = W_2$, we obtain

$$\rho(W_1 \vee_1 W_2) \vee_2 \rho(W_1 \wedge_1 yW_2) = \rho(\mathbf{1}) \vee_2 \rho(\mathbf{0}) = 1,$$

which is not less than or equal to

$$\rho(W_1) \vee_2 \rho(W_2) = \max\,(r_1, r_2).$$

The proof is similar for lattice L_S. □

11.8.4 Global Ranking

By taking the direct product of lattices L_{Web} and L'_{Web}, a new lattice is obtained. We have the following possibilities:

(a) $L_{Web} = L_W$ and $L'_{Web} = L_{W'}$.
(b) $L_{Web} = L_W$ and $L'_{Web} = L_S$.
(c) $L_{Web} = L_S$ and $L'_{Web} = L_{S'}$.

Figure 11.18 shows an example for the direct product $L_W \times L_S$ between a page lattice L_W and a site lattice L_S.

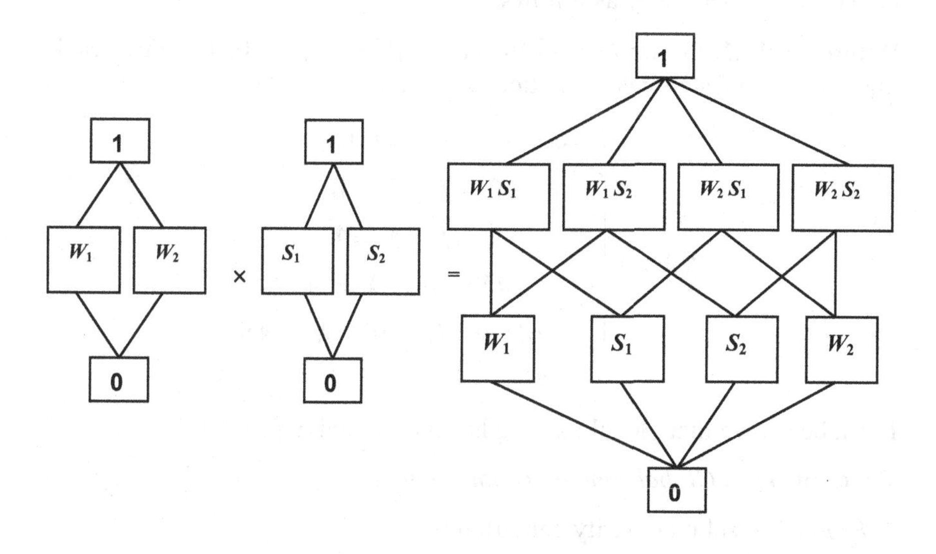

Fig. 11.18. Direct product of a page lattice and a site lattice (example).

Figure 11.19 shows an example for the direct product $L_W \times L_{W'}$ between two page lattices.

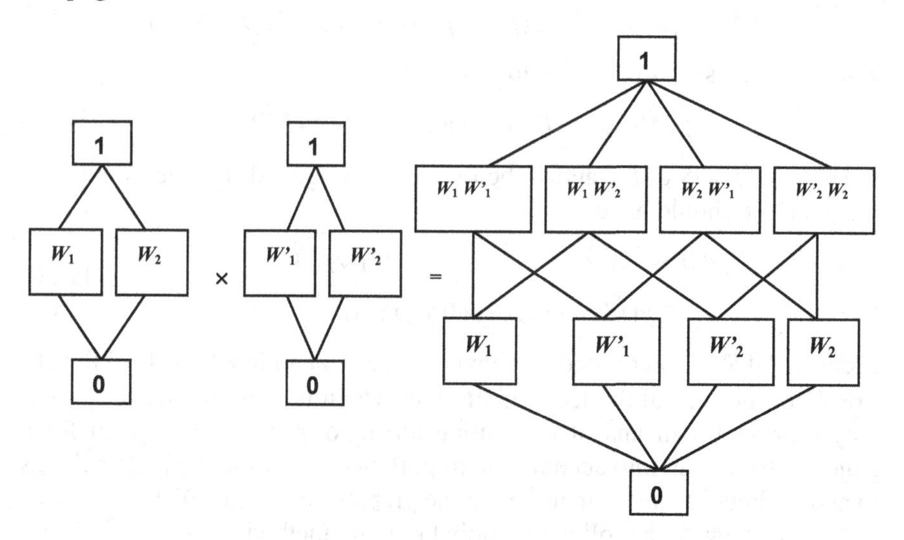

Fig. 11.19. Direct product of two page lattices (example).

Taking into account the meet and join operations in the direct product lattice, we can define a global ranking for the direct product of two Web lattices in a general way as follows:

Definition 11.2. Let $\rho: L_{Web} \rightarrow [0; 1]$ and $\rho': L'_{Web} \rightarrow [0; 1]$ be two rankings. A *global ranking* is the lattice-lattice function

$$\gamma: L_{Web} \times L'_{Web} \rightarrow [0; 1],$$

$$\gamma(X) = \begin{cases} 0 & if \quad X = \mathbf{0} \\ 1 & if \quad X = \mathbf{1} \\ \rho(X) & if \quad X = A \\ f(\rho(A), \rho'(B)) & if \quad X = AB \end{cases}.$$ □

It can be shown that global ranking has the following property:

Theorem 11.4. *Global ranking is not submodular.*

Proof. The submodularity condition is

$$\gamma(x \vee_1 y) \vee_2 \gamma(x \wedge_1 y) \leq \gamma(x) \vee_2 \gamma(y).$$

Let $\rho(AB) = r_1$ and $\rho(BC) = r_2$. If we take $x = AB$ and $y = CD$, we obtain

$$\gamma(AB \vee_1 BC) \vee_2 \gamma(AB \wedge_1 BC) = \gamma(\mathbf{1}) \vee_2 \gamma(B) =$$

$$\rho(\mathbf{1}) \vee_2 \rho'(B) = \max(\rho(\mathbf{1}), \rho'(B)) = \max(1, \rho'(B)) = 1,$$

which is not less than or equal to

$$\gamma(AB) \vee_2 \gamma(BC) = \max(\gamma(AB), \gamma(BC)).$$ □

Let $W \in L_W$, $S \in L_S$, and W be on site S. As global ranking is not submodular, we should have

$$\gamma(W \vee_1 S) \vee_2 \gamma(W \wedge_1 S) > \gamma(W) \vee_2 \gamma(S)$$
$$\gamma(WS) > \max(\rho(W), \rho'(S)). \qquad (11.50)$$

Element "WS" is interpreted as viewing page W at a global level (of the entire Web), not just at the level of site S to which it belongs. Hence, $\gamma(WS)$ may represent a means of computing an importance at Web level for a page W also taking into account the importance of the site the page belongs to and without having to manipulate the graph of the entire Web.

We can suggest the following global ranking method:

Global Ranking Method

1. Let L_W and L_S be two Web lattices ($W \in L_W$, $\exists S \in L_S$: $W \in S$).
2. Use a method to produce rankings ρ_W and ρ'_S.
3. Compute the global importance $\gamma(W)$ for every page W using the global ranking function γ for the direct product lattice $L_W \times L_S$:

$$\gamma(X) = f(X, a, b),$$

where a and b are real parameters.

From a practical point of view, the really useful case is when X is equal to a page and a site, i.e., $X = AB$, where A is a page and B is a site. In such a case, we can use the following form for $\gamma(X)$:

$$\gamma(X) = \gamma(AB) = a\rho(A) + b\rho'(B), \quad A \in L_W, B \in L_S. \tag{11.51}$$

i.e., we can utilize a linear (weighted) combination of importances (or some other form that can be defined and fine-tuned experimentally).

Aberer and Wu (2003) reported experimental results for a version of the global ranking method (the exact values of parameters a and b were not reported; these were adjusted and tuned experimentally). The ranked hit lists they obtained were compared with each other using the Spearman footrule equation:

$$\sum_{i=0}^{n} | R_0(i) - R_1(i) |, \tag{11.52}$$

where $R_0(i)$ and $R_1(i)$ denote the rank of page (site) i. The results obtained showed that using global ranking yielded rankings that were at least as good as those using a link analysis method globally. The rankings were also at least as good as those obtained by the Google search engine.

The global ranking method has the following advantages over traditional methods:

1. Global ranking can be obtained from local rankings (e.g., some link analysis method is applied to rank the pages on a site and to rank sites, and then a ranking for pages can be obtained without manipulating larger graphs).
2. Owing to point 1, both running time and disk space (memory) is saved. This is very important because computation costs are prohibitively

high (the size of the Web is increasing very rapidly, approximately exponentially). For example, one does not have to apply a traditional link analysis method for the entire Web.

3. The Global ranking method allows for computing rankings in a distributed fashion, and thus better scalability can be achieved.
4. The ranking function for the site lattice L_S need not be computed very frequently because intersite links change at a slower rate than interpage links.
5. If interpage links change, then only the affected page lattice L_W has to be recomputed.
6. Algebraic equivalences and other mathematical properties can be exploited in order to find alternative rankings.
7. Relevance feedback from users can be more easily included and taken into account because recomputation refers only to the affected Web lattice (not to the whole Web), or it can used to fine-tune the global ranking function γ.
8. The information provided by hub sites can be used to enhance the global ranking function γ.

11.8.5 Structure-Based Ranking

The ranked hit list returned by a Web search engine is typically very long. This may frustrate the user if he/she wants to assess all the corresponding pages. Further, the hit list fails to indicate how a relevant page is related to the site to which it belongs. Similarly, at the page level, the structure of the page could help the user better assess its content or relevance.

A solution to these problems is described in what follows. We first prove that:

Theorem 11.5. *Any tree can be transformed into a lattice.*

Proof. Let $T = (V, E)$ denote a tree (i.e., a connected and acyclic graph). Let $V = \{root, v_1, v_2, \ldots, v_n\}$ denote the vertices of tree T.

A lattice $(L, \wedge, \vee)$ can be constructed in the following way:

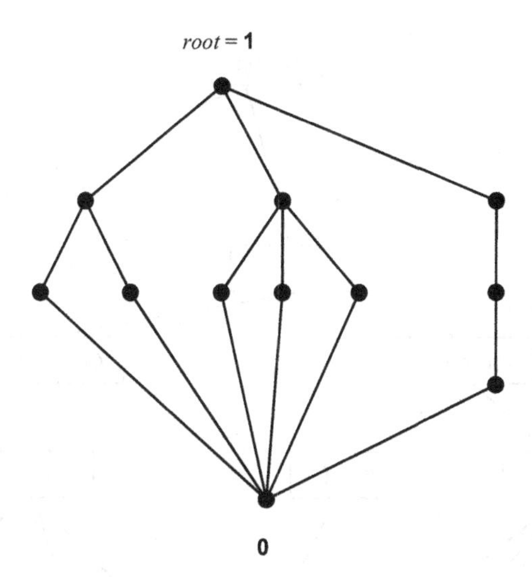

- The elements of the lattice are $L = V \cup \{\mathbf{0}\}$, where $\mathbf{0}$ denotes the null element.
- $root = \mathbf{1}$.
- The meet $v_i \wedge v_j = \mathbf{0}$, if v_i and v_j are leaves ($i \neq j$).
- The meet $v_i \wedge v_j$ is obtained by following paths downward to the first vertex where the two paths intersect [if v_i and v_j are not both leaves ($i \neq j$)].
- The join $\vee$ of any two elements is given by following the paths upward to the first vertex where they intersect. □

Next, we prove that:

Theorem 11.6. *Any document can be transformed into a lattice.*

Proof. Any document has a well-defined structure consisting of sections organized into a hierarchy. For example: document, part(s), chapter(s), part(s), paragraph(s), sentence(s).

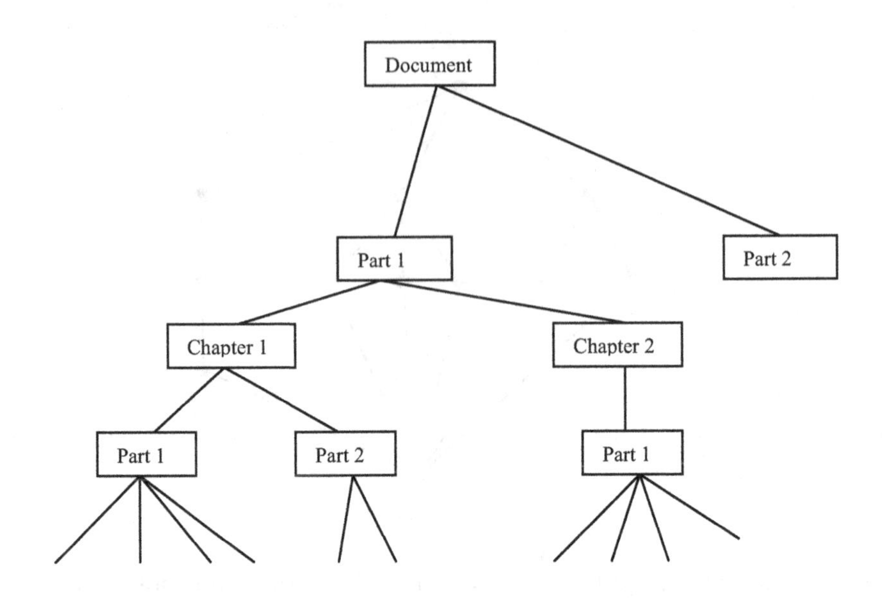

Formally, the hierarchy of sections is a tree. Thus (Theorem 11.5), the structure of any document can be transformed into a lattice. □

A graphical representation of the document tree, called DocBall, is proposed in Vegas et al. (2007) (**Fig. 11.20**):

- DocBall consists of concentric rings.
- The innermost (central) ring corresponds to the root, i.e., to the document (as a whole) being considered.
- The outermost ring is divided into sections corresponding to the leaves of the document (i.e., to the basic structural elements).
- The remaining rings are divided into sections s according to the remaining tree vertices. Every section is defined as a triple $s = (l, a, b)$, where l denotes the level of the section, a is the angular degree where the section begins (the origin of DocBall is at 12 o'clock), and b is the angular degree where the section ends.

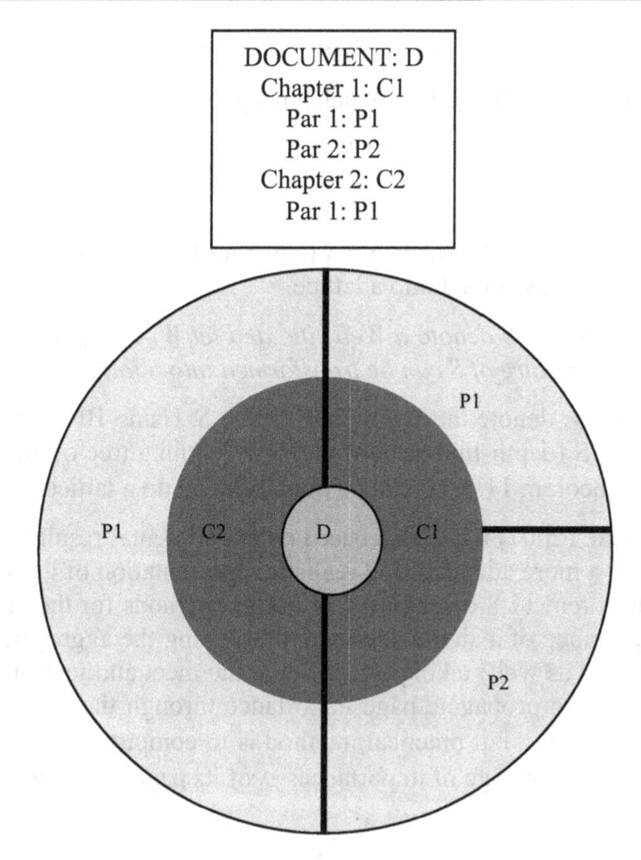

Fig. 11.20. DocBall representation of a document.

A section $s_j = (l_j, a_j, b_j)$ is said to be the *ancestor* of section $s_i = (l_i, a_i, b_i)$, which is expressed as $s_j \leq s_i$, if $a_i \leq a_j$ and $b_j \leq b_i$.

It can be shown that there is a connection between DocBall and concept lattices:

Theorem 11.7. *The notions of ancestor (in DocBall) and subconcept (in concept lattice) are equivalent to each other.*

Proof. Let $s = (l, a, b)$ denote a section in a DocBall, and $C = (A, B)$ a concept in a concept lattice. With the correspondences

- $s \leftrightarrow C$
- $b \leftrightarrow A$
- $a \leftrightarrow B$
- $\leq \leftrightarrow \subseteq$

the definition of the notion of ancestor, i.e.,

$$s_j \leq s_i \Leftrightarrow (a_i \leq a_j \text{ and } b_j \leq b_i),$$

and the definition of the notion of subconcept,

$$C_j \subseteq C_i \Leftrightarrow (B_i \subseteq B_j \text{ and } A_j \subseteq A_i),$$

are equivalent.

It can be shown that a Web site S (e.g., any URL ending with the symbol '/') can be transformed into a lattice.

Theorem 11.8. *Let S denote a Web site, and let $W_1, \ldots, W_n$ be the pages on it. Then, the structure of S can be transformed into a lattice.*

Proof. Let G_S denote the Web graph of site S. Using BFS (breadth-first-search) or DFS (depth-first-search) yields a spanning tree of graph G_S. According to Theorem 11.5, G_S can be transformed into a lattice.

Vegas et al. (2007) noted that based on experimental results, BFS yields a tree that is a more adequate and realistic representation of the structure of the site. Theorem 11.8 can be used to design methods for the computation of the importance of a single site (and, thus, using the aggregated method, for Web pages as well) taking into account the meet and join operations in the site lattice to propagate page importance through the lattice up to the site. A very simple, but practical, method is to compute the importance ρ_S of the site as the average of importances ρ_i of its pages W_i as follows:

$$\rho_S = \frac{1}{n} \sum_{i=1}^{n} \rho_i \, . \tag{11.53}$$

11.9 P2P Retrieval

11.9.1 P2P Network

A *peer-to-peer* (briefly P2P) network is a computer network in which each computer (workstation) has equal "rights" (capabilities) and "duties" (responsibilities), as opposed to the client/server network (in which some computers, the servers, are dedicated to serve other computers, the clients). In other words, one may say that in a P2P network every computer is both a server and a client at the same time.

On the Internet, a P2P network allows a group of computers (and thus their users as well) running the same network software to connect with one another and access one another's information. One widely used P2P network software is Gnutellanet.[7] The network program stores the IP addresses of the participating computers, and thus all users who are online connect to one another. (Another P2P network software is Napster.)

One of the advantages of a P2P network is that, e.g., the employees of an organization (company) can share data (e.g., files) without the need to have a central server. One of the disadvantages of a P2P network is that major producers of content (e.g., record companies) are concerned about illegal sharing of copyrighted material (e.g., music) over a P2P network.

11.9.2 Information Retrieval

Each peer has a repository of information (texts, audio, video, etc.). When a query is initiated at one of the peers, it is sent to the other peers. These peers generate hit lists from their own repositories in response to the query they receive, and then send them to the peer that initiated the query.

Information retrieval in a P2P network has a number of characteristics:

- There is no central repository (as, e.g., in a traditional IR system),
- A large quantity of data can be added to and deleted from the network in an ad hoc manner,
- Computers enter and leave the network in an ad hoc way.

The way in which peers generate hit lists varies, e.g., using keyword match (Boolean retrieval with AND-ed query) or some other method.

What is really specific to P2P retrieval is the way in which the query is sent over the network (i.e., from the querying computer to peer computers). Several of the methods that are utilized (Kalogeraki et al. 2002) are described briefly below.

11.9.2.1 BFS Method

In the BFS method the querying node (computer) sends the query to all its neighbor peers. When a neighbor receives the query, it sends it to other peers, generates its own hit list from its repository, and sends it to the querying peer. The BFS method is simple. However, it is very resource demanding: the query is sent along all paths, so a low bandwidth node can

[7] http://www.gnutella.com.

considerably reduce retrieval time. (Flooding can be avoided or prevented by assigning a TTL, time-to-live, parameter to the query.)

11.9.2.2 RBFS Method

In the RBFS (random breadth first search) method, the querying peer does not send the query to all its neighbors, only to randomly selected ones, e.g., to half of them selected at random. This, of course, has the advantage of being faster than the BFS method, but important nodes may never be reached (disadvantage).

11.9.2.3 ISM Method

In the ISM (intelligent search mechanism) method, each peer keeps a profile of its neighboring peers. The profile contains the queries (typically, the last N queries) and the corresponding hit lists that the neighboring peers have answered. At the same time, every peer produces a ranking of its neighboring peers in order to decide to which ones a new query should be forwarded. Ranking is produced using a similarity measure S_i between itself and peer i (the peer of interest). The similarity measure is computed as follows. Given a new query q originated at a peer, the cosine measure (or other measure) c_{ij} is calculated between q and all queries j in the profile of peer i. Every c_{ij} is multiplied by the corresponding number n_j of hits to query j, and then the products are added up, i.e., $S_i = \sum_j c_{ij}^{\alpha} n_i$, where α is a parameter that allows increasing the weight of most similar queries (and should be tuned manually). The querying peer ranks its neighbors i on their scores S_i, and sends the query to those highly ranked. The ISM method works well when peers store specialized information. As search may get locked in a cycle, it is recommended that a few randomly chosen peers (apart from those chosen based on S_i) also receive the query.

11.9.2.4 >RES Method

In the >RES (the most results in past) method, the querying peer sends the query to the peers that returned the highest number of hits for the last M queries.

11.9.2.5 HDS Method

The HDS (high degree seeking) method exploits the power law property of the peer graph. First an arbitrary node is chosen, and then a node with a degree higher than the current node. Once the highest-degree node has been found, a node having second highest degree will be chosen, and so on. The query is iteratively sent to all the nodes in a neighborhood of the current node until a match is found. This broadcasting is costly in terms of bandwidth. If every node keeps adequate information (e.g., file names) about its first and second neighbors, then HDS proves useful. As storage is likely to remain less expensive than bandwidth, and since network saturation is a weakness of P2P, HDS can be an efficient alternative to exhaustive searching. Adamic et al. (2003) showed that the expected degree $E(\alpha, n)$ of the richest neighbor of a node having degree n is given by

$$E(\alpha, n) = \frac{n(\alpha-2)}{(1-N^{2/\alpha-1})^n} \sum_{x=0}^{\lfloor N^{1/\alpha} \rfloor} x(1+x)^{1-\alpha}(1-(x+1)^{2-\alpha})^{n-1}, \tag{11.54}$$

where N denotes the number of peers and α is the power law exponent. **Figure 11.21** shows simulation results for the ratio $E(\alpha, n)/n$ for different values of α.

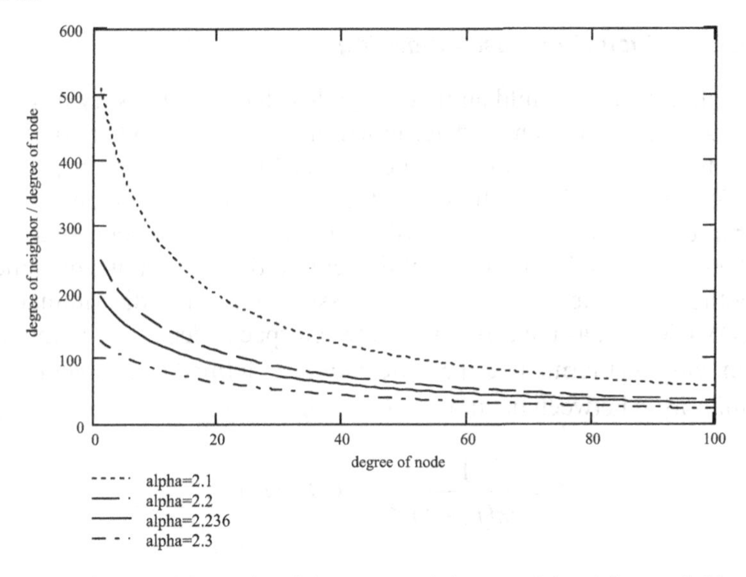

Fig. 11.21. Simulation of the ratio of the expected degree of the richest neighbor of a node with degree n for different values of the power law exponent alpha. The total number of nodes is equal to 100,000,000.

It can be seen that for a power law exponent between 2 and 2.3, the chance of finding a richer neighbor is higher than the degree of the node itself within a relatively large interval of degree values, which means that HDS can be applied nontrivially. (As the exponent is fairly robust,

HDS-based crawling may have a theoretical justification in the golden section (see Section 11.1).

11.9.3 Lattice-Based Indexing

In order for a peer to produce a hit list in response to a query q, a similarity measure between q and documents stored in its repository should be computed. Toward this end, the technology presented in Chapter 4 can be used to compute term weights. Extracting terms from documents in the repository can yield a number of terms that may be too large for a P2P network (superfluous terms, strict bound on the size of posting lists, high bandwidth consumption).

In order to ameliorate problems caused by too many terms, one can use a document-based indexing method based on lattices (Skobeltsyn et al. 2007).

11.9.3.1 Document-Based Indexing

Initially, the peers build an index together that contains very discriminative terms. Such terms can be determined in several ways. One possibility is to use the term discrimination model (TDM) introduced in Salton et al. (1974, 1975a). TDM is based on the underlying assumption that a "good" term causes the greatest possible separation of documents in space, whereas a "poor" term makes it difficult to distinguish one document from another. Each term of interest is assigned a term discrimination value (TDV) defined as the difference between space "densities" before and after removing that term. The space density Δ is defined as an average pairwise similarity ρ between documents $D_1,\ldots,D_m$:

$$\Delta = \frac{1}{m(m-1)} \sum_{\substack{i,j=1 \\ i \neq j}}^{m} \rho(D_i, D_j) . \qquad (11.55)$$

Alternatively, space density Δ can be computed—faster—as the average similarity between documents and a centroid document (defined as one in which terms have frequencies equal to their average frequencies across the collection of documents). Let Δ_{bk} and Δ_{ak} denote the space densities before and after removing term t_k, respectively. Then, the TDV_k of term t_k is defined as $\text{TDV}_k = \Delta_{bk} - \Delta_{ak}$. The best discriminators generally have positive TDVs, whereas the worst discriminators usually have negative TDVs.

Terms having TDVs around zero do not modify space density significantly when used as index terms. TDV can be used to decide which terms should be used as index terms. Based on experimental reults, terms with average document frequencies (between approximately $m/100$ and $m/10$) usually have positive TDVs, and can be used directly for indexing purposes. Terms whose document frequency is too high generally have negative TDVs, and are the worst discriminators. Too rare or specific terms have TDVs near zero and should not be used directly as index terms.

Another method to compute term discrimination values is based on the notion of entropy and given in Dominich et al. (2004):

$$H = -\sum_{j=1}^{m} p_j \log_2 p_j$$

$$p_j = \frac{\rho_j}{\sum_{k=1,\ldots,m} \rho_k} \quad j = 1,\ldots, m, \tag{11.56}$$

where m is the number of documents and ρ_k is the similarity between a term of interest and document k. The higher the entropy H of a term, the better its discrimination power.

11.9.3.2 Query Lattice-Based Indexing

The index will be gradually enriched by query terms at retrieval time using the following method (**Fig. 11.22**).

1. When a query $q = \{t_1,\ldots,t_n\}$ (where $t_1,\ldots,t_n$ are the query terms) originates at a peer, the peer generates the Boolean algebra of query terms $t_1,\ldots,t_n$, i.e., $(\wp(\{t_1,\ldots,t_n\}), \subseteq)$.
2. Then, the peer starts exploring the query lattice from the supremum $\mathbf{1} = \{t_1,\ldots,t_n\}$ in decreasing subset size order to single terms.
3. For every node in the lattice, the querying peer requests hit lists from other peers.
4. If the node is indexed in a peer, then the nodes whose join that node is will be skipped.
5. If two nodes are not indexed in any peer, then their meet will still be sent out to peers for retrieval.
6. After exploring the entire query lattice, the querying peer makes the union of all the hit lists received, and reranks the documents against the query (using some ranking method).

7. When a peer recognizes that a query term (or lattice node) is a new term (i.e., it is not in the index), then it sends an indexing request to other peers that can include it in their indexes. Thus, that term (or lattice node) can be used in retrieval for a new query in future. This on-demand indexing mechanism can also, or only, be performed for popular terms (i.e., for terms that are considered to be popular in some sense, e.g., they have been used frequently).

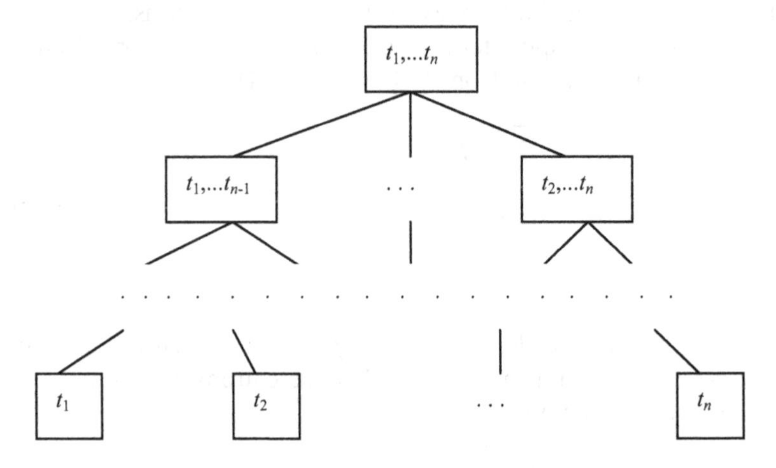

Fig. 11.22. Query lattice used in P2P indexing.

11.10 Exercises and Problems

1. Let M denote a Web graph. Represent it as an array, an adjacency list, or using some other method. Observe which representation method requires the least amount of memory.

2. Let W denote a portion of the Web and M the Web graph of W. Modify W so as to contain weighted links.

3. According to experimental results accumulated so far, the Web power law (for page degree distribution) is characterized by a robust exponent value around 2.5. Check the validity of this result by performing your own experiment.

4. Show that the connectivity method (Section 11.2) yields equal rank values when the Web graph W is a bipartite graph (hubs and authorities) with vertices having equal degrees.

5. Modify the connectivity method so as to obtain useful ranking for a Web graph W as in Exercise 4.

6. Given a Web graph *W*. Calculate the importance of pages using: (i) the mutual citation method, and (ii) the PageRank method. Analyze the differences in the importance values obtained.

7. You are given a Web portion containing only pages with zero outdegree. Compute link importance for these pages.

8. Let *W* denote: (i) a citation graph of authors; (ii) a partnership graph (vertex: partner, and there is an arc from partner p_i to partner p_j if p_i is using a product made by p_j); (iii) a "who knows who" graph. Compute the importance of authors, partners, and persons using several link analysis methods. Discuss the results obtained.

9. Prove that in the associative interaction method there will always be at least one reverberative circle.

10. Given a Web graph *W*. Apply the similarity merge method to compute the combined importance Ψ, also using a weighted version. Discuss the results obtained and experiment with several weightings.

11. Show that in the belief network method importance Ψ reduces to the normalizing parameter if the cosine measure is equal to 1.

12. Show that function $\Psi = Pe^L$ is also a combined importance function that can be used in the aggregated method. Experiment with this combined importance function using a Web portion of your choice.

Solutions to Exercises and Problems

Chapter 2

2.4.1. It is a proposition if either the value true (T) or false (F) can be assigned to it for certain. But is this sure? Discussion.

2.4.2. It depends on what is being asserted, i.e., on what the emphasis is on: is it on sun or on shining?

2.4.3. Identify the logical structure of the sentence. Let P = "John is liar," and Q = "Peter is a truth teller." Take into account the definition of implication $P \Rightarrow Q$.

2.4.4. Identify the logical structure of the sentence ($P \Rightarrow Q$). Use formal implication to write the sentence and then also use conjunction.

2.4.5. Formulate a logical expression using the terms given. Pay attention to negating what you are not interested in (namely Web ranking method). For example, waltz AND rock-and-roll AND NOT(Web ranking method).

2.4.6. You obtain two posets (one on the date and the other on the title length). What you should observe as an interesting and practically important fact is that the two posets are different (in general).

2.4.7. For example, you can order the toys by color, by shape, by function, by dimension (according to childrens' needs, pedagogical considerations, storing requirements).

2.4.8. Denote the number of occurrences of term T_i in the entire set B of Web pages (or books) by $1, 2, \ldots, n_i$. There are n_i classes (why?). If this partitioning is performed for every term (i.e., $i = 1, 2, \ldots, m$), there will be corresponding partitions. In general, different partitions are obtained.

2.4.9. Because any element in a set occurs at most once, the collection $B_1,\ldots,B_n$ forms a set only if every page (book) occurs only once. As this is usually not true (there are books or Web pages that are held in several copies), they do not generally form a set.

Chapter 3

3.12.1. Verify the validity of conditions (3.15) in Definition 3.2.

3.12.2. We introduce the smallest element, **0**, and the largest element, **1**. Thus, the set $\{\mathbf{0}, t_1, t_2, t_3, \mathbf{1}\}$ is obtained. Case (i):

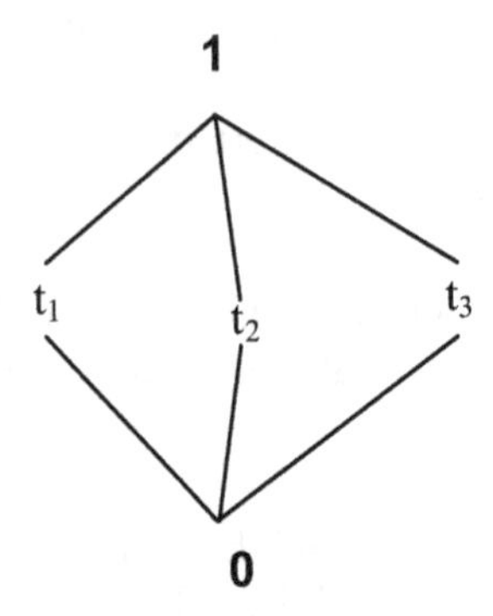

In a similar manner, cases (ii)–(iv).

3.12.3. Sufficiency: $X \vee (Y \wedge Z) = X \vee (Z \wedge X) = X$, $(X \vee Y) \wedge (X \vee Z) = (X \vee X) \wedge (Z \vee Y) = X \wedge (Z \vee X) = X$. Necessity can be proved in a similar manner.

3.12.4. Show that the distributive law Eq. (3.15) does not hold.

3.12.5. Let P denote the set of all distinct publications (e.g., their ISBN numbers). Use the notion of powerset ordered by set inclusion.

3.12.6. The thesaurus may be viewed as being a poset. A poset is a lattice if any two of its elements have a supremum and an infimum. Analyze and discuss when these properties hold.

3.12.7. See problem 3.12.6.

3.12.8. In general, no. For example, antisymmetry need not be satisfied. Check when the axioms of lattice are satisfied.

3.12.9. Yes. It is finite, so it is complete, etc.

3.12.10. Yes (subset of the Cartesian product $W \times W$). Let $w_i \rightarrow w_j \Leftrightarrow w_i \leq w_j$. Does any pair of pages have a supremum and infimum?

3.12.11. Check the validity of the corresponding definitions.

Chapter 4

4.11.1. Identify the terms first and then count the occurrence of each. Approximate C and α. Discussion.

4.11.2. Measure running time using built-in functions of the programming language you use. Store the term-document matrix using different formats, e.g., adjacency list, binary coding (for Boolean matrix, etc.). Observe whether economical storage and matrix operations are directly or indirectly proportional to each other. Discussion.

4.11.3. First do a literature search and follow that by your own experience. Discuss the Web characteristics learned or discovered.

4.11.4. The problem is straightforward. However, if, e.g., the precision-recall graph you obtain is not decreasing, then (very probably) something must have gone wrong.

Chapter 5

5.5.1. Use the method given in Section 5.3.2.

5.4.2. Verify the distributive law, or use Theorem 5.4.

5.5.3. Analyze several concept lattices.

5.5.4. See Section 5.1.1.

5.5.5. Verify the distributive law, or use Theorem 5.4 or Exercise 5.5.2.

Chapter 6

6.4.1. Identify terms, e.g., route planner, Europe, car. Build a Boolean expression using the terms.

6.4.2. Identify terms, e.g., opening hours, chair, museum, United Kingdom, car seat, chair of organization. Build a Boolean expression using the terms.

6.4.3. Both the Boolean method and the selection operator are based on Boolean expressions. Observe the data on which they operate.

6.4.4. Use the technology described in Chapter 4 to select index terms T, and then build the lattice $\wp(T)$. Experiment with different queries.

6.4.5. See problem 6.4.4.

6.4.6. See problems 6.4.4 and 6.4.5. Pay attention to identifying terms (they are special medical terms, so use domain knowledge, e.g., medical dictionary being assisted by a neuroradiologist).

Chapter 7

7.13.1. Check the validity of axioms defining a metric.

7.13.2. Let $D = \{d_1,\ldots,d_i,\ldots,d_n\}$, $d_i \in \wp(T)$. Take $|\,|d_i| - |d_j|\,|$, where $|d_i|$ and $|d_j|$ denote lengths (number of terms, cardinality) of documents.

7.13.3. Verify the validity of the axioms of a metric.

7.13.4. Verify the validity of the axioms of a linear space. (The answer is no.)

7.13.5. Verify the validity of the axioms of a metric.

7.13.7. E_n is a Euclidean space; hence it is also a Hilbert space. The closed subspaces of any Hilbert space form a lattice, and because E_n is finite, this lattice is modular.

7.13.8. For example, col(W). Use the Gram-Schmidt procedure starting from the columns of W. Use Example 7.11.

7.13.9. Verify the definition of a linear operator (Section 7.5). Verify whether it is self-adjoint (Section 7.10).

Chapter 8

8.7.1. You should note different rankings depending on the weighting scheme and similarity measure used.

8.7.2. Comment on the hit lists obtained as to their relevance to your request.

8.7.3. Observe whether length normalization has an effect on ranking.

8.7.4. Use Definition 8.1. Note that this also depends on the choice of terms you use.

8.7.5. Use Theorem 8.4 and Definition 8.2.

Chapter 9

9.9.1. Write it, e.g., for the three- or two-dimensional case first. Then, generalize.

9.9.2. Use Eq. (9.4) for each column of W.

9.9.3. Use Eq. (9.6).

9.9.4. Use the corresponding definitions.

9.9.5. Apply technologies described in Chapter 4 and use the precision-recall graph method.

9.9.6. Same as problem 9.9.5.

9.9.7. Same as problem 9.9.6.

9.9.8. For example, use a distribution function for term occurrences across documents.

Chapter 10

10.7.1. $f_i / \Sigma_i f_i$.

10.7.2. Use the definition of probability measure (Section 10.1) and Eq. (10.9).

Chapter 11

11.9.2. See Section 11.1 (in *M*, the entry is equal to a weight instead of 1).

11.9.3. See Section 11.1.1, and Chapter 4, for fitting the power law curve.

11.9.4. See the connectivity method (Section 11.2).

11.9.5. When retrieving, using the Boolean method, matching pages, determine the frequency (number of occurrences) of query terms, and use these in ranking.

11.9.7. Importance can be calculated in several ways. For example, if pages have nonzero indegree, this can be used to rank them. Another way is to apply the PageRank method with dangling pages.

11.9.8. Define matrix *M* for the method chosen, and then apply the method.

11.9.9. See Section 11.6.2.

11.9.10. The weighted version is

$$\Psi = \sum_i a_i \frac{s_i - s_{\min}}{s_{\max} - s_{\min}},$$

where a_i is a weight reflecting the importance of method *i*.

11.9.11. Use the equation for the belief network method.

11.9.12. Show that Assumptions 1–4 are satisfied.

References

(Aberer and Wu 2003) Aberer, K., and Wu, J.: A framework for decentralized ranking in Web information retrieval. In: *APWeb*, ed by Zhou, X., Zhang, Y., and Orlowska, M. E., *Lecture Notes in Computer Science*, vol 2642 (Springer, Berlin/Heidelberg/New York, 2003), pp 213–226

(Adamic 2003) Adamic, L. A.: Zipf, Power-laws, and Pareto—a ranking tutorial. http://ginger.hpl.hp.com/shl/papers/ranking/ranking.html (2003) Cited Dec. 18, 2003

(Adamic and Huberman 2000) Adamic, L. A., and Huberman, B. A.: Power-law distribution of the World Wide Web. *Science*. **287**, 2115a (2000)

(Albert 2000) Albert, R. (2000). *Power Laws on Internet and World Wide Web*. Ph.D. dissertation, University of Notre Dame

(Baeza-Yates 2003) Baeza-Yates, R.: Information retrieval in the Web: beyond current search engines. *International Journal of Approximate Reasoning*. **34**, 97–104 (2003)

(Baeza-Yates and Ribeiro-Neto 1999) Baeza-Yates, R., and Ribeiro-Neto, B.: *Modern Information Retrieval* (Addison Wesley Longman, 1999)

(Barabási et al. 2000) Barabási, A-L., Albert, R., and Jeong, H.: Scale-free characteristics of random networks: the topology of the world-wide web. *Physica A*. **281**, 69–77 (2000)

(Belew 2000) Belew, K. K.: *Finding Out About* (Cambridge University Press, Cambridge, UK, 2000)

(Berry and Browne 1999) Berry, W. M., and Browne, M.: Understanding Search Engines (SIAM, Philadelphia, 1999)

(Birkhoff and von Neumann 1936) Birkhoff, G., and von Neumann, J.: The logic of quantum mechanics. *Annals of Mathematics*. **37**(4), 823–843 (1936)

(Blair 2006) Blair, D.: *Back to the Rough Ground: Wittgenstein Philosophy of Language* (Springer, Berlin/Heidelberg/New York, 2006)

(Bollmann-Sdorra and Raghavan 1993) Bollmann-Sdorra, P., and Raghavan, V. V.: On the delusiveness of adopting a common space for modelling information retrieval objects: are queries documents? *Journal of the American Society for Information Science*. **44**(10), 579–587 (1993)

(Borlund 2003) Borlund, P.: The IIR evaluation model: a framework for evaluation of interactive information retrieval systems. *Information Research*. **8**(3), 1–66 (2003)

(Brin and Page 1998) Brin, S., and Page, L.: The anatomy of a large-scale hypertextual Web search engine. In: *Proceedings of the 7th World Wide Web Conference* (Brisbane, Australia, April 14–18, 1998), pp 107–117

(Broder et al. 2000) Broder, A., Kumar, R., Maghoul, F., and Raghavan, P.: Graph structure in the Web. *Computer Networks*. **33**, 309–320 (2000)

(Burris and Sankappanavar 2000) Burris, S., and Sankappanavar, H. P. *A Course in Universal Algebra* (The Millennium Edition, 2000)

(Calado et al. 2003) Calado, P., Ribeiro-Neto, B., and Ziviani, N.: Local versus global link information in the Web. *ACM Transactions of Inormation Systems*. **21**(1), 1–22 (2003)

(Carpineto and Romano 2005) Carpineto, C., and Romano, G.: Using concept lattices for text retrieval and mining. In: *Formal Concept Analysis*, ed by Ganter, B., et al., *Lecture Notes in Artificial Intelligence*, vol 3626 (Springer, Berlin/Heidelberg/New York, 2005), pp 161–170

(Carrière and Kazman 1997) Carrière, S. J., and Kazman, R.: WebQuery: searching and visualizing the Web through connectivity. *Computer Network and ISDN Systems*. **29**, 1257–1267 (1997)

(Cheung and Vogel 2005) Cheung, K. S. K., and Vogel, D.: Complexity reduction in lattice-based information retrieval. *Information Retrieval*. **8**, 285–299 (2005)

(Chu and Rosenthal 1996) Chu, H., and Rosenthal, A. U.: Search engines for the World Wide web: a comparative study and evaluation methodology. In: *Proceedings of the 59th Annual Meeting of the American Society for Information Science* (Baltimore, 1996), pp 127–135

(Clark and Willett 1997) Clark, S. J., and Willett, P.: Estimating the recall performance of Web search engines. *ASLIB Proceedings*. **49**(7), 184–189 (1997)

(Cooper 1968) Cooper, W. S.: Expected search length: a single measure of retrieval effectiveness based on the weak ordering action of retrieval systems. *American Documentation.* **19**, 30–41 (1968)

(Cooper 1971) Cooper, W. S.: A definition of relevance for information retrieval. *Information Storage and Retrieval.* **7**(1), 19–37 (1971)

(Cooper 1990) Cooper, G. F.: The computational complexity of probabilistic inference using Bayesian belief network. *Artificial Intelligence.* **42**(2–3), 393–405 (1990)

(Crestani et al. 1998) Crestani, F., Lalmas, M., and van Rijsbergen, C. J.: *Information Retrieval: Uncertainty and Logics* (Springer, Berlin/Heidelberg/New York, 1998)

(Croft and Lafferty 2003) Croft, W. B., and Lafferty, J.: *Language Modelling for Information Retrieval* (Springer, Berlin/Heidelberg/New York, 2003)

(Croft and Ponte 1998) Croft, W. B., and Ponte, J.: A language modeling approach to information retrieval. In: *Proceedings of 21st Annual International Conference on Research and Development in Information Retrieval ACM SIGIR* (Aug. 24–28, Melbourne, Australia, 1998), pp 275–281

(Cummins and O'Riordan 2006) Cummins, R., and O'Riordan, C.: Evolving local and global weighting schemes in information retrieval. *Information Retrieval.* **9**, 311–330 (2006)

(De Wilde 1996) De Wilde, P.: *Neural Network Models* (Springer, Berlin/Heidelberg/New York, 1996)

(Deerwester et al. 1990) Deerwester, S., Dumais, S., Furnas, G., Landauer, T., and Harshman, R.: Indexing by latent semantic analysis. *Journal of the American Society for Information Science.* **41**, 391–407 (1990)

(Dewey 1929) Dewey, J.: *Experience and Nature* (Open Court, La Salle, IL, 1929)

(Dominich 2001) Dominich, S.: *Mathematical Foundations of Information Retrieval* (Kluwer, Dordrecht/Boston/London, 2001)

(Dominich 2002) Dominich, S.: Paradox-free formal foundation of vector space model. In: *Proceedings of the ACM SIGIR Workshop Mathematical/Formal Methods in Information Retrieval MF/IR '02* (University of Tampere, Tampere, Finland, Aug. 12–15, 2002), pp 43–48

(Dominich 2003) Dominich, S.: Connectionist interaction information retrieval. *Information Processing and Management.* **39**(2), 167–194 (2003)

(Dominich and Kiezer 2005) Dominich, S., and Kiezer, T.: Zipf-törvény, kis világ és magyar nyelv. (Zipf law, small world, and Hungarian language). *Alkalmazott Nyelvtudomány*, **1-2**, 5–24 (2005) (in Hungarian).

(Dominich and Kiezer 2007) Dominich, S., and Kiezer, T.: A measure theoretic approach to information retrieval. *Journal of the American Society of Information Science and Technology.* **58**(8), 1108–1122 (2007)

(Dominich et al. 2004) Dominich, S., Góth, J., Kiezer, T., and Szlávik, Z.: Entropy-based interpretation of retrieval status value-based retrieval. *Journal of the American Society for Information Science and Technology.* **55**(7), 613–627 (2004)

(Dominich et al. 2005) Dominich, S., Góth, J., Horváth, M., and Kiezer, T.: 'Beauty' of the World Wide Web: cause, goal, or principle. In: *Proceedings of the European Conference on Information Retrieval ECIR '05.* Lecture Notes in Computer Science, vol. 3408 (Springer, Berlin/Heidelberg/New York, 2005), pp 67–81

(Dominich et al. 2006) Dominich, S., Tuza, Zs., and Skrop, A.: Formal theory of connectionist Web retrieval. In: *Soft Computing in Web Retrieval*, ed by Crestani, F. et al. (Springer, Berlin/Heidelberg/New York, 2006), pp 161–194

(Doob 1994) Doob, J. L.: *Measure Theory* (Springer, Berlin/Heidelberg/New York, 1994)

(Egghe 2000) Egghe, L.: The distribution of n-grams. *Scientometrics.* **47**(2), 237–252 (2000)

(Faloutsos et al. 1999) Faloutsos, M., Faloutsos, P., and Faloutsos, Ch.: On power-law relationship of the Internet topology. In: *Proceedings of the ACM SIGCOMM* (Cambridge, MA, 1999), pp 251–262

(Farahat et al. 2006) Farahat, A., Lofaro, T., Miller, J. C., Rae, G., and Ward, L. A.: Authority rankings from HITS, PageRank, and SALSA: existence, uniqueness, and effect of initialisation. *SIAM Journal of Computing.* **27**(4), 1181–1201 (2006)

(Feldman and Ballard 1982) Feldman, J. A., and Ballard, D. H.: Connectionist models and their properties. *Cognitive Science.* **6**, 205–254 (1982)

(Feynman et al. 1964) Feynman, R. P., Leighton, R. B., and Sands, M.: *The Feynman Lectures on Physics* (Addison-Wesley, Reading, MA, 1964)

(Fox and Shaw 1994) Fox, E. A., and Shaw, J. A.: Combination of multiple searches. In: *The 2nd Text Retrieval Conference*, ed by Harman, D. K. (NIST Publication 500215, Washington, DC, US Government Printing Office. 1994), pp 243–252

(Fuhr 1992) Fuhr, N.: Probabilistic models in information retrieval. *The Computer Journal*. **35**(3), 243–255 (1992)

(Garfield 1955) Garfield, E.: Citation indexes for science. *Science*. 108 (1955)

(Garfield 1972) Garfield, E.: Citation analysis as a tool in journal evaluation. *Science*. 471–479 (1972)

(Geller 1978) Geller, N.: On the citation influence methodology of Pinski and Narin. *Information Processing and Management*. **14**, 93–95 (1978)

(Godin et al.1989) Godin, R., Gecsei, J., and Pichet, C.: Design of a browsing interface for information retrieval. In: *Proceedings of the 12th International Conference on Research and Development in Information Retrieval ACM SIGIR* (Cambridge, MA, 1989), pp 32–39

(Godin et al. 1998) Godin, R., Missaoui, R., and April, A.: Experimental comparison of navigation in a Galois lattice with conventional information retrieval methods. *International Journal of Man-Machine Studies*. **38**, 747–767 (1998)

(Grinbaum 2005) Grinbaum, A.: Information-theoretic principle entails orthomodularity of a lattice. *Foundations of Physics Letters*. **18**(6), 563–572 (2005)

(Grossberg 1976) Grossberg, S.: Adaptive pattern classification and universal recoding: I. Parallel development and coding of neural feature detectors. *Biological Cybernetics*. **23**, 121–134 (1976)

(Grossman and Frieder 2004) Grossman, D. A., and Frieder, O.: *Information Retrieval: Algorithms and Heuristics* (Springer, Berlin/Heidelberg/New York, 2004)

(Guilmi et al. 2003) Guilmi, C., Gaffeo, E., and Gallegati, M.: Power law scaling in the world income distribution. *Economics Bulletin*. **15**(6), 1–7 (2003)

(Gwizdka and Chignell 1999) Gwizdka, J., and Chignell, M.: Towards information retrieval measures for evaluation of Web search engines. Technical Report 99-01 (University of Toronto, 1999)

(Hays 1966) Hays, D. D.: *Readings in Automatic Language Processing* (American Elsevier, New York, 1966)

(Heine 1999) Heine, M. H.: Measuring the effects of AND, AND NOT and OR operators in document retrieval systems using directed line segments. In: *Proceedings of the Workshop on Logical and Uncertainty Models for Information Systems*, ed by Crestani, F., and Lalmas, M. (University College, London, July 5, 1999), pp 55–76

(Hopfield 1984) Hopfield, J. J.: Neurons with graded response have collective computational properties like those of two-states neurons. In: *Proceedings of the National Academy of Sciences.* **81**, 3088–3092 (1984)

(Ingwersen and Jarvelin 2005) Ingwersen, P., and Jarvelin, K.: *The Turn: Integration of Information Seeking and Retrieval in Context* (Springer, Berlin/Heidelberg/New York, 2005)

(James 1988) James, W.: Psychology—briefer course. In: *Neurocomputing: Foundations of Research*, ed by Anderson, J. A., and Rosenfeld, E. (MIT Press, Cambridge, MA, 1988), pp 253–279

(Jauch 1968) Jauch, J. M.: *Foundations of Quantum Mechanics* (Addison-Wesley, Reading, MA, 1968)

(Kahng et al. 2002) Kahng, B., Park, Y., and Jeong, H.: Robustness of the indegree exponent for the World Wide Web. *Physical Review.* **66**(046107), 1–6 (2002)

(Kalogeraki et al. 2002) Kalogeraki, V., Gunopulos, D., and Zeinalipour-Yazti, D.: A local search mechanism for peer-to-peer networks. In: *Proceedings of the 11th Conference on Information Knowledge Management CIKM '02* (Nov. 4–9, McLean, VA, 2002), pp 300–307

(Kim and Compton 2004) Kim, M., and Compton, P.: Evolutionary document management and retrieval for specialized domains on the web. *International Journal of Human-Computer Studies.* **60**, 201–241 (2004)

(Kiyosi 2000) Kiyosi, I.: Encyclopedic Dictionary of Mathematics, Second edition, The MIT Press, Cambridge-Massachusetts-London (2000)

(Kleinberg 1999) Kleinberg, J. M.: Authoritative sources in a hyperlinked environment. *Journal of the ACM*. **46**(5), 604–632 (1999)

(Kneale and Kneale 1962) Kneale, K., and Kneale, M.: *The Development of Logic* (Oxford University Press, Oxford, 1962)

(Knuth 2005) Knuth, K. H.: Lattice duality: the origin of probability and entropy. *Neurocomputing*. **67**, 245–274 (2005)

(Koester 2005) Koester, B.: Conceptual knowledge processing with Google. In: *Proceedings of the LWA 2005, Lernen Wissensentdeckung Adaptivitat* (Saarbrücken, 2005), pp 178–183

(Koester 2006a) Koester, B.: Conceptual knowledge retrieval with FooCA: improving Web search engine results with contexts and concept hierarchies. In: *Lecture Notes in Artificial Intelligence*, vol. 4065 (Springer, Berlin/Heidelberg/New York, 2006a), pp 176–190

(Koester 2006b) Koester, B.: *FooCA Web information retrieval with formal concept analysis* (Verlag Allgemeine Wissenschaft, Mühltal, 2006b)

(Kolmogorov 1956) Kolmogorov, A.: *Foundation of the Theory of Probability* (Chelsea, New York, 1956)

(Korfhage 1997) Korfhage, R. R.: *Information Storage and Retrieval* (Wiley, New York, 1997)

(Kowalski 1997) Kowalski, G.: *Information Retrieval Systems: Theory and Implementation* (Kluwer, Dordrecht/Boston/London, 1997)

(Kumar et al. 1998) Kumar, R., Raghavan, P., Rajagopalan, S., and Tomkins, A.: Trawling the web for emerging cyber-communities. In: *Proceedings of the WWW8 Conference* (1998) http://www8.org/w8-papers

(Kurtz 1991) Kurtz, M.: *Handbook of Applied Mathematics for Engineers and Scientists* (McGraw-Hill, New York, 1991)

(Lánczos 1970) Lánczos, K.: *Space through the Ages* (Academic Press, London, 1970)

(Langville and Meyer 2005) Langville, A. N., and Meyer, C. D.: A survey of eigenvector methods for information retrieval. *SIAM Review*. **47**(1), 135–161 (2005)

(Langville and Meyer 2006) Langville, A. N., and Meyer, C. D.: *Google's PageRank and Beyond* (Princeton University Press, Princeton, NJ, 2006)

(Leighton and Srivastava 1999)	Leighton, H. V., and Srivastava, J.: First twenty precision among World Wide Web search services. *Journal of the American Society for Information Science*. **50**(10), 882–889 (1999)
(Le Quan Ha et al. 2003)	Le Quan Ha, Sicilia-Garcia, E. I., Ming, J., and Smith, F. J.: Extension of Zipf's law to word and character N-grams for English and Chinese. *Computational Linguistics and Chinese Language Processing*. **1**, 77–102 (2003)
(Lempel and Moran 2001)	Lempel, R., and Moran, S.: SALSA: the stochastic approach for link-structure analysis. *ACM Transactions on Information Systems*. **19**(2), 131–160 (2001)
(Lesk 2007)	Lesk, M.: The Seven Ages of Information Retrieval. http://www.ifla.org/VI/5/op/udtop5/udtop5.htm, Cited July 2007
(Luhn 1966)	Luhn, H. P.: Keyword-in-context index for technical literature (KWIC index). In: *Readings in Automatic Language Processing*, ed by Hays, D. D. (American Elsevier, New York, 1966), pp 159–167
(Maron and Kuhns 1960)	Maron, M. E., and Kuhns, J.-L.: On relevance, probabilistic indexing and information retrieval. *Journal of the Association for Computing Machinery*. **7**, 219–244 (1960)
(Martin and Reissner 1961)	Martin, W. T., and Reissner, E.: *Elementary Differential Equations* (Addison-Wesley, Reading, MA, 1961)
(Meadow et al. 1999)	Meadow, C. T., Boyce, B. R., and Kraft, D. H.: *Text Information Retrieval Systems* (Academic Press, San Diego, CA, 1999)
(Messai et al. 2006)	Messai, N., Devignes, M-D., Napoli, A., and Smail-Tabbone, M.: BR-explorer: an FCA-based algorithm for information retrieval. In: *Proceedings of the 4th International Conference on Concept Lattices and Their Applications*, Oct. 30 (Hammamet, Tunisia 2006) http://hal.inria.fr/inria-0010391/en/
(Metzler and Croft 2004)	Metzler, D., and Croft, W. D.: Combining the language model and inference model approaches to retrieval. *Information Processing and Management*. **40**, 735–750 (2004)
(Mooers 1959)	Mooers, C. N.: A mathematical theory of language symbols in retrieval. In: *Proceedings of the International Conference on Scientific Information* (Washington, DC, 1959), pp 1327–1352

(Ng et al. 2001) Ng, A. Y., Zheng, A. X., and Jordan, M. I.: Link analysis, eigenvectors and stability. In: *Proceedings of the 17th International Joint Conference on Artificial Intelligence IJCAI-01* (2001) http://robotics.stanford.edu/~ang/papers/ijcai01-linkanalysis.pdf (cited 2007)

(Nielson 1993) Nielson, J.: *Usability Engineering*. (Academic Press, San Diego, CA, 1993)

(Over et al. 2007) Over, D. E., Hadjichristidis, C., Evans, J. St. B. T., and Sloman, S. A.: The probability of causal conditionals. *Cognitive Psychology*. **54**(1), pp 62–97 (2007) (in press)

(Pearl 1988) Pearl, J.: *Probabilistic Reasoning in Intelligent Systems: Networks and Plausible Inference* (Morgan Kaufmann, Palo Alto, CA, 1988)

(Pennock et al. 2002) Pennock, D. M., Flake, G. W., Lawrence, S., Glover, E., and Giles, L.: Winners don't take all: characterizing the competition for links on the Web. In: *Proceedings of the New York Academy of Sciences*. **99**(8), pp 5207–5211 (2002)

(Piziak 1978) Piziak, R.: Orthomodular lattices and quantum physics. *Mathematics Magazine*. **51**(5), 299–303 (1978)

(Ponte and Croft 1998) Ponte, J. M., and Croft, W. B.: A language modeling approach to information retrieval. In: *Proceedings of the 21st Annual International Conference on Research and Development in Information Retrieval ACM SIGIR* (Melbourne, 1998), pp 275–281

(Priss 2000) Priss, U.: Lattice-based information retrieval. *Knowledge Organisation*. **27**(3), 132–142 (2000)

(Rajapakse and Denham 2006) Rajapakse, R. K., and Denham, M.: Text retrieval with more realistic concept matching and reinforcement learning. *Information Processing and Management*. **42**, 1260–1275 (2006)

(Recski 1989) Recski, A.: *Matroid Theory and its Applications in Electric Network Theory and in Statics* (Springer Verlag and Akadémiai Kiadó, Budapest, 1989)

(Rédei 1996) Rédei, M.: Why John von Neumann did not like the Hilbert space formalism of quantum mechanics. *Studies in the History and Philosophy of Modern Physics*. **27**, 493–510 (1996)

(Robertson 1977) Robertson, S.: The probability ranking principle in IR. *Journal of Documentation*. **33**, 294–304 (1977)

(Robertson 2002) Robertson, S. E.: On Bayesian models and event spaces in IR. *Proceedings of the ACM SIGIR MF/IR Workshop on Mathematical/Formal Methods in Information Retrieval* (Tampere, Finland, 2002) http://www.dcs.vein.hu/CIR Cited 2007

(Robertson and Sparck-Jones 1977) Robertson, S. E., and Sparck-Jones, K.: Relevance weighting of search terms. *Journal of the American Society for Information Science*. **27**(3), 129–146 (1977)

(Robertson et al. 1982) Robertson, S. E., Maron, M. E., and Cooper, W. S.: Probability of relevance: a unification of two competing models for document retrieval. *Information Technology: Research and Development*. **1**, 1–21 (1982)

(Salton 1966) Salton, G.: Automatic phrase matching. In: *Readings in Automatic Language Processing*, ed by Hays, D. D. (American Elsevier, New York, 1966), pp 169–189

(Salton 1971) Salton, G.: *The SMART Retrieval System—Experiment in Automatic Document Processing* (Prentice-Hall, Englewood Cliffs, NJ, 1971)

(Salton 1986) Salton, G.: Another look at automatic text-retrieval systems. *Communications of the ACM*. **29**(7), 648–656 (1986)

(Salton and Buckley 1988) Salton, G., and Buckley, C.: Term-weighting approaches in automatic text retrieval. *Information Processing and Management*. **24**(5), 513–523 (1988)

(Salton and Lesk 1968) Salton, G., and Lesk, M.E.: Computer evaluation of indexing and text processing. *Journal of the Association for Computing Machinery*. **15**(1), 8–36 (1968)

(Salton and McGill 1983) Salton, G., and McGill, M.: *Introduction to Modern Information Retrieval* (McGraw Hill, New York, 1983)

(Salton et al. 1974) Salton, G., Yang, C. S., and Yu, C. T.: *Contribution to the Theory of Indexing. Information Processing 74* (North Holland, Amsterdam, 1974), pp 584–590

(Salton et al. 1975a) Salton, G., Wong, A., and Yang, C. S.: A vector space model for automatic indexing. *Communications of the ACM*. **18**(11), 613–620 (1975a)

(Salton et al. 1975b) Salton, G., Yang, C. S., and Yu, C. T.: Theory of term importance in automatic text analysis. *Journal of the American Society for Information Science*. **26**(1), 33–44 (1975b)

(Savoy and Debois 1991) Savoy, J., and Desbois, D.: Information retrieval in hypertext systems: an approach using Bayesian networks. *Electronic Publishing*. **4**(2), 87–108 (1991)

(Shafi and Rather 2005) Shafi, S. M., and Rather, R. A.: Precision and Recall of Five Search Engines for Retrieval of Scholarly Information in the Field of Biotechnology. Webology. **2**(2) (2005). http://www.webology.ir/2005/v2n2/a12.html

(Shannon and Weaver 1949) Shannon, C. E., and Weaver, W.: *The Mathematical Theory of Communication* (University of Illinois Press, Urbana, IL, 1949)

(Shiodde and Batty 2000) Shiode, N., and Batty, M. (2000). *Power Law Distribution in Real and Virtual Worlds.* http://www.isoc.org/inet2000/cdproceeedings/2a/2a_2.htm (Cited Dec. 18, 2003)

(Silva et al. 2004) Silva, I. R., de Souza, J. N., and Santos, K. S.: Dependence among terms in vector space model. In: *Proceedings of the International Database Engineering and Applications Symposium* (2004), pp 97–102

(Silverstein et al. 1998) Silverstein, C., Henzinger, M., Marais, J., and Moricz, M.: *Analysis of a very large Altavista query log* (Technical Report 1998-014, COMPAQ Systems Research Center, Palo Alto, CA, 1998)

(Simmonds 1982) Simmonds, J. G.: *A Brief on Tensor Analysis* (Springer, Berlin/Heidelberg/New York, 1982)

(Skobeltsyn et al. 2007) Skobetltsyn, G., Luu, T., Zarko, I. P., Rajman, M., and Aberer, K.: Query-driven indexing for peer-to-peer text retrieval. In: *Proceedings of the WWW 2007 Conference*. (May 8–12, Banff, Alberta, 2007), pp 1185–1186

(Smith and Devine 1985) Smith, F. J., and Devine, K.: Storing and retrieving word phrases. *Information Processing and Management*. **21**(3), 215–224 (1985)

(Song and Croft 1999) Song, F., and Croft, W. B.: A general language model for information retrieval. In: *Proceedings of CIKM '99* (Kansas City, MO, 1999), pp 316–321

(Spink and Cole 2005) Spink, A., and Cole, C.: *New Directions in Cognitive Information Retrieval* (Springer, Berlin/Heidelberg/New York, 2005)

(Stalnaker 1968) Stalnaker, R.: A theory of conditionals. *American Philosophical Quarterly Monograph Series*. **2**, 98–112 (1968)

(Strogatz 2001) Strogatz, S. H.: Exploring complex networks. *Nature*. **410**, 268–276 (2001)

(Su 1994) Su, L. T.: The relevance of recall and precision in user evaluation. *Journal of the American Society of Information Science*. **45**(3), 207–217 (1994)

(Su et al. 1998) Su, L. T., Chen, H. L., and Dong, X. Y.: Evaluation of Web-based search engines from an end-user's perspective: a pilot study. In: *Proceedings of the 61st Annual Meeting of the American Society for Information Science* (Pittsburgh, PA, 1998) pp 348–361

(Tait 2005) Tait, J.: *Charting a New Course: Natural Language Processing and Information Retrieval. Essays in Honour of Karen Sparck-Jones* (Springer, Berlin/Heidelberg/New York, 2005)

(Tang and Sun 2003) Tang, M. C., and Sun, Y.: Evaluation of Web-based search engines using user-effort measures. *LIBRES*. **13**(2), (2003) (electronic journal)

(Thelvall and Vaughan 2004) Thelwall, M., and Vaughan, L.: New versions of PageRank employing alternative Web document models. *ASLIB Proceedings*. **56**(1), 24–33 (2004)

(Thorndike 1937) Thorndike, E. L.: On the number of words of any given frequency of use. *Psychological Record*. **1,** 399–406 (1937)

(Tsikrika and Lalmas 2004) Tsikrika, T., and Lalmas, M.: Combining evidence for Web retrieval using the inference network model: an experimental study. *Information Processing and Management*. **40**(5), 751–772 (2004)

(Turtle and Croft 1991) Turtle, H., and Croft, W. B.: Evaluation of an inference network-based retrieval model. *ACM Transactions on Information Systems*. **9**(3), 187–222 (1991)

(Ullman 1980) Ullman, J. D.: *Principles of Database Systems* (Computer Science Press, New York, 1980)

(Van Rijsbergen 1979) Van Rijsbergen, C. J.: *Information Retrieval* (Butterworth, London, 1979)

(Van Rijsbergen 1992) Van Rijsbergen, C. J.: Probabilistic retrieval revisited. *The Computer Journal*. **35**(3), 291–298 (1992)

(Van Rijsbergen 2004) Van Rijsbergen, C. J.: *The Geometry of IR* (Cambridge University Press, Cambridge, 2004)

(Vegas et al. 2007) Vegas, J., Crestani, F., and de la Fuente, P.: Context representation for web search results. *Journal of Information Science*. **33**(1), 77–94 (2007)

(Wang and Klir 1991) Wang, Z., and Klir, G. J.: *Fuzzy Measure Theory* (Plenum Press, New York, 1991)

(Weiss 1995) Weiss, M.A.: *Data Structures and Algorithm Analysis* (Benjamin/Cummings, New York/Amsterdam, 1995)

(Wellish 1991) Wellish, H.: *Indexing from A to Z* (H. W. Wilson 1991)

(Widdows 2003) Widdows, D.: Orthogonal negation in vector space for modelling word-meanings and document retrieval. In: *Proceedings of the 41st Annual Meeting of the Association for Computational Linguistics* (Sapporo, Japan, 2003), pp 136–143

(Widdows 2004) Widdows, D.: *Geometry and Meaning* (CSLI, Stanford, 2004)

(Widdows and Peters 2003) Widdows, D., and Peters, S.: Word vectors and quantum logic—experiments with negation and disjunction. In: *Proceedings of Mathematics of Language*, ed by Oehrle, R. T., and Rogers, J. (Bloomington, IN, 2003), pp 141–154

(Wille 2005) Wille, R.: Formal concept analysis as mathematical theory of concepts and concept hierarchies. In: *Formal Concept Analysis,* ed by Ganter, B., et al. *Lecture Notes in Artificial Intelligence LNAI* vol. 3626 (Springer, Berlin/Heidelberg/New York, 2005), pp 1–33

(Wolff 1993) Wolff, K. E.: A first course in formal concept analysis. In: *SoftStat'93 Advances in Statistical Software*, ed by Faulbaum, F., **4**, 429–438 (1993)

(Wong and Raghanavan 1984) Wong, S. K. M., and Raghavan, V. V.: Vector space model of information retrieval—a re-evaluation. In: *Proceedings of the 7th ACM SIGIR International Conference on Research and Development in Information Retrieval* (Kings College, Cambridge, UK, 1984), pp 167–185

(Wong et al. 1985) Wong, S. K. M., Ziarko, W., and Wong, P. C. N.: Generalized vector space model in information retrieval. In: *Proceedings of the 8th ACM SIGIR International Conference on Research and Development in Information Retrieval* (ACM Press, New York, 1985), pp 18–25

(Yang and Chen 1996) Yang, A., and Chen, P. P.: Efficient data retrieval and manipulation using Boolean entity lattice. *Data and Knowledge Engineering*. **20**, 211–226 (1996)

(Yu et al. 1989) Yu, C. T., Meng, W., and Park, S.: A framework for effective retrieval. *ACM Transactions on Database Systems*. **14**(2), 147–167 (1989)

(Yule 1924) Yule, G. U.: A mathematical theory of evaluation, based on the conclusions of Dr. J.Willis FRS. *Philosophical Transactions*. 213–221 (1924)

(Zimmerman 1996) Zimmerman, H.-J.: *Fuzzy Set Theory—and Its Applications* (Kluwer, Norwell-Dordrecht, 1996)

(Zipf 1949) Zipf, G.: *Human Behavior and the Principle of Least Effort* (Addison-Wesley, Reading, MA, 1949)

Index

The Information Retrieval Series

Gerald Kowalski. ***Information Retrieval Systems:*** **Theory and Implementation**. ISBN 0-7923-9926-9

Gregory Grefenstette. ***Cross-Language Information Retrieval.*** ISBN 0-7923-8122-X

Robert M. Losee. ***Text Retrieval and Filtering:*** **Analytic Models of Performance**. ISBN 0-7923-8177-7

Fabio Crestani, Mounia Lalmas, and Cornelis Joost van Rijsbergen. ***Information Retrieval: Uncertainty and Logics:*** **Advanced Models for the Representation and Retrieval of Information**. ISBN 0-7923-8302-8

Ross Wilkinson, Timothy Arnold-Moore, Michael Fuller, Ron Sacks-Davis, James Thom, and Justin Zobel. ***Document Computing:*** **Technologies for Managing Electronic Document Collections**. ISBN 0-7923-8357-5

Marie-Francine Moens. ***Automatic Indexing and Abstracting of Document Texts.*** ISBN 0-7923-7793-1

W. Bruce Croft. ***Advances in Informational Retrieval:*** **Recent Research from the Center for Intelligent Information Retrieval**. ISBN 0-7923-7812-1

Gerald J. Kowalski and Mark T. Maybury. ***Information Storage and Retrieval Systems:*** **Theory and Implementation**, Second Edition. ISBN 0-7923-7924-1

Jian Kang Wu, Mohan S. Kankanhalli, Joo-Hwee Lim, Dezhong Hong. ***Perspectives on Content-Based Multimedia Systems***. ISBN 0-7923-7944-6

George Chang, Marcus J. Healey, James A.M. McHugh, Jason T.L. Wang. ***Mining the World Wide Web:*** **An Information Search Approach**. ISBN 0-7923-7349-9

James Z. Wang. ***Integrated Region-Based Image Retrieval.*** ISBN 0-7923-7350-2

James Allan. ***Topic Detection and Tracking:*** **Event-based Information Organization**. ISBN 0-7923-7664-1

W. Bruce Croft, John Lafferty. ***Language Modeling for Information Retrieval.*** ISBN 1-4020-1216-0

Yixin Chen, Jia Li and James Z. Wang. ***Machine Learning and Statistical Modeling Approaches to Image Retrieval***. ISBN 1-4020-8034-4

David A. Grossman and Ophir Frieder. ***Information Retrieval:*** **Algorithms and Heuristics**. Second edition; ISBN 1-4020-3003-7; PB: ISBN 1-4020-3004-5

John I. Tait. ***Charting a new Course: Natural Language Processing and Information Retrieval.*** ISBN 1-4020-3343-5

Udo Kruschwitz. ***Intelligent Document Retrieval:*** **Exploiting Markup Structure**. ISBN 1-4020-3767-8

Peter Ingwersen, Kalervo Jrvelin. ***The Turn:*** **Integration of Information Seeking and Retrieval in Context**. ISBN 1-4020-3850-X

Amanda Spink, Charles Cole. ***New Directions in Cognitive Information Retrieval.*** ISBN 1-4020-4013-X

James G. Shanahan, Yan Qu, Janyce Wiebe. ***Computing Attitude and Affect in Text:*** **Theory and Applications**. ISBN 1-4020-4026-1

Marie-Francine Moens. ***Information Extraction, Algorithms and Prospects in a Retrieval Context.*** ISBN 1-4020-4987-0

Maristella Agosti. ***Information Access through Search Engines and Digital Libraries.*** ISBN 978-3-540-75133-5

Jin Zhang. ***Visualization for Information Retrieval.*** ISBN 978-3-540-75147-2

Sándor Dominich. ***The Modern Algebra of Information Retrieval.***
ISBN 978-3-540-77658-1